Earliest Life on Earth:
Habitats, Environments and Methods of Detection

Suzanne D. Golding • Miryam Glikson
Editors

Earliest Life on Earth: Habitats, Environments and Methods of Detection

Springer

Editors

Suzanne D. Golding
School of Earth Sciences
University of Queensland
Brisbane
Australia
s.golding1@uq.edu.au

Miryam Glikson
School of Earth Sciences
University of Queensland
Brisbane
Australia
miryam7@bigpond.com

ISBN 978-90-481-8793-5 e-ISBN 978-90-481-8794-2
DOI 10.1007/978-90-481-8794-2
Springer Dordrecht Heidelberg London New York

Library of Congress Control Number: 2010934231

Cover illustration: Top picture: Thermally degraded and compressed primary microbial cell constituting the predominant insitu, parallel to bedding organic matter (OM) in 3.5 billion year old rock formation in Western Australia. Observed and recorded by Miryam Glikson using TEM.
Below 1 (from left to right): Primary microbial cells, not thermally degraded, preserved in fluid inclusions within quartz crystals in 3.5 billion year old rock formation in Western Australia. Observations using Transmission Electron Microscopy (TEM) by Miryam Glikson.
2. OM coating quartz grains (represented by voids), part of major insitu OM in 3.5 billion year old rock formation in Western Australia. TEM observations and recording by Miryam Glikson.
3. Laboratory produced OM in simulated hydrothermal conditions, seen here to coat quartz grains (black). Note the close resemblance to similar OM occurrence in the 3.5 B.Y old rocks. Experimental work, observations and recording of synthesized OM using TEM by Bradley De Gregorio.

Printed on acid-free paper

Springer is part of Springer Science+Business Media (www.springer.com)

Preface

This volume integrates the latest findings on earliest life forms, identified and characterised in some of the oldest sedimentary rocks and seafloor hydrothermal systems on Earth and includes a number of papers from authors with alternate perspectives on the evidence of earliest life. Emphasis is placed on the integration of microanalytical methods with observational techniques and experimental simulations because of the challenges involved in study of organic matter in such ancient rocks. The opening section focuses on submarine vent systems that the majority of researchers postulate served as the cradle of life on Earth. In the following sections evidence for life in strongly metamorphosed rocks such as those in Greenland is evaluated and early ecosystems identified in the well preserved Barberton and Pilbara successions in Southern Africa and Western Australia. This volume will be of great value to graduate students and researchers interested in the origin and record of earliest life on Earth.

The quest for evidence of the earliest life forms on Earth has occupied geologists and palaentologists in earnest for the past 50 years after a somewhat rocky start in the late nineteenth and early twentieth century dogged by controversy surrounding the theory of evolution. It was recognized early that the first forms of life had to be primitive microbial forms with the ability to utilize inorganic compounds and convert them into organic matter and were thus unlikely to leave fossil traces as we know them from geologically younger rocks. Stromatolites are a potential exception to this generalisation and the most abundant biogenic structures identified in Archaean age rocks because of their distinctive morphologies. Unfortunately, laminated sedimentary structures can form in a variety of ways including seafloor precipitation such that the biogenicity of early Archaean stromatolites (>3.0 Ga) is highly debated due to the general lack of biosedimentary fabrics and organic remains of microfossils. In this volume new research on stromatolites and other microbially induced sedimentary structures are presented and evaluated in the context of previous work. Brasier documents the history of the search for Precambrian life and the numerous diversions along the way, which has led to the recognition that morphological complexity can develop abiologically in natural systems. Webb and Kamber show that high precision, high resolution trace element analysis coupled with other geochemical evidence may allow distinction between competing abiogenic and biogenic hypotheses for the origin of sedimentary structures inferred to be stromatolites or other microbialites on the basis of morphology and geological setting. A third contribution by Noffke summarizes her work on microbially induced sedimentary structures that result from the biotic-physical interaction of microbial mats with the sedimentary dynamics in sandy marine environments and are well represented in the 2.9 Ga Pongola Supergroup of South Africa.

Seafloor hydrothermal systems are postulated as the cradle of life on Earth because their chemistry favours the synthesis of reduced carbon compounds and present day thermophilic microbes such as archaea and sulfate reducing bacteria inhabiting deep sea vent environments sit at the base of the tree of life. Evidence of hydrothermal activity is widespread in the well preserved 3.5 to 3.2 Ga rocks of the Pilbara Craton of Western Australia and the Barberton Greenstone Belt in Southern Africa, which comprise a number of volcanic packages overlain in turn by thin units of clastic and chemical sedimentary rocks. This book focuses on such environments and the carbonaceous matter that occurs in both clastic and chemical sedimentary rocks of the Pilbara Craton and Barberton Greenstone Belt, most notably in thin chert units and transgressive bodies (chert dykes) formed by precipitation on the seafloor and replacement of other rock types. Golding et al. review the literature on the multiple sulfur isotope geochemistry of seafloor hydrothermal systems and sediments and present new data for the 3.24 Ga Sulphur Springs massive sulfide deposit and the 3.5 Ga Dresser Formation of the Pilbara Craton, which are respectively the earliest well preserved and documented black smoker and epithermal style seafloor hydrothermal systems on Earth. Hofmann's study of the Barberton Greenstone Belt in South Africa describes the Archean hydrothermal systems and their potential significance as a habitat for early life. Williams and her colleagues report their experimental work on the role of clay minerals particularly the smectite clays in the abiotic synthesis of organic molecules in seafloor hydrothermal systems. This is suggested as a possible mechanism for the beginning of life and the primordial birth of the first replicating cell.

Another group of papers focuses on the remains of organic matter preserved in the least thermally altered early Archean rocks of the Pilbara Craton and Barberton Greenstone Belt and demonstrates a possible relationship between co-existing abiotic and biological carbonaceous materials. Glikson et al. summarize their comprehensive studies of organic matter concentrates from these ancient rocks using sophisticated electron microscopical methodologies and geochemistry, as well as a comparison of this material with the remains of a modern hyperthermophilic microbe cultured under simulated hydrothermal conditions and thermally stressed. In another contribution De Gregorio and colleagues evaluate a variety of nano scale methods for establishing the biogenicity of ancient organic matter and associated putative microfossils and stromatolites. Furthermore, this group of researchers conducted laboratory Fischer-Tropsch-Type (FFT) experiments that produced abiotic carbonaceous matter, which they compared to carbonaceous matter obtained from ancient rocks, enhancing the distinction between biological and abiologically sourced organic matter.

The remaining two chapters in this volume discuss the Archean nitrogen and carbon isotope records and their use as tools in the search for evidence of early life.

Pinti and Hashizume show that nitrogen isotopes can be used for modeling the interplay of changing microbial metabolisms through early Earth history and related broadly to environmental changes including the progressive oxygenation of the Earth. On the other hand McCollom argues that carbon isotopes alone are insufficient to establish a biological origin for ancient organic matter because carbonaceous matter produced in FTT experiments can have similarly ^{13}C-depleted carbon isotopic compositions. The comparison of FTT carbonaceous matter synthesized under simulated hydrothermal conditions with thermally stressed microbial cultures and undisputedly ancient biogenic carbonaceous matter is a common thread of several chapters in this book, and an avenue for future research.

Finally, as editors of this volume, we would like to thank all the authors and co-authors and reviewers for their contributions and Springer Publishing Editor Petra van Steenbergen and her assistant Cynthia de Jonge for initially suggesting the topic for this book and their consistent support during its writing.

Suzanne D. Golding and Miryam Glikson

Contents

Contributors

Ms. Kim A. Baublys
School of Earth Sciences, University of Queensland, Brisbane, Queensland
4072, Australia

Prof. Martin D. Brasier
Department of Earth Sciences, University of Oxford, Parks Road, Oxford
OX1 3PR, UK
e-mail: martinb@earth.ox.ac.uk

Dr. Brandon Canfield
Department of Chemistry and Biochemistry, Arizona State University, Tempe,
Arizona 85287, USA

Dr. Bradley T. De Gregorio
NASA Johnson Space Center, Mail Code KT,
2101 NASA Parkway Houston, TX 77058, USA
e-mail: bradley.t.degregorio@nasa.gov

Dr. Jeffrey M. Dick
Department of Chemistry and Biochemistry, Arizona State University, Tempe,
Arizona 85287, USA

Dr. Lawrence J Duck
School of Earth Sciences, University of Queensland, Brisbane, Queensland
4072, Australia

Dr. Christopher R. Glein
School of Earth and Space Exploration, Arizona State University,
Tempe, Arizona 85287, USA

Dr. Miryam Glikson
P.O. Box 553, Palmwoods,
Queensland 4555, Australia
e-mail: miryam7@bigpond.com

Prof. Suzanne D. Golding
School of Earth Sciences, University of Queensland, Brisbane, Queensland 4072, Australia
e-mail: s.golding1@uq.edu.au

Dr. Hilairy E. Hartnett
School of Earth and Space Exploration & Department of Chemistry and Biochemistry, Arizona State University, Tempe, Arizona 85287, USA

Dr. Ko Hashizume
Department of Earth & Space Sciences, Graduate School of Science, Osaka University, Toyonaka, Osaka 560-0043, Japan
e-mail: kohash@ess.sci.osaka-u.ac.jp

Dr. Arthur H. Hickman
Geological Survey of Western Australia, 100 Plain Street, East Perth, Western Australia 6004, Australia

Dr. Axel Hofmann
Palaeoproterozoic Mineralization Research Group, Department of Geology, University of Johannesburg, Johannesburg 2006, South Africa
e-mail: hofmannaxel@hotmail.com

Prof. John R. Holloway
School of Earth and Space Exploration & Department of Chemistry and Biochemistry, Arizona State University, Tempe, Arizona 85287, USA

Prof. Balz S. Kamber
Department of Earth Sciences, Laurentian University, 935 Ramsey Lake Road, Sudbury, Ontario P3E 2C6, Canada

Dr. Thomas M. McCollom
CU Centre for Astrobiology & Laboratory for Atmospheric and Space Physics, University of Colorado, Campus Box 392, Boulder, CO 80309-0392, USA
e-mail: mccollom@lasp.colorado.edu

Prof. Nora Noffke
Department of Ocean, Earth & Atmospheric Sciences, Old Dominion University, Norfolk, VA 23529, USA
e-mail: nnoffke@odu.edu

Prof. Daniele L. Pinti
GEOTOP and Département des Sciences de la Terre et de l'Atmosphère, Université du Québec à Montréal, CP 8888, Centre-Ville, Montréal, Quebec H3C 3P8, Canada
e-mail: pinti.daniele@uqam.ca

Dr. Ahmed I. Rushdi
COGER, College of Agricultural and Food Sciences, King Saud University, P.O. Box 2460, Riyadh 11452, Saudi Arabia

Prof. Thomas G. Sharp
School of Earth and Space Exploration, Arizona State University,
P.O. Box 871404, Tempe, Arizona 85287-1404, USA

Prof. Everett L. Shock
School of Earth and Space Exploration & Department of Chemistry
and Biochemistry, Arizona State University, Tempe, Arizona 85287, USA

Prof. Bernd R.T. Simoneit
COGER, College of Agriculture and Food Science, King Saud University,
P.O. Box 2460, Riyadh 11452, Saudi Arabia
Oregon State University, Corvallis, OR, USA

Prof. Lynda B. Williams
School of Earth and Space Exploration, Arizona State University, Tempe,
Arizona 85287, USA
e-mail: lynda.williams@asu.edu

Dr. Gregory E. Webb
School of Natural Resource Sciences, Queensland University of Technology,
GPO Box 2434, Brisbane, Queensland 4001, Australia
e-mail: ge.webb@qut.edu.au

Ms. Robyn E. Webb
Centre for Microscopy and Microanalysis, University of Queensland,
Brisbane, Queensland 4072, Australia

Dr. Elisa Young
School of Earth Sciences, University of Queensland, Brisbane,
Queensland 4072, Australia

Dedication to John F. Lindsay

22 January 1941–20 June 2008

John Lindsay was a visiting scientist at the Lunar and Planetary Institute, Houston, Texas up until his death on 20 June 2008 at the age of 67. John obtained BSc and MSc degrees in geology from the University of New England in northern NSW in 1962 and 1964, respectively, and was awarded a PhD in geology from the Ohio State University in 1968. He joined the Apollo Program in Houston with NASA in 1969 and was involved in planning of the lunar missions and scientific training of astronauts. John had worked in Antarctica during his graduate program at Ohio State University that provided the impetus for organisation of the first Mars analogue expedition to the Antarctic Dry Valleys in 1972. John received a NASA Achievement Award for his work on the Apollo Lunar Program and the United States Polar Medal for Antarctic Service. These early years with NASA set the tone for much of John's professional career where he used modern analogues to understand the early history of the Earth and the Earth's extreme environments as analogues for other planetary environments.

During the 1970s and early 1980s John held academic positions at La Trobe University in Victoria and the Marine Science Institute at the University of Texas in Galvaston and was employed by the Exxon Production Research Company in Houston. He joined the then Bureau of Mineral Resources in Canberra as a principle research scientist in 1984 and worked on Australia's resource rich sedimentary basins over the next 15 years. John continued to publish extensively and received the Australian Institute of Cartographers Award for his work on the Amadeus Basin of central Australia. It was during this period that John became convinced of the importance of global tectonics in early biosphere evolution and the need for a more scientific approach to the study of early life on Earth, which he explored in collaboration with the University of Oxford's Professor Martin Brasier.

John returned to the United States in 2002. He worked first with the Astrobiology Group at the NASA Johnson Space Centre and in 2004 accepted a position as visiting scientist at the Lunar and Planetary Institute where he had worked so many years before. John continued his research on the origins of life on Earth and concluded that prebiotic processes dominate the Paleoarchean record of the Pilbara Craton of Western Australia and the Kaapvaal Craton of southern Africa. This was a controversial view at odds with the collective view of many earth scientists that life developed prior to 3.8 Ga and was widespread by 3.5 Ga. Nevertheless, the work of several authors in this volume supports John's view and further suggests these strata may preserve a record of the prebiotic-biotic transition. During his career John published some 200 scientific papers, a text book on lunar stratigraphy and sedimentology and a number of book chapters and showed a generation of earth and planetary scientists that there is no substitute for careful documentation and critical analysis of the rock record.

Introduction

Miryam Glikson and Suzanne D. Golding

Abstract The major questions as to the origin of life on Earth, when, where, how and what are the subject of this chapter and book. The exact timing cannot be determined as a review of previous work shows that carbonaceous matter with recognizable source-related structures has not been found prior to 3.5 Ga. As to where, scientists are in agreement that the first life forms developed and emerged in the marine environment where carbonaceous compounds derived from mantle venting and potentially catalytic minerals were available. Modern analogues can be found today in seafloor hydrothermal vent environments where diverse chemoautotrophic microbial communities occur that include methanogens and methanotrophs. Similar unicellular microbial forms have been observed and studied from 3.5 Ga rocks in Western Australia and South Africa. Interestingly, side by side with recognizable structures of microbial cells, amorphous carbonaceous matter (CM) is prominent as coating of the mineral matrix, and may record the co-existence of primeval abiotic CM and some of the earliest biologically-formed cells. These findings place the early Archean as the transition or first step from the abiotic world to the biological world. Between 3.5 and 3.2 Ga the microbial communities have further developed and evolved to form the South African Barberton and the West Australian Sulphur Springs assemblages of filamentous, coccoid and spindle shaped forms comparable to present-day primitive microbes. On the other hand, additional criteria are required in rocks older than 3.0 Ga to determine the biogenicity of sedimentary structures inferred to be stromatolites in the absence of biosedimentary fabrics and organic remains of microfossils.

Keywords Archaean carbonaceous matter • Oldest microbial remains • Abiotic-formed carbonaceous matter • Stromatolite

M. Glikson (✉)
P.O. Box 553, Palmwoods, Queensland 4555, Australia
e-mail: miryam7@bigpond.com

S.D. Golding
School of Earth Sciences, University of Queensland, Brisbane, Queensland 4072, Australia

S.D. Golding and M. Glikson (eds.), *Earliest Life on Earth: Habitats, Environments and Methods of Detection*, DOI 10.1007/978-90-481-8794-2_1,

The origin of life on Earth and the nature of the earliest life forms have been topics of great interest to scientists and the wider community since the publication of Charles Darwin's *On the Origin of Species* in 1859. Darwin recognized that the lack of fossils prior to the Cambrian Period posed a problem for his theory of evolution (Schopf 2009), and appreciated that the earliest life forms were most likely to be primitive microbes (O'Malley 2009). On 1 February 1871, in a letter to Joseph Dalton Hooker, his closest scientific collaborator, he speculated that life may have originated in "some warm little pond with all sorts of ammonia and phosphoric salts, light, heat, electricity etc. present (so) that a protein compound was chemically formed, ready to undergo still more complex changes" (Pereto et al. 2009). Notwithstanding these early insights and the pioneering work of the American paleontologist Charles Doolittle Walcott in the late nineteenth and early twentieth centuries, it was not until the 1950s that the inorganic synthesis of amino acids in the famous Miller-Urey experiments (Miller 1953; Miller and Urey 1959) and the discovery of structurally preserved microorganisms in the 1.9 Ga Gunflint Formation of Ontario (Tyler and Baghoorn 1954) stimulated a new generation of scientists to focus on the origin and record of earliest life on Earth. Scientists today are in agreement that the first forms of life on Earth were simple anaerobic chemoautotrophic microbes that are found today in niches of extreme marine habitats such as seafloor hydrothermal systems that existed on Earth as soon as liquid water accumulated more than 4.2 billion years ago (Martin et al. 2008).

While microbial life may have existed during the Hadean Eon (4.38–3.85 Ga), all evidence is likely to have been erased by the 3.9–3.8 Ga late heavy bombardment of Earth coupled with crustal recycling (Kamber et al. 2001; Abramov and Mojzsis 2009; Jorgensen et al. 2009). The earliest Archean rocks of Greenland and Canada have also been heavily deformed and metamorphosed, which has destroyed their geological context and microstructural characteristics that might have provided support for a biogenic origin of the graphite in these rocks (Fedo and Whitehouse 2002; Van Zuillen et al. 2002; Moorbath 2005). Attention has therefore shifted to the well preserved, 3.5–3.2 Ga, volcano-sedimentary successions of the Pilbara Craton of Western Australia and the Barberton Greenstone Belt of South Africa and Swaziland. Nevertheless, the bacterial fossils of the 1.9 Ga Gunflint Formation (Barghoorn and Tyler 1965) still constitute the gold standard evidence of Precambrian life with many discredited reports of stromatolites and microfossils in Archean strata older than 3.0 Ga (e.g. Brasier et al. 2002, 2004). Furthermore fossil-like mineralized forms such as have been recorded from Archean rocks and some meteorites have also been produced in laboratory experiments in recent years (Garcia-Ruiz et al. 2003). Mineral precipitates mimicking biological structures are not uncommon as end products of cooling saturated solutions of calcite, silica, iron and other metals (Horodyski 1981). Mineralized fossil-like forms have been reported, for example, from the Martian meteorite ALH84001 (McKay et al. 1996) and from the early Archean of the Pilbara in Western Australia (e.g. Schopf and Packer 1987; Schopf 1993). The characteristics of these forms and their structures are the uniformity in shape and precision in repetitive sub-structures. The biological invalidity of these pseudo-fossils has been outlined in a detailed paper by Brasier et al. (2006) and as expressed in the same document "are the self-organizing structures that do not pass the test for biogenicity".

Among the fossil-like forms reported over the last 40 years from outcrops in the early Archean rocks of the Pilbara of Western Australia are finely laminated light and dark undulating, wrinkled and folded mineralized structures regarded as fossil remains of stromatolite bioherms such as are found today in some shallow marine environments. However, some of these structures resemble stromatolites in the domal part only, which when found as a separate fragment may give the impression of a stromatolite morphology. Photosynthesizing cyanobacteria are the dominant microbes that build these structures by depositing layer upon layer upwards towards the light in the photic zone of a relatively shallow marine habitat. Some of the early Archean structures are wavy and folded downwards as if building by these microorganisms was upside-down at times, which is impossible of course. When following these structures for a distance towards ground level the laminations straighten out and become parallel fine light and darker bands. Such structures are a typical occurrence today around hot springs as sinter deposits and also forming from evaporite precipitation (Lowe 1994). The same study suggested the domal structures in the 3.5–3.2 Ga Warrawoona Group rocks in Western Australia and the 3.5–3.3 Ga Onverwacht Group in South Africa resulted from "soft sediment deformation of originally flat layers".

These types of structures may also form where hydration of a mineral has occurred such as the anhydrite in the Dead Sea varved marls (Fig. 1) where similar

Fig. 1 The Dead Sea varves consist of annual couplets or triplets of fine detritus, aragonite and gypsum/anhydrite that are generally flat lying in the subsurface. The hydration of anhydrite at the surface results in expansion that distorts bedding and produces a similar macro-morphology to some stromatolites. Outcrop photograph taken a few 100 m west of the Dead Sea shoreline by Miryam Glikson

distortions are evident. In these instances expansion of a mineral occurs when absorbing water resulting in folding and wrinkling. In the Dead Sea marls the anhydrite was converted to gypsum upon reaching the surface. In the case of silica and iron precipitates, hydration of the silica and/or oxidation of the iron may bring on changes in volume resulting in contortions and distortion in the form of wrinkles and folds, as well illustrated in Lindsay et al. (2005) and documented from studies of terrestrial hot springs (Jones and Renaut 2003). In a similar vein, Grotzinger and Rothman (1996) presented a purely abiotic model for growth of stromatolitic features during times when precipitation at the sea floor was perhaps common. Conophyton-like occurrence studied by Batchelor et al. (2004) from the much younger 1.7 Ga Dungaminnie Formation in the Northern Territory, Australia throws light on these striking bioherms. Although superficially similar structures have been documented recently from the Pilbara of Western Australia (Allwood et al. 2006, 2007), it is yet to establish their internal structure to be able to confirm a similarity to true Conyphyton, as well as sub-surface occurrences in drill core. Sedimentary structures whether directly or indirectly microbially linked are generally easier to reconcile in sediments younger than 3.2 Ga. A concise analysis of such occurrences is presented by Noffke (this volume).

True bioherm building stromatolites most probably came later with the majority of Archean stromatolite reports from 3.0 to 2.5 Ga strata, most notably in sediments of the Transvaal Supergroup and Insuzi Groups of South Africa and Zimbabwe and the Fortescue Group of Western Australia (Beukes and Lowe 1989; Kamber and Webb 2001; Kamber et al. 2004; Schopf et al. 2007). This is not to say that anoxygenic photosynthesizing mat-forming cyanobacteria (e.g. Oscillatoria; Cohen et al. 1975) have not been around earlier. Indeed some of Pflug's (1966) findings from the Barberton closely resemble the cyanobacterium Oscillatoria. These were among the oldest identifiable structures of microorganisms comparable to known life forms. The rocks that yielded these microbial remains were described and documented by Pflug from the 3.2 Ga shales of the Fig Tree Group in the Barberton Greenstone Belt of South Africa. The assemblage contains also various coccoid forms suggestive of sulfate reducers and methanogens. Coccoid and spindle-shaped microbes closely resembling the forms documented by Pflug have also been recorded by Westall et al. (2001) and Westall (2004) from the Barberton Greenstone Belt. Westall notes "colonial association of filamentous and coccoid forms which may form thin coatings or thick mats on the surfaces of rocks". Noffke et al. (2003, 2006) documented microbial mats from the 3.2 Ga Moodies Group, South Africa.

3.2 Ga also seems to be a turning point in the appearance of fossil remains of microbial communities in carbonaceous materials that can be compared to known groups living in today's specialized habitats. In the Pilbara's 3.24 Ga Sulphur Springs black smoker deposit Duck et al. (2007) recorded the remains of filamentous microbial mats and spherical forms where both the individual filaments and spherical entities closely resemble the Barberton microbial remains recorded by Pflug and Westall from rocks of a comparable age. Going further back in time, a gap of about 200 million years is evident, where no carbonaceous material (CM) has been preserved in the rock record until the ca. 3.46–3.43 Ga Strelly Pool and

Apex Basalt Formations in the Pilbara of Western Australia where preserved CM has been studied. The 3.46 Ga Apex Basalt has been studied both by observational as well as analytical methods by a number of scientists. Observational studies documented the imprints of microbial filaments on rock surfaces (Schopf 1993; Schopf et al. 2007). However, studies of the Apex Chert CM (De Gregorio and Sharp 2003; De Gregorio et al. 2005) and the Strelley Pool CM (De Gregorio and Sharp 2006) using observational and microanalytical techniques also showed filamentous CM coating quartz grains. Based on their comparative studies of abiotically synthesized CM and biogenic remains, De Gregorio and his colleagues suggested the possibility of co-existence of abiotic CM and biogenic material in the same rocks. Binet et al. (2003) in an attempt to clarify the source of CM in early Archean Dresser Formation cherts using EPR concluded that the CM may have a biogenic origin. Philippot et al. (2006) reported presence of both abiotic and biological indicators in their study of the 3.49 Ga Dresser Formation CM. Interestingly, Hofmann and Bolhar (2007) in their extensive and detailed study of the early Archean Barberton Greenstone Belt cherts that contained CM noted that the association of CM was not always with ultramafic rocks that can account for serpentinization processes as its source. Some cherts according to the same study were associated with felsic rocks. Their concluding comments were that it is unlikely that all CM in the Barberton Greenstone Belt is abiogenically formed. Hofmann and Bolhar suggest the presence of biogenic and abiogenic CM at the same time.

Our studies of Apex Basalt CM (Glikson et al. this volume) using argillite-CM core that has been intruded by peridotite at >1,000°C found optically isotropic graphite (in contrast to anisotropic standard graphite), which when observed at high magnification using High Resolution TEM (HRTEM) revealed nano-tubes, fullerenes and clusters of spheres enclosed within a perfectly formed wall closely resembling such as documented from meteorites by Harris and Vis (2003) and Nakamura-Messenger et al. (2006). There could be little doubt that the perfect spherical structures were formed abiotically at very high temperature. Whether the graphite and fullerenes were formed from primary biogenic material is not possible to establish but the possibility cannot be entirely ruled out. At such high temperatures no primary features may be retained as the CM solidified and crystallized from a liquid or vapor phase. On the other hand, our study of CM from the 3.49 Ga Dresser Formation in the Pilbara of Western Australia and the 3.47 Ga Hooggenoeg Formation in the Barberton Greenstone Belt in South Africa found primary CM within fluid inclusions in quartz grains (Glikson et al. 2008). Spherical and oval typical cell structures ranging from 0.3–0.6 μm across were apparently released when demineralization of the host rock was carried out in order to concentrate the CM. The hollow vesicular cells must have been entombed within inclusions trapped in quartz crystals. Their entombment has preserved them without there having been subjected to the severe thermal degradation as have the rest of the colonies. Thermally degraded microbial remains within the same rocks were compared to thermally stressed hypothermophile archaea cultures from today's deep sea hydrothermal environment (Glikson et al. 2008). Thermal degradation morpho-textures and structures when compared were found to be identical. The degradation process and its resulting morphological structures were indistinguishable between the microbes subjected to simulated thermal stress and those recovered from the Dresser and Hooggenoeg Formations. The living cultures were dying between 80°C and 100°C and began degradation and disintegration at 132°C.

The main question arising is how can we distinguish analytically and/or visually between the natural presumed biologically formed remains in earliest Archean rocks and abiotically synthesized CM. Thermal degradation ruptures and tears apart cells, which only occasionally retain their outline as observed in both the simulated microbial cultures and the fossil remains in the Dresser Formation (Glikson et al. 2008). Therefore the two possibilities, namely the presence of abiotic and/or biological CM have to be considered and evaluated. Abiotic formation of organic compounds in hydrothermal systems has been shown to occur during serpentinization processes through hydration of ultramafic silicates (Berndt et al. 1996; Shock and Schulte 1998; Horita and Berndt 1999; Sherwood Lollar 2004) resulting in methane and hydrogen release. Simoneit (1995) in his extensive studies of CM in active hydrothermal systems noted the probability of potential synthesis reactions. The same study noted, however, that any such CM would be masked in the present hydrothermal environments by the vast amount of degradation products from biological precursors as recently confirmed by Bassez et al. (2009) from analysis of peridotites from active seafloor hydrothermal sites. On the other hand Lindsay et al. (2005) suggest the abiotic output may overwhelm the signatures of primitive life in early Archean hydrothermal systems.

Laboratory simulation of processes involved in abiogenic synthesis or Fischer-Tropsch-type reactions were conducted under hydrothermal conditions by several groups (McCollom et al. 2001; Rushdi and Simoneit 2005; McCollom and Seewald 2006). Such processes involve the reduction and polymerization of CO or CO_2 emanating from seafloor vents with H_2 dissociated from water and varied metal catalysts. An array of organic compounds resulted from the experiments including lipids, most of which could be building blocks for assemblage of a primordial cell and/or a substrate for metabolic processes of earliest microbial communities. McCollom and Seewald's (2006) experiments showed predominance of homologous n-alkanes. The common factor in these experiments is temperature, which is always above 100°C that is also approaching the limit of life (Russell 2003; Glikson et al. 2008; Nitchke and Russell 2009). However, this does not necessarily exclude the building blocks for primary cell assemblage being produced at >100°C by chemical reactions in hydrothermal systems. Thermodynamic calculations by Amend and Shock (1998) concluded that "autotrophic synthesis of all 20 protein-forming amino acids was energetically favored at 100°C moderately reduced submarine hydrothermal solutions".

Carbonaceous matter delivered to Earth by meteorites has been discussed and suggested as a source of pre-biotic building blocks for earliest life forms. Meteorites contain a variety of organic compounds delivered to Earth in large quantities during intense bombardment prior to 3.8 Ga (Chyba et al. 1990; Rollinson 2007; McCollom, this volume). Amino acids in significant quantities have been reported from several meteorites such as the Orgueil, Murchison, Allenede and Tagish Lake although many of the amino acids identified are rare or non-existent on Earth

(Cronin and Pizzarello 1997; Pizzarello et al. 2005). Some of the extra-terrestrial amino acids have been suggested over the years as the building blocks of the first life forms on Earth. However, even though they may have formed in cool conditions (Rollinson 2007), they would have experienced high temperature incoming to Earth. The high temperature that the amino acids were subjected to would render them chemically non-reactive, like graphite the inert end product of CM thermal maturation. A very high temperature effect on the meteorite amino acids is supported by the significant enrichment in $\delta^{13}C$ (Pizzarello et al. 2003).

Although gaseous generation of hydrocarbons such as methane are known to form abiotically, studies of hydrothermal petroleum forming today showed generally that the petroleum originated from thermal diagenesis of microorganisms (Venkatasan et al. 2003; Simoneit et al. 2004, 2008). According to these studies "A search for molecular evidence for abiogenic compounds (in hydrocarbons) was negative, except methane". Similar conclusions were reported from serpentinization processes by Horita and Berndt (1999) and Sherwood Lollar (2004). The most likely environment for the assembly of a cell, the prototype of a chemoautotrophic microbe is the early Archean seafloor where venting of hot fluids containing CO, CO_2, H_2S and metals was a common product from degassing of the mantle (Shock et al. 1995; Shock and Schulte 1998). The self organization into vesicles has been studied and demonstrated in laboratory experiments using lipids, in simulated hydrothermal conditions (e.g., Hazen 2006; Simoneit et al. 2007). Hazen (2006) suggested "lipid self-organization proceeds spontaneously at a critical concentration of amphiphilic molecules, but vesicle formation may be affected dramatically by the presence of minerals". The question arising from these experiments is what is the ideal mineral for this kind of process?

Hanczyc et al. (2003) demonstrated experimentally the encapsulation, growth and division of a cell compartment. The same study used lipids for membrane formation at a specific pH. It was shown that montmorillonite accelerated the conversion of the lipidic matter into vesicles. Montmorillonite is also suggested by Ferris (2005) as playing a central role in the evolution of life by its adsorptive capacity and catalytic properties. Earlier studies of Lahav (1975) and Lahav et al. (1978) suggested montmorillonite as a key factor in the interaction of water and amino acids in the prebiotic synthesis. Recently, Williams et al (2005; and this volume) simulated seafloor hydrothermal conditions in reacting montmorillonite, saponite and illite clays with dilute methanol as a carbon source. Various aromatic and polycyclic aromatic hydrocarbons were produced in smectite clays, but not in illitic clays. Williams et al.'s experiments showed that organic molecules up to C20 were formed, and the smectite protected them from thermal destruction.

The major questions as to the beginning of life on Earth, when, where, how and what are the subject of this book. The exact timing cannot be determined since no carbonaceous matter with any source-related recognizable structures has been found prior to 3.5 Ga. Carbonaceous matter found in 3.8 Ga rocks in Greenland is graphite (Fedo and Whitehouse 2002; Van Zuillen et al. 2002), which by definition does not retain any primary structures as a result of precipitation from vaporized carbon. Processes leading to the first microbial forms of life may have initiated between 3.8 and 3.5 Ga or even earlier. As to where, scientists are in agreement that the first life forms developed and emerged in the marine environment where fundamental organic compounds emanating from mantle venting and/or atmospheric reactions were available. Such seafloor hydrothermal environments can be compared to those at today's spreading centers. Black smokers were suggested by Wachtershauer (2007) as the chemoautotrophic origin of life where the volcanic iron-sulfur system enabled the formation of a pioneer microorganism. The latter is postulated to have formed from an inorganic substrate of a catalyctic metal centre promoting C-fixation in an encasing superstructure. Martin et al. (2008) on the other hand suggests off-axis white smoker vents as the cradle of life forms due to their ultramafic substrate, which closely resembles the material that erupted onto the early Earth ocean floor. Off-axis white smokers such as the modern Lost City hydrothermal field contain high concentrations of dissolved H_2, CH_4 and other low molecular hydrocarbons, the basic materials for the metabolism of anaerobic chemoautotrophs such as archaea. These compounds lodged within protective clay sheets or sulfide bubbles may have slowly encapsulated into vesicular cell-like entities. To quote Hanczyc et al. (2003) "These experiments constitute a proof-of-principle demonstration that vesicle growth and division can result from simple physico-chemical forces, without any complex biochemical machinery". It is interesting though that experimentally produced vesicles or proto-cells are unnaturally perfect in shape, the perfect spheres, both in the above mentioned experiments as well as earlier ones (Yanagawa and Kojima 1985). On the other hand, the vesicular cell-like structures released from fluid inclusions in quartz crystals in the 3.5 Ga Dresser Formation rocks, Pilbara, Western Australia (Glikson et al 2008) vary a lot in outline with hardly two cells being exactly the same in shape. There is no absolute perfection in natural biological forms, whether pollen, spores, microbial cells or other living things, as worded in Velasco-Santos et al. (2003) "nature does not have a strict control of conditions as synthetic routes".

References

Abramov O, Mojzsis S (2009) Microbial habitability of the Hadean Earth during the late heavy bombardment. Nature 459:419–422

Allwood AC, Walter MR, Kamber BS et al (2006) Stromatolite reef from the early Archean era of Australia. Nature 441:714–718

Allwood AC, Walter MR, Burch IW et al (2007) 3.43 billion-year-old stromatolite reef from the Pilbara Craton of Western Australia: ecosystem-scale insights to early life on Earth. Precambr Res 158(3–4):198–227

Amend JP, Shock EL (1998) Energetics of amino acid synthesis in hydrothermal ecosystems. Science 281:1859–1662

Barghoorn ES, Tyler SA (1965) Microorganisms from the Gunflint chert. Science 147:563–577

Bassez M-P, Takano Y, Ohkouchi N (2009) Organic analysis of peridotite rocks from the Ashadze and Logatchev hydrothermal sites. Int J Mol Sci 10:2986–2998

Batchelor MT, Burne RV, Henry BI (2004) The case for biotic morphogenesis of coniform stromatolites. Phys A-Stat Mech Appl 337:319–326

Berndt ME, Allen DE, Seyfried WE (1996) Reduction of CO_2 during serpentinization of olivine at 300°C and 500 bar. Geology 24:351–354

Beukes NJ, Lowe DR (1989) Environmental control on diverse stromatolite morphologies in the 300o Ma Pongola Supergroup, South Africa. Sedimentology 36:383–397

Binet L, Skrzypezak A, Gouvier D (2003) Origin of the organic matter in the oldest Archean cherts; an EPR insight. Lunar and Planetary Science Conference XXXIV, Houston

Brasier MD, Green OR, Jephcoat AP et al (2002) Questioning the evidence for Earth's oldest fossils. Nature 416:76–81

Brasier M, Green O, Lindsay J et al (2004) Earth's oldest (3.5Ga) fossils and the 'early Eden hypothesis': questioning the evidence. Orig Life Evol Biosph 34:257–269

Brasier M, MCLoughlin N, Green O et al (2006) A fresh look at the fossil evidence for early Archean cellular life. Phil Trans R Soc B 361:887–902

Chyba CF, Thomas PJ, Brookshaw L et al (1990) Cometary delivery of organic molecules to the early Earth. Science 249:366–373

Cohen Y, Jorgensen BB, Paden E et al (1975) Sulphide-dependent anoxygenic photosynthesis in the cyanobacterium Oscillatoria limnetica. Nature 257:489–492

Cronin JR, Pizzarello S (1997) Enantiomeric excesses in meteorite amino acids. Science 275:951–955

De Gregorio BT, Sharp TG (2003) Determining the biogenicity of microfossils in the Apex chert, Western Australia, using Transmission Electron Microscopy. Lunar and Planetary Science Conference XXXIV, Houston

De Gregorio BT, Sharp TG (2006) Possible abiotic formation of kerogen-like carbon in the Strelley Pool chert. Lunar and Planetary Science Conference XXXVII, Houston

De Gregorio BT, Sharp TG, Flynn GJ (2005) A comparison of the structure and bonding of carbon in Apex chert kerogenous material and Fischer-Tropsch-type carbons. Lunar and Planetary Science Conference XXXVI, Houston

Duck LJ, Glikson M, Golding SD et al (2007) Microbial remains and other carbonaceous forms from the 3.24 Ga sulphur springs black smoker deposit, Western Australia. Precambr Res 154:205–220

Fedo CM, Whitehouse MJ (2002) Metasomatic origin of quartz-pyroxene rock, Akilia, Greenland, and implications for Earth's earliest life. Science 296:448–1452

Ferris JP (2005) Mineral catalysis and pre-biotic synthesis: montmorillonite-catalyzed formation of RNA. Elements 1:145–149

Garcia-Ruiz JM, Hyde ST, Carnerup AM et al (2003) Self-assembled silica-carbonate structures and detection of ancient microfossils. Science 302:1194–1197

Glikson M, Duck LJ, Golding SD et al (2008) Microbial remains in some earliest Earth rocks: comparison with a potential modern analogue. Precambr Res 164:187–200

Grotzinger JP, Rothman DH (1996) An abiotic model for stromatolite morphogenesis. Nature 383:423–425

Hanczyc MM, Fujikawa SM, Szostak JW (2003) Experimental models of primitive cellular compartments. Encapsulation, growth and division. Science 302:618–622

Harris PJF, Vis RD (2003) High-resolution transmission electron microscopy of carbon and nanocrystals in the Allende meteorite. Proc R Soc Lond A 459:2069–2076

Hazen RM (2006) Mineral surfaces and the prebiotic selection and organization of biomolecules. Am Miner 91:1715–1729

Hofmann A, Bolhar R (2007) Carbonaceous cherts in the Barberton Greenstone Belt and their significance for the study of early life in the Archean record. Astrobiology 7:355–388

Horita J, Berndt ME (1999) A biogenic methane formation and isotopic fractionation under hydrothermal conditions. Science 285:1055–1057

Horodyski RJ (1981) Pseudomicrofossils and altered microfossils from a middle Proterozoic shale, Belt Supergroup, Montana. Precambr Res 16:143–154

Jones B, Renaut RW (2003) Hot spring and geyser sinter – the integrated conduct of precipitation, replacement and deposition. Can J Earth Sci 40:1549–1569

Jorgensen UG, Appel PWU, Hatsukawa Y (2009) The Earth-Moon system during the late heavy bombardment period- geochemical support for impacts dominated by comets. Icarus 204:368–380

Kamber BS, Webb GE (2001) The geochemistry of late Archean microbial carbonate: implications for ocean chemistry and continental erosion history. Geochim Cosmochim Acta 65:2509–2525

Kamber BS, Moorbath S, Whitehouse MJ (2001) The oldest rocks on Earth; time constraints and geological controversies. Geol Soc Spec Publ 290

Kamber BS, Bolhar R, Webb GE (2004) Geochemistry of late Archean stromatolites from Zimbabwe: evidence for microbial life in restricted epicontinental seas. Precambr Res 132:379–399

Lahav N (1975) Simultaneous existence of different environments in aqueous clay systems and its possible role in prebiotic synthesis. J Mol Evol 5(3):243–247

Lahav N, White D, Chang S (1978) Peptide formation in the pre-biotic Earth: thermal condensation of glycine in fluctuating clay environments. Science 201:67–69

Lindsay JF, Brasier MD, McLoghlin N et al (2005) The problem of deep carbon- an Archean paradox. Precambr Res 143:1–22

Lowe D (1994) Abiological origin of described Stromatolites older than 3.2 Ga. Geology 22:287–390

Martin W, Baross J, Kelley D et al (2008) Hydrothermal vents and the origin of life. Nat Rev 6:805–814

McCollom TM, Seewald JS (2006) Carbon isotope composition of organic compounds produced by abiotic synthesis under hydrothermal conditions. Earth Planet Sci Lett 243:74–84

McCollom TM, Seewald JS, Simoneit BRT (2001) Reactivity of monocyclic aromatic compounds under hydrothermal conditions. Geochim Cosmochim Acta 65:455–468

McKay DS, Gibson EK Jr, Thomas-Keprta KL et al (1996) Search for past life on Mars: possible relic biogenic activity in Martian meteorite ALH84001. Science 273:924–930

Miller SL (1953) A production of amino acids under possible primitive Earth conditions. Science 117:528–529

Miller SL, Urey HC (1959) Organic compound synthesis on the primitive Earth. Science 130:245–251

Moorbath S (2005) Dating earliest life. Nature 434:155

Nakamura-Messenger K, Messenger S, Keller P et al (2006) Organic globules in the Tagish Lake meteorite remnants of the protosolar disc. Science 314:1439–1440

Nitchke W, Russell M (2009) Hydrothermal focusing of chemical and chemiosmotic energy, supported by delivery of catalytic Fe, Ni, Mo/W, Co, S and Se, forced life to emerge. J Mol Evol 69:481–496

Noffke N, Hazen RM, Nhleko N (2003) Earth's earliest microbial mats in a siliciclastic marine environment (2.9 Ga Mozaan Group, South Africa). Geology 31:673–676

Noffke N, Ericson KA, Hazen RM et al (2006) A new window into early Archean life: microbial mats in Earth's oldest siliciclastic tidal deposits (3.2 Ga Moodies Group, South Africa). Geology 34:253–256

O'Malley MA (2009) What did Darwin say about microbes, and how did microbiology respond. Trends Microbiol 17:341–347

Pereto J, Bada JL, Lazcano A (2009) Charles Darwin and the origin of life. Orig Life Evol Biosph 39:395–406

Pflug HD (1966) Structured organic remains from the Fig Tree series of the Barberton mountainland. Univ Witwatersrand Econ Geol Res Unit Info Circular 28:1–17

Philippot P, Lepot K, Thomas C et al (2006) Abiotic versus biotic carbonaceous matter in 3525-Myr-old hydrothermally altered sub-seafloor sediments (Pilbara Craton, Western Australia). Abstract EGU 6, Vienna

Pizzarello S, Huang Y, Fuller M (2003) The carbon isotopic distribution of the Murchison amino acids. Lunar and Planetary Science Conference XXXIV, Houston

Pizzarello S, Huang Y, Becker L et al (2005) The organic content of the Tagish Lake meteorite. Geochim Cosmochim Acta 69:599–605

Rollinson H (2007) Early Earth systems: a geochemical approach. Wiley-Blackwell Publishing, Oxford

Rushdi AI, Simoneit BRT (2005) Abiotic synthesis of organic compounds from carbon disulfide under hydrothermal conditions. Astrobiology 5(6):749–768
Russell MJ (2003) Origin and evolution of life: clues from ore deposits. Appl Earth Sci 112:177–178
Schopf JW (1993) Microfossils of the early Archean Apex chert: new evidence of the antiquity of life. Science 260:640–646
Schopf JW (2009) The hunt for Precambrian fossils: an abbreviated genealogy of the science. Precambr Res 173:4–9
Schopf JW, Packer BM (1987) Early Archean (3.3 to 3.5 billion year-old) microfossils from Warrawoona Group, Australia. Science 237:70–73
Schopf JW, Kudryavtsev AB, Czaja AD (2007) Evidence of Archean life: stromalites and microfossils. Precambr Res 158:141–155
Sherwood Lollar B (2004) Life's chemical kitchen. Science 134:272–273
Shock EL, Schulte MO (1998) Organic synthesis during fluid mixing in hydrothermal systems. J Geophys Res 103:28513–28517
Shock EL, McCollom T, Schulte MD (1995) Geochemical constrains on chemolithotrophic reactions in hydrothermal systems. Orig Life Evol Biosph 25:141–159
Simoneit BRT (1995) Evidence for organic synthesis in high temperature aqueous media, facts and prognosis. Orig Life Evol Biosph 25:119–140
Simoneit BRT, Lein AY, Perespkyn VI et al (2004) Composition and origin of hydrothermal petroleum and associated lipids in the sulfide deposits of the Rainbow field (Mid-Atlantic Ridge at 36°N). Geochim Cosmochim Acta 68:2275–2294
Simoneit BRT, Rushdi AI, Deamer DW (2007) Abiotic formation of acyglycerols under simulated hydrothermal conditions and self-assembly properties of such lipid products. Adv Space Res 40:1649–1656
Simoneit BRT, Deamer DW, Komparichako V (2008) Characterization of hydrothermally generated oil from Uzon caldera, Kamtchatka. Appl Geochem 24:303–309
Tyler SA, Baghoorn ES (1954) Occurrence of structurally preserved plants in Precambrian rocks of the Canadian shield. Science 119:606–608
Van Zuillen MA, Lepland A, Arrhenius G (2002) Reassessing the evidence for the earliest traces of life. Nature 418:627–630
Velasco-Santos C, Martinez-Hernandez AL, Consultchi A et al (2003) Naturally produced carbon nanotubes. Chem Phys Lett 373:272–276
Venkatasan MI, Ruth E, Rao PS et al (2003) Hydrothermal petroleum in the sediments of the Andaman Backarc Basin, Indian Ocean. Appl Geochem 18:845–861
Wachtershauer G (2007) On the chemistry and evolution of the pioneer organism. Chem Biodivers 4:584–602
Westall F (2004) Early life on Earth: the ancient fossil record. In: Ehrenfreud P, Irvine WM, Owen T et al (eds) Astrobiology, future perspectives. Kluwer, Dordrecht, The Netherlands
Westall F, de Wit MJ, Dann J et al (2001) Early Archean fossil bacteria and biofilms in hydrothermally-influenced sediments from the Barberton Greenstone belt, South Africa. Precambr Res 106:93–116
Williams LB, Canfield B, Voglesonger KM et al (2005) Organic molecules formed in "primordial womb". Geol Soc Am 33(11):913–916
Yanagawa H, Kojima K (1985) Thermophilic microspheres of peptide-like polymers and silicates formed at 250°C. J Biochem 97:913–916

Part I
Submarine Hot Springs and Venting Environments – Cradle of Life

Earliest Seafloor Hydrothermal Systems on Earth: Comparison with Modern Analogues

Suzanne D. Golding, Lawrence J. Duck, Elisa Young, Kim A. Baublys, Miryam Glikson, and Balz S. Kamber

Abstract Recent developments in multiple sulfur isotope analysis of sulfide and sulfate minerals provide a new tool for investigation of ore-forming processes and sources of sulfur in Archean hydrothermal systems, with important implications for the Archean sulfur cycle, the origin and impact of various microbial metabolisms and the chemistry of surface waters. In the current study we show that most of the sulfides and sulfates in the 3.49 Ga Dresser Formation and 3.24 Ga Panorama Zn–Cu field of Western Australia have non zero $\Delta^{33}S$ values that indicate variable proportions of seawater sulfate and elemental sulfur of UV-photolysis origin were incorporated into the deposits. Our results show that the multiple sulfur isotope systematics of the Dresser Formation sulfides and sulfates mainly reflect mixing between mass independently fractionated sulfur reservoirs with positive and negative $\Delta^{33}S$. Pyrite occurring with barite is depleted in ^{34}S relative to the host barite that has been interpreted as evidence for microbial sulfate reduction. We note, however, that the reported quadruple sulfur isotope systematics of pyrite-barite pairs are equally permissive of a thermochemical origin for this pyrite, which is consistent with inferred formation temperatures for the chert-barite units in excess of 100°C. The variably positive $\Delta^{33}S$ anomalies of the Panorama VHMS deposits, disequilibrium relations among sulfides and sulfates and general trend of increasing sulfide $\Delta^{33}S$ with stratigraphic height in individual ore systems most likely reflects temperature evolution and fluid mixing through the life of the hydrothermal system. The absence of sulfides with significant negative $\Delta^{33}S$ anomalies suggests that volcanic sulfur, not seawater sulfate, was the dominant sulfur source for the Panorama mineral system. The data presented here require Paleoarchean seawater to be at least locally sulfate bearing.

S.D. Golding (✉), L.J. Duck, E. Young, K.A. Baublys, and M. Glikson
School of Earth Sciences, University of Queensland, Brisbane, Queensland 4072, Australia
e-mail: s.golding1@uq.edu.au

B.S. Kamber
Department of Earth Sciences, Laurentian University, 935 Ramsey Lake Road, Sudbury, Ontario P3E 2C6, Canada

S.D. Golding and M. Glikson (eds.), *Earliest Life on Earth: Habitats, Environments and Methods of Detection*, DOI 10.1007/978-90-481-8794-2_2,

Keywords Archean sulfur cycle • Seafloor hydrothermal systems • Multiple sulfur isotopes • Thermochemical sulfate reduction • Microbial sulfate reduction • Dresser formation

1 Introduction

Submarine hydrothermal systems provide physically and chemically diverse habitats for microbial and other life and are chemically reactive environments that support abiotic synthesis of carbonaceous compounds (e.g. Kelley et al. 2005; Fiebig et al. 2007; McCollom and Seewald 2007; Proskurowksi et al. 2008). The widespread occurrence of hydrothermal systems in the early Precambrian has led to the suggestion that such environments may have constituted the cradle of life on Earth (e.g. Corliss et al. 1981; Simoneit 1995; Henley 1996; Shock 1996; Stetter 1996; Summons et al. 1996; Walter 1996; Shock and Schulte 1998; Holm and Charlou 2001; Russell 2003; Williams et al. 2005), largely because the chemistry of the H_2/CO_2 redox pair in seafloor hydrothermal systems favours the synthesis of reduced carbon compounds, the essential constituents of life (Shock 1990, 1996; Martin and Russell 2007). However, it is also generally accepted that the emergence of life in a hydrothermal vent environment would have been constrained by temperature and most likely required long lived hydrothermal activity (Poole et al. 1999; Ertem 2004; Russell et al. 2005; Russell and Hall 2006; Martin et al. 2008). As such, a low to mid temperature (≤110°C), freely convecting (non-magma driven), metalliferous seafloor hydrothermal seep or vent system like the Lost City hydrothermal field, located 30°N and some 15 km west of the Mid-Atlantic ridge (MAR) (Früh-Green et al. 2003; Kelley et al. 2005) is a most likely milieu for emergence of life on Earth (Russell et al. 2005; Russell and Hall 2006; Martin and Russell 2007; Russell 2007; Martin et al. 2008). In such environments, the serpentinisation of ophiolite complexes driven by subsurface seawater circulation, produces a self-sustaining exothermic reaction (Kelley et al. 2005), which, in the case of Lost City, has been in operation for at least 30,000 years, and may last for millions of years (Früh-Green et al. 2003; Kelley et al. 2005; Russell and Hall 2006; Martin et al. 2008). Alkaline systems such as Lost City may be representative of very early Earth (Hadean) hydrothermal environments (Kelley et al. 2005), where the eruption of Mg-rich komatiitic lavas was common (e.g. Macleod et al. 1994, and references therein). As such, these environments present as probably the most likely environment for the emergence of life given the thermodynamic and geochemical constraints imposed upon such a process (e.g. Russell et al. 2005; Russell and Hall 1997, 2006). However, the record of such ancient ophiolitic hydrothermal systems in the rock record is greatly limited; hence, the main geological focus of the search for early life is those ancient hydrothermal deposits and sedimentary sequences that have survived destruction over time.

A variety of morphological, petrological, mineralogical and geochemical approaches have been taken in the search for earliest life on Earth with recent attention focused on relatively well preserved, ca. 3.5–3.2 Ga sedimentary rocks and seafloor hydrothermal systems in the Pilbara Craton in Western Australia and the

Kaapvaal Craton in South Africa (e.g. Schopf 1993, 2006; Ueno et al. 2001, 2004, 2006; Tice and Lowe 2004; Lindsay et al. 2005; Allwood et al. 2006, 2007; Orberger et al. 2006; Westall and Southam 2006; Duck et al. 2007; Hofmann and Bolhar 2007; Glikson et al. 2008; Pinti et al. 2009). This is necessary because the oldest known sedimentary rocks circa 3.8 Ga in age, including those in Greenland in the Isua Supracrustal Belt and on Akilia Island, have experienced tectonism and high grade metamorphism, which has modified mineralogy and destroyed microstructural characteristics that might provide support for a biogenic origin of the graphite in these rocks (Fedo and Whitehouse 2002; Van Zuilen et al. 2002). While there is general agreement that life existed in the Palaeoarchean, much of the evidence is equivocal (Brasier et al. 2002, 2004, 2005; Lindsay et al. 2005), and it is possible that abiogenic carbon from Fischer-Tropsch-type synthesis (FTS) (McCollom and Seewald 2007) and/or atmospheric sources (Pavlov et al. 2001) dominated seafloor hydrothermal systems and sediments during the transition of the early Earth from geochemical to biogeochemical cycles.

In previous work we have reported the results of work on well preserved carbonaceous matter (CM) in sedimentary rocks and seafloor mineral deposits from the Warrawoona, Sulphur Springs and Gorge Creek Groups of the Pilbara Craton and the Onverwacht Group of the Kaapvaal Craton (Duck et al. 2004, 2005, 2007; Glikson et al. 2008, 2010) and multiple sulfur isotope evidence for dual sulfur sources in the 3.24 Ga Sulphur Springs massive sulfide deposit and the 3.49 Ga Dresser Formation of the Pilbara Craton (Golding and Young 2005; Golding et al. 2006). In the current study we use stable and radiogenic isotope geochemistry to understand the genesis of the hydrothermal ore deposits at Sulphur Springs and in the Dresser Formation, which are arguably and respectively the earliest well preserved and documented black smoker and epithermal style seafloor hydrothermal systems on Earth and potential habitats for early life.

2 Sulfur Isotopes in Seafloor Hydrothermal Systems and Sediments

This section provides a brief overview of our current understanding of the sulfur isotope geochemistry of seafloor hydrothermal systems and sediments and the use of multiple sulfur isotopes to constrain the early origins and activity of sulfur-based microbial life. Although the modern sulfur cycle is dominated by microbial transformations of seawater sulfate, magmatic, hydrothermal and atmospheric processes may have played a more significant role in the early Earth sulfur cycle (e.g. Canfield and Raiswell 1999; Farquhar et al. 2000; Farquhar and Wing 2003; Strauss 2003; Canfield 2005). Normal terrestrial cycling fractionates sulfur isotopes according to relative mass difference so that variations in $\delta^{33}S$ and $\delta^{36}S$ follow those in $\delta^{34}S$. The mass difference between ^{33}S and ^{32}S is about half the mass difference between ^{34}S and ^{32}S, which is why observed shifts in $\delta^{33}S$ are about half those in $\delta^{34}S$ and define the mass dependent fractionation (MDF) line ($\delta^{33}S \cong 0.515 \times \delta^{34}S$; Hulston and Thode 1965; Farquhar et al. 2000). Deviations from the mass dependent line (i.e. the $\Delta^{33}S$ record expressed as $\Delta^{33}S = \delta^{33}S - 0.515 \times \delta^{34}S$) have been identified in many Archean rocks, as well as modern sulfate aerosols and volcanic horizons resulting from Plinean eruptions (e.g. Farquhar et al. 2000; Mojzsis et al. 2003; Ono et al. 2003; Romero and Thiemens 2003; Savarino et al. 2003). Experimental studies indicate that atmospheric gas-phase reactions (SO_2–SO photolysis) produce two main products with different mass independent fractionation (MIF), i.e., elemental sulfur aerosols with positive $\Delta^{33}S$ and sulfate aerosols with negative $\Delta^{33}S$ (Farquhar et al. 2000, 2001; Farquhar and Wing 2003; Thiemens 2006). These atmospheric sources of oxidized and reduced sulfur may be incorporated into sulfide and sulfate minerals provided the atmospheric level of oxygen is sufficiently low (i.e., $<10^{-5}$ times the present atmospheric level; Pavlov and Kasting 2002). Based on the occurrence of sulfur MIF and other geochemical proxies, the majority of researchers conclude that the early atmosphere was anoxic to neutral and evolved to an oxic atmosphere between 2.45 and 2.32 Ga (e.g. Farquhar et al. 2000, 2001, 2007; Kasting 2001; Bekker et al. 2004).

Significant mass-dependent sulfur isotope fractionation accompanies microbial utilization of sulfur, particularly sulfate reduction that is the most important process for anaerobic mineralization of organic matter in marine sediments (Ohmoto and Goldhaber 1997). This is why pyrite in modern marine sediments exhibit very negative $\delta^{34}S$ values with a fractionation of up to 70 per mil relative to modern seawater sulfate. Accordingly sulfides in the Precambrian sedimentary rock record with variable $\delta^{34}S$ values and/or a substantial fractionation relative to the sulfur isotope composition of coeval seawater sulfate have been interpreted to form by microbial sulfate reduction (e.g. Kakegawa and Ohmoto 1999; Kakegawa et al. 1999; Towe 2000; Shen et al. 2001; Habicht et al. 2002). These authors have attributed the relatively small fractionation of sulfide relative to sulfate in many Archean settings to higher rates of seawater sulfate reduction in response to limiting sulfate concentrations and/or higher temperatures. Different mass dependent sulfur isotope fractionations accompany biogeochemical processes such as sulfate and elemental sulfur reduction, disproportionation of elemental sulfur and oxidation of sulfide and elemental sulfur. Many of these transformations can also occur inorganically and result in a spectrum of fractionations between sulfide and sulfate, which is why it is not always possible to interpret the ancient sulfur isotope record unambiguously even with multiple sulfur isotopes. The three main reservoirs of biologically useful sulfur in the Archean are oceanic sulfate with positive $\delta^{34}S$ and negative $\Delta^{33}S$, elemental sulfur with positive $\Delta^{33}S$ and volcanic SO_2 and H_2S with near zero $\delta^{34}S$ and $\Delta^{33}S$ (Farquhar and Wing 2003). Thus microbial sulfate reduction produces sulfide with negative $\Delta^{33}S$, whereas sulfur reducers produce sulfide with positive $\Delta^{33}S$. Both sulfides may have similar $\delta^{34}S$ values so this distinction is only possible for pre 2.4 Ga sulfides using multiple sulfur isotopes. Significantly, virtually all sedimentary and some hydrothermal sulfides older than 2.8 Ga have positive $\Delta^{33}S$ (Fig. 1), which is inconsistent with a significant role for microbial sulfate reduction and may suggest an early origin for organisms that reduce or disproportionate elemental sulfur (Golding et al. 2006; Philippot et al. 2007). From 2.8 to 2.5 Ga, sedimentary and some hydrothermal sulfides exhibit a wide range of positive and negative $\Delta^{33}S$, which is good evidence for a sulfur cycle that included microbial sulfur and sulfate reduction (Partridge et al. 2008). On the other hand, Watanabe et al. (2009) have shown experimentally using amino acid powders and sulfate that mass independently fractionated sulfur isotopes can be

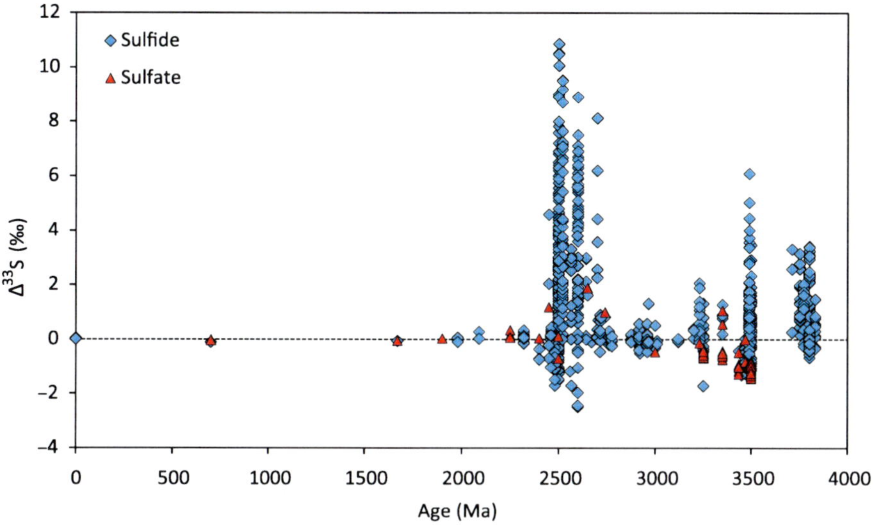

Fig. 1 A plot of $\Delta^{33}S$ versus age for sedimentary and hydrothermal sulfides and sulfates shows that the majority of sulfides older than ~2.8 Ga exhibit positive $\Delta^{33}S$ that implicates an elemental sulfur source of UV-photolysis origin irrespective of whether their precipitation occurred inorganically or was microbially mediated. Thermochemical reduction of mass dependently fractionated sulfate may also have produced small positive $\Delta^{33}S$ anomalies in some sulfides (cf. Watanabe et al. 2009). Barites and some sulfides have negative $\Delta^{33}S$ anomalies that we attribute to an ultimate seawater sulfate source. Additional data sources are Farquhar et al. (2000, 2007), Hu et al. (2003), Mojzsis et al. (2003), Ono et al. (2003, 2007, 2009), Bekker et al. (2004), Whitehouse et al. (2005), Cates and Mojzsis (2006), Ohmoto et al. (2006), Papineau and Mojzsis (2006), Bao et al. (2007), Philippot et al. (2007), Kamber and Whitehouse (2007), Kaufman et al. (2007), Partridge et al. (2008), Ueno et al. (2008) and Shen et al. (2009)

produced during thermochemical sulfate reduction at 150–200°C. The sulfur-MIF anomalies produced were relatively small but potentially complicate interpretation of multiple sulfur isotopes of ancient sedimentary rocks affected by hydrothermal alteration.

Multiple sulfur isotope studies of ancient rocks have focused on Mesoarchean and Neoarchean sedimentary successions that have been least affected by metamorphism and tectonism, particularly those in Western Australia and southern Africa (e.g. Ono et al. 2003; Bekker et al. 2004; Farquhar et al. 2007; Kaufman et al. 2007; Partridge et al. 2008). However, bedded and discordant barite deposits from Western Australia and South Africa have similar $\delta^{34}S$ values consistent with a common hydrothermal origin (Runnegar et al. 2001; Huston and Logan 2004; Bao et al. 2007), and there is wide agreement that there was a high level of hydrothermal activity in the Archean. Notwithstanding controversy as to whether the common ancestor to extant life forms was thermophilic (Doolittle 1999; Galtier et al. 1999), Archean seafloor hydrothermal environments would have supported the evolution of chemolithautotrophs and thermophiles and mesophiles that left isotopic and other biosignatures in the rock record (cf. Ueno et al. 2006). Furthermore, convective circulation of seawater is a fundamental part of seafloor hydrothermal systems, which provides a mechanism for imparting atmospheric sulfur MIF signatures to hydrothermal sulfides whatever the process mediating their precipitation. Thus multiple sulfur isotopes in Archean seafloor hydrothermal systems and sediments provide the basis for investigation of sulfur sources, the origin and impact of various microbial metabolisms and the chemistry of surface waters, with important implications for the Archean sulfur cycle and the composition of the atmosphere. In this context, Philippot et al. (2007) undertook a multiple sulfur isotope study of different sulfide generations in the Dresser Formation of Western Australia and concluded that the negative $\delta^{34}S$ and positive $\Delta^{33}S$ of microscopic sulfides in growth zoned barite crystals provides evidence for the involvement of organisms that disproportionated elemental sulfur of atmospheric origin. More recently, Ueno et al. (2008) and Shen et al. (2009) have shown that the majority of barite-pyrite pairs in Dresser Formation barites have $\Delta^{33}S$-$\Delta^{36}S$ systematics consistent with that expected for microbial sulfate reduction.

3 Geologic Setting of Dresser Formation and Sulphur Springs Deposit

The 3.80–2.83 Ga Pilbara Craton of Western Australia (Fig. 2a) contains some of the oldest, best-preserved supracrustal rocks on Earth (Van Kranendonk 1998; Van Kranendonk et al. 2002). The craton is essentially an aggregate of granite-greenstone terranes overlain by sedimentary basins with six major elements recognized in the outcropping northern part of the craton. These are rare fragments of 3.80–3.55 Ga early crust, the 3.53–3.17 Ga East Pilbara Terrane, the 3.27–3.06 Ma West Pilbara Superterrane, the 3.20–2.89 Ga Kurrana Terrane, the 3.02–2.93 Ma De Grey Superbasin and the 2.89–2.83 Ga Split Rock Supersuite of late- to post-tectonic granitic intrusions (Van Kranendonk et al. 2004). The hydrothermal systems sampled for the present study occur within the 3.53–3.17 East Pilbara Terrane.

The Dresser Formation outcrops only in the North Pole Dome locality of the Panorama Greenstone Belt, approximately 160 km southeast of Port Hedland and 40 km west of Marble Bar (Fig. 2a). The Dresser is an upper formation of the Talga Talga Subgroup of the Warrawoona Group (Fig. 2b), which is the oldest lithostratigraphic group of the Pilbara Supergroup in the East Pilbara Terrane. It comprises a 1,000 m thick sequence of interbedded black and variously coloured chert, massive barite, minor felsic and mafic/ultramafic volcaniclastics and pillowed to massive basalts (Buick and Dunlop 1990; Nijman et al. 1999; Van Kranendonk et al. 2002, 2006b). Up to five lenticular members of bedded chert ± barite ± carbonate ± jasper, interbedded with pillow basalts have been recognized in the Dresser Formation. These stratiform units are thought to have been fed by a set of chert-barite veins that were emplaced within, and immediately above, listric normal growth faults, which were active during deposition of the cherts and carbonates (Nijman et al. 1998, 1999; Van Kranendonk 2006; Van Kranendonk et al. 2006a). The dominant lithology is extensively carbonate altered and silicified volcanics (Barley 1993), which are metamorphosed to prehnite-pumpellyite facies in the North Pole Dome area (Nijman et al. 1999; Terabayashi et al. 2003; Van Kranendonk and Pirajno 2004).

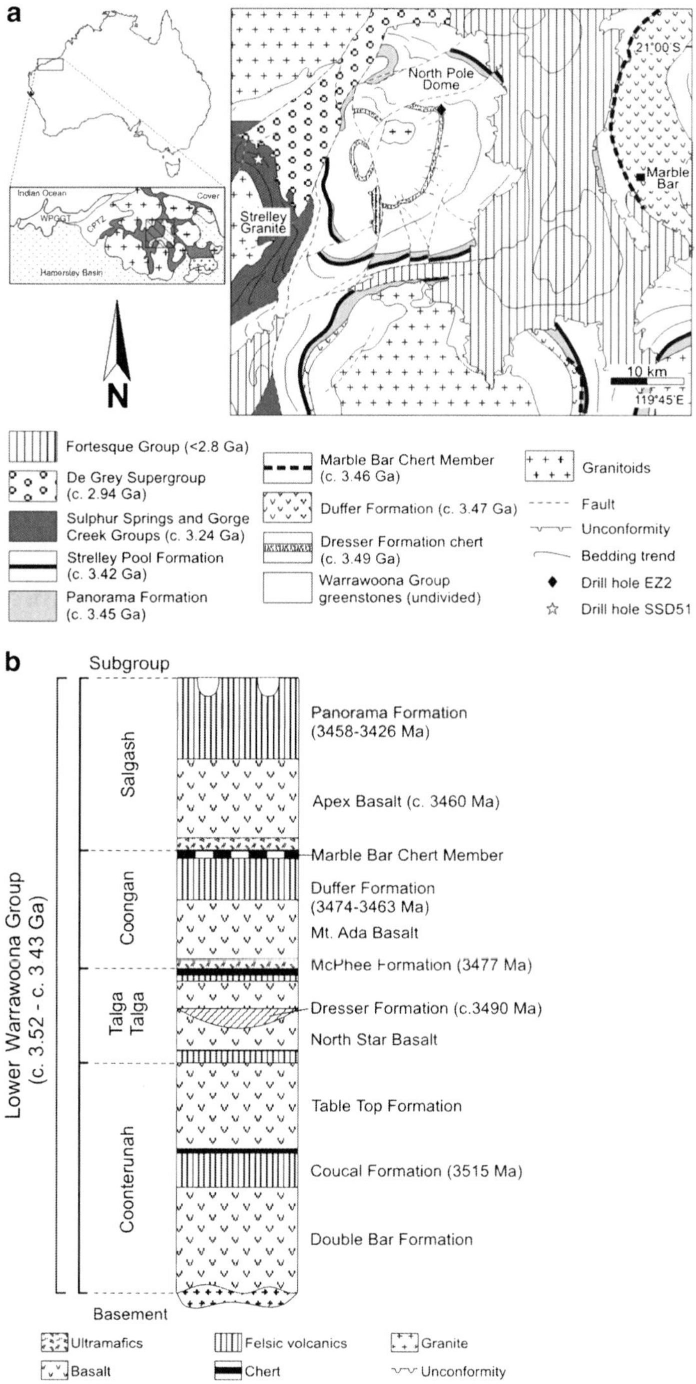

Fig. 2 (**a**) Geology of the North Pole Dome and Panorama Zn–Cu district, Pilbara Craton, (**b**) Stratigraphic column for Dresser samples

The Dresser Formation has been previously dated to circa 3,490 ± 10 Ma, using two Pb–Pb model ages for galena in presumed syngenetic barite (Thorpe et al. 1992). Recent SHRIMP U–Pb zircon dating of a felsic volcaniclastic sandstone from the top of the lowermost chert unit of the formation indicate a maximum depositional age of 3,481 ± 3.6 Ma (Van Kranendonk et al. 2008). The monzogranite exposed in the core of the North Pole Dome has been dated at 3,459 ± 18 Ma using conventional U–Pb zircon methods (Thorpe et al. 1992). New Sensitive High Resolution Ion Microprobe-Reverse Geometry (SHRIMP-RG) zircon geochronology of felsic porphyry dykes shows these intrusions have similar ages to the main granitoid body and the overlying Panorama Formation (Harris et al. 2009).

The 3.24 Ga Sulphur Springs volcanic hosted massive sulfide (VHMS) deposit is situated in the Panorama Zn–Cu field in the Soanesville Greenstone Belt, approximately 100 km southeast of Port Hedland and 50 km west of Marble Bar (Fig. 2a). The prospect was discovered in 1984 by PhD student Harry Wilhelmij, who reported finding sulfate encrustations on rocks in a then unnamed creek bed (Morant 1995). Sulphur Springs have since been described as Earth's oldest black smoker (Vearncombe et al. 1995; Morant 1998), and is interpreted to have formed at a water depth of around 1,000 m in an Archean setting equivalent to a modern back-arc basin (Vearncombe et al. 1995, 1998). A lack of any major tectonic overprint and very low metamorphic grade (prehnite-pumpellyite facies or below) has resulted in an exceptional level of structural and textural preservation (Morant 1995; Vearncombe 1995; Vearncombe et al. 1995; Duck et al. 2007).

The Sulphur Springs VHMS deposit is hosted within the upper part of the Kangaroo Caves Formation that is the upper stratigraphic unit of the 3,238–3,235 Ma Sulphur Springs Group and overlain by turbidites of the Gorge Creek Group (Morant 1995; Buick et al. 2002) (Fig. 3). The 1.5 km-thick Kangaroo Caves Formation is composed of massive, pillowed, pillow-brecciated and hyaloclastic andesitic volcanics, and dacite, rhyodacite and rhyolite sills and lavas (Brauhart 1999; Van Kranendonk 2000). VHMS mineralisation at Sulphur Springs and elsewhere in the district occurs mostly at or immediately below the marker chert, a regionally extensive silicified siltstone unit at the top of the Kangaroo Caves Formation. (Morant 1995; Vearncombe et al. 1995; Brauhart et al. 1998; Buick et al. 2002). Strata dip east such that the current exposure constitutes a cross section through a VHMS district and its underlying subvolcanic intrusion (Brauhart et al. 1998, 2000). Mineralisation consists of stratabound lenses of pyrite, sphalerite, chalcopyrite, galena, tennantite, arsenopyrite, barite, pyrrhotite and quartz, with underlying stringer zones comprising veins of quartz, pyrite, chalcopyrite, minor sphalerite and carbonate (Vearncombe et al. 1995; Barley 1998).

4 Sampling and Analytical Methods

Samples of the Dresser Formation used in this study were collected during the 2003 and 2005 field seasons from surface outcrops and diamond drill cores intersecting the massive barite deposits, which outcrop along the eastern margin of the North

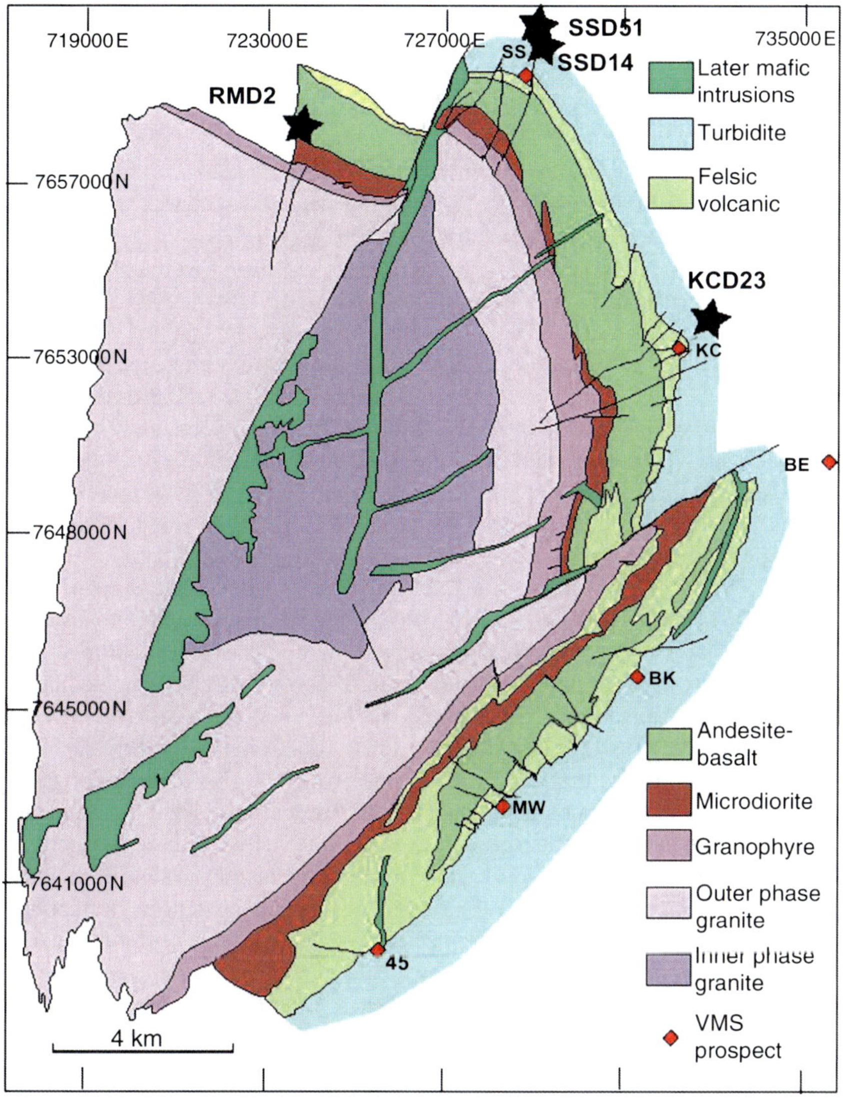

Fig. 3 Simplified local geology of the Panorama Zn–Cu district, showing locations of the VMS prospects and diamond drill holes SSD14, SSD51, RMD2 and KCD23 (adapted from Brauhart et al. 1998). The VMS prospects occur mainly at the top of the volcanic Kangaroo Caves Formation that is overlain by the turbidites of the Gorge Creek Group and intruded by the synvolcanic Strelley Granite (SS = Sulphur Springs; KC = Kangaroo Caves)

Pole Dome (Fig. 2a). The cores were drilled during the 1980s as part of the delineation of the barite deposits and are stockpiled at a location near the old Dresser mining camp (50K 753337E, 7666430N UTM), south of Panorama Homestead. Surface materials consisted primarily of massive barite, bedded chert and interbedded chert-barite samples, some of which showed obvious signs of iron-sulfide weathering.

The subsurface samples are from diamond holes EZ1 (50K 753940E, 7665686N UTM) and EZ2 (50K 753790E, 7665956N UTM) and associated core trays (Glikson et al. 2008). The drill core samples are predominantly fine-grained, carbonaceous matter (CM)-rich siliceous material variably cut by chalcedonic veins and late-stage carbonate veins. Diamond drill cores PDP2B and PDP2C from the recent Pilbara Drilling Project Provide context for interpretation of the geological setting of these samples that comprise the immediate footwall of the massive barite deposits in the case of EZ1 and EZ2 (Van Kranendonk et al. 2008).

Samples from the Sulphur Springs and Gorge Creek Groups were collected during the 2003 field season from diamond drill cores SSD14 (AMG 729043.07E, 7659869.47N), SSD51 (AMG 729103E, 7660264N) and RMD2 (AMG 723451.617E, 7658609.885N) at the Sipa Resources Port Hedland core storage facility (Fig. 3) (Duck et al. 2007). Samples of diamond drill core KCD23 also drilled by Sipa Resources Limited were subsequently obtained from the University of Western Australia Geology Museum. SSD14 commenced within the immediate hanging wall sedimentary strata of the Gorge Creek Group, intersected 4.5 m of massive sulfide mineralisation at some 220 m and a 45 m lens of combined massive sulfide and stringer zone mineralisation at some 251 m, before terminating at 354 m within the underlying felsic volcanics of the Kangaroo Caves Formation of the Sulphur Springs Group. SSD51 extends 826.2 m through the Paddy Market Formation and Corboy Formations of the Gorge Creek Group into the Kangaroo Caves Formation and intersected the marker chert horizon at some 730 m. It comprises a volcano-sedimentary sequence rich in carbonaceous cherts, ironstones and silicified shales with only minor sulfide occurrences. RMD2 was drilled in close proximity to the Strelley Granite through the lower part of the Kangaroo Caves Formation. This core comprises variably altered intermediate to mafic volcanics with lesser cherts, siltstones and arenites and is heavily intruded by dacite sills. Several semi massive sulfide horizons were intersected at some 144 m. KCD23 commenced within hanging wall turbidites of the Gorge Creek Group, intersected the marker chert and massive sulfide mineralisation between 480.6 and 504.1 m, several zones of massive to stringer sulfide mineralisation in the underlying dacite sill and terminated in dacite at 601 m.

Pyrite samples for Pb–Pb dating and sulfur isotope analysis were sourced from selected Dresser black chert samples. An attempt was made to limit the pyritic material selected to a first stage or earliest primary phase by visually isolating those areas of each selected sample not affected by cross-cut veining. Once selected, each rock sample was coarse crushed in a stainless steel percussion mortar. Sulfide-rich materials were then handpicked from the crush and transferred to 50 ml Falcon tubes prior to immersion in cold HF for a period of 48 h, with a 5 min agitation in an ultrasonic bath every 2 h. Tubes were then subjected to centrifugation to concentrate the denser pyritic material. The pyrite free supernatant fraction was decanted and the bottom residue treated with cold HCl to liberate any carbonate and fluoride complexes. The tubes again underwent centrifugation to concentrate the denser pyritic material, and the supernatant was discarded. Each sample was then thoroughly rinsed in DI water up to eight times, and the resultant pyritic slurry was dried overnight in a low temperature oven. Each sample concentrate was subsequently added to a 250 ml conical flask of DI water and agitated to separate out the last of the carbonaceous material and concentrate the pyritic material at the base of the flask. The liquid fraction was decanted off and retained for repeat phases if needed. This filtration step was repeated until the dominant colour at the bottom of the 250 ml conical flask was that of a brassy-yellow, or until no further material remained in suspension. The pyritic fraction was then decanted into ceramic crucibles and the concentrate oven dried at 40°C. Those samples found to be of low sulfide content were repeated to increase concentration or discarded as non-usable if the final sulfide concentration was too low. Samples displaying foreign matter inclusions (e.g. barite) were hand picked clean under magnification. Macroscopic sulfide laminates and barite crystals in stratiform barite-sulfide deposits at North Pole and sulfides in Sulfur Springs samples were generally sufficiently coarse grained to be prepared by hand picking of a coarse crush or micro-drilling.

Lead isotope analyses on extracted pyrite crystals were conducted at the University of Queensland Radiogenic Isotope Laboratory (UQ-RIL). Pyrite samples were boiled in de-ionised water, then twice evaporated to dryness in open Savillex beakers using 4 ml ultra-pure HF, in order to remove any silica traces and to maximise Pb yield. Samples were then digested at 145°C for 48 h in a 1:1 HNO_3/HF mixture, converted to HBr form, and Pb separated using standard HCl–HBr chemistry, via double-pass through miniaturised anion-exchange columns. Pb-isotope ratios were acquired on a VG Sector54 instrument in static mode at pyrometer-controlled 1,350–1,450°C. Mass bias was estimated using NIST SRM 981. Corrected results are reported at 2σ confidence level (Table 1).

Table 1 Lead isotope data for Dresser pyrite samples

Sample	$^{206}Pb/^{204}Pb$	$^{207}Pb/^{204}Pb$	$^{208}Pb/^{204}Pb$	$^{207}Pb/^{206}Pb$	$^{208}Pb/^{206}Pb$	Age (Ma)	Error (±Ma)
EZ1	12.219	13.845	32.031	1.13307	2.62129	3,312	62
153.20m	±0.005	±0.006	±0.012	±0.00010	±0.00022		
EZ2a	11.803	13.573	31.699	1.15005	2.68585	3,506	53
180.50m	±0.008	±0.009	±0.022	±0.00015	±0.00032		
EZ2b	11.704	13.531	31.661	1.15245	2.69624	3,533	51
199.50m	±0.002	±0.002	±0.005	±0.00005	±0.00015		
EZ4	11.795	13.621	31.784	1.15484	2.69484	3,493	29
Core	±0.006	±0.007	±0.019	±0.00017	±0.00039		
L6-B	11.969	13.745	31.869	1.14840	2.66278	3,509	55
Surface	±0.004	±0.004	±0.011	±0.00009	±0.00020		
DF-7	11.482	13.635	31.771	1.15147	2.68291	3,567	60
Core	±0.004	±0.005	±0.013	±0.00007	±0.00019		
DF-11	11.988	13.735	31.874	1.14570	2.65880	3,508	65
Core	±0.005	±0.004	±0.010	±.00010	±0.00021		
DF-16	11.962	13.728	31.827	1.14761	2.66061	3,515	60
Core	±0.004	±0.004	±0.011	±0.00009	±0.00017		
DF-17	11.909	13.752	31.896	1.15480	2.67860	3,501	42
Core	±0.002	±0.004	±0.011	±0.00011	±0.00045		

Oxygen isotope analyses ($\delta^{18}O$) were undertaken on a Micromass 602E mass spectrometer housed at the University of Queensland Stable Isotope Geochemistry Laboratory (UQ-SIGL), which uses the dynamic gas flow method to compare unknown sample gas with reference gas. Samples comprising a minimum 20 mg of crushed, hand picked quartz were reacted overnight at 650°C with bromine pentafluoride (BrF_5) in nickel vessels in order to liberate oxygen (Clayton and Mayeda 1963). The liberated gas was converted to CO_2 by reaction with an internally heated carbon rod. Results are reported in per mil relative to the V-SMOW standard, with analytical uncertainties better than ±0.2‰ (1σ) (Table 2).

Multiple sulfur isotope analyses ($\delta^{34}S$, $\delta^{33}S$) were carried out at the UQ-SIGL using a Carlo Erba NA1500 elemental analyser coupled in continuous flow mode to an Isoprime mass spectrometer (EA-CF-IRMS) (Baublys et al. 2004). Analytical precision was determined through replicate analyses of international standards NBS127 and IAEA-S3 run as unknowns, while international standards IAEA-S1 and IAEA-S2 were used to set up the calibration for each analytical run. Results are reported in per mil relative to V-CDT and have an analytical uncertainty better than ±0.2‰ (1σ) (Tables 2 and 3).

5 Geochronology

The Pb-isotope compositions of the pyrites extracted from the samples are plotted in conventional common Pb space, i.e. $^{207}Pb/^{204}Pb$ versus $^{206}Pb/^{204}Pb$ (Fig. 4). The isotope ratio ranges for the samples from the Dresser Formation are: $^{206}Pb/^{204}Pb$ = 11.740–12.219; $^{207}Pb/^{204}Pb$ = 13.531–13.845; and $^{208}Pb/^{204}Pb$ = 31.661–32.031. The variability in isotope ratio is not random but the data define a tight linear array (Fig. 4). The slope of this array is steeper than the Geochron and implies an origin as a binary mixing line. In other words, the spread in isotope composition is not primarily caused by an age range but by a different source origin of the Pb incorporated into the pyrites, which can be successfully described with the depleted (MORB) mantle source and a global average sediment mixture (EMIX). Those samples from the EZ drill holes that plot close to the MORB curve of Kramers and Tolstikhin (1997) are interpreted to have incorporated almost pure mantle Pb (Kamber et al. 2003). In contrast, the Pb-isotope results for sample DF17 plot almost exactly onto the EMIX curve, indicating that the Pb incorporated into this sample potentially originated from the weathering/recycling of non-juvenile sediment, i.e. some older exposed continental crustal component (Green et al. 2000). The remaining Pb samples from the dataset (DF7, DF11, DF16, and L6B) have a mixed MORB/EMIX signature. For these, a 50:50 MORB-EMIX origin was used in the model age calculation.

The model age calculation was performed by fitting the $^{206}Pb/^{204}Pb$ and the $^{207}Pb/^{204}Pb$ temporal evolutions for the MORB-source mantle, the EMIX (both from Kramers and Tolstikhin 1997) and the 50:50 mixture versus time using polynomial fits. For each sample, two model ages were thus obtained by comparison of the

Table 2 Stable isotope data for Dresser sulfide, sulfate and quartz samples (WR = whole rock sample; V = vein sample)

Sample	Lithology	Sulfur species	$\delta^{34}S‰$	$\delta^{33}S‰$	$\Delta^{33}S‰$	Quartz variety	$\delta^{18}O‰$
EZ1	Black chert	Pyrite	3.47	2.07	0.29	WR	13.7
153.20m			3.65	1.98	0.11		
EZ2a	Black chert	Pyrite	2.98	1.81	0.27		
180.50m			3.07	2.81	1.24		
EZ2s	Black chert	Pyrite	2.84	1.11	−0.35		
198.20m			2.64	1.15	−0.21		
EZ2b	Black chert	Pyrite	1.83	1.75	0.81	WR	15.1
199.50m			2.55	1.00	−0.31		
			2.24	2.53	1.38		
EZ4	Black chert	Pyrite	2.49	3.19	1.91		
Core			1.17	2.05	1.45		
L6-B	Massive barite	Pyrite	−3.93	−2.90	−0.88		
Surface			−3.76	−2.83	−0.89		
		Barite	4.35	0.87	−1.37		
			4.13	0.88	−1.25		
			4.09	0.66	−1.45		
DFNP-WP38	Massive barite	Barite	4.78	1.63	−0.83		
Surface			4.64	1.13	−1.26		
			4.42	1.18	−1.09		
DFNP-WP38A	Massive barite	Barite	3.52	0.68	−1.14		
Surface			4.56	1.18	−1.16		
			4.25	1.49	−0.78		
DFNP-WP41A	Massive barite	Barite	4.80	1.51	−0.96		
Surface			4.91	1.41	−1.11		
			4.00	1.20	−0.86		
DFNP-WP45	Massive barite	Barite	4.99	1.68	−0.89		
Surface			4.68	1.39	−1.02		
			4.80	1.39	−1.08		
DF-2	Black chert	Pyrite	−8.93	−4.83	−0.22	WR	14.8
Core			−0.27	0.47	0.61		
			0.84	0.72	0.29		
			−2.45	−0.78	0.48		
DF-5	Black chert					WR	14.5
Core							
DF-7	Black chert	Pyrite	0.63	1.14	0.82		
Core			0.43	0.73	0.51		
DF-10	Black chert					WR	15.0
Core						V	14.9
DF-11	Black chert	Pyrite	1.69	1.12	0.25	WR	14.6
Core			0.95	0.20	−0.29		
			−4.31	−2.45	−0.23		
DF12	Black chert					WR	13.5
Core						V	14.6
DF-16	Black chert	Pyrite	3.93	2.81	0.79	WR	15.0
Core			4.82	2.62	0.14	V	14.8
DF-17	Black chert	Pyrite	2.28	1.53	0.36	WR	15.5
Core			1.03	1.43	0.90	V	16.1
			1.35	1.07	0.37		

Table 3 Multiple sulfur isotope data for sulfides and sulfates of the Panorama VHMS district (HW = hangingwall strata of deposit; MS = massive sulfide mineralization; FW = footwall strata of deposit)

Sample	Host lithology	Deposit	Sulfur species	$\delta^{34}S$‰	$\delta^{33}S$‰	$\Delta^{33}S$‰
SSD14-EY1	Fe-carbonate	Sulphur	Pyrite	−0.01	1.95	1.95
135.1 m	altered siltstone	Springs		0.09	2.13	2.08
		(HW)		0.27	2.29	2.15
				0.06	2.09	2.06
SSD14-EY2/1	Fe-carbonate	Sulphur	Pyrite	−0.49	0.54	0.79
147.9 m	altered siltstone	Springs		−0.28	0.71	0.85
		(HW)		−0.67	0.96	1.31
				−0.64	0.88	1.21
SSD14-EY2/2			Chalcopyrite-	1.06	0.91	1.46
147.9 m			pyrite	−0.42	1.10	1.32
				−0.62	0.91	1.23
				−0.39	1.32	1.53
				0.59	1.26	0.96
				−0.15	1.09	1.16
SSD14-EY2/3			Pyrite;	0.14	0.90	0.83
147.9 m			minor sphalerite	0.46	1.08	0.84
				−0.76	0.38	0.77
				−0.25	0.98	1.12
				0.54	1.01	0.73
SSD14-EY3	Silicified siltstone	Sulphur	Pyrite	2.44	1.66	0.40
192.3 m		Springs				
		(HW)				
SSD14-EY4	Silicified	Sulphur	Pyrite	0.78	2.44	2.03
210.6 m	volcanoclastic	Springs		0.93	2.31	1.84
		(HW)		1.01	2.33	1.80
SSD14-EY5/1	Massive sulfide	Sulphur	Chalcopyrite;	−0.49	−0.19	0.06
270.5 m		Springs	minor pyrite	−0.12	−0.02	0.04
		(MS)		−0.04	0.13	0.15
				0.58	0.51	0.21
				0.14	0.46	0.39
				0.57	0.45	0.16
SSD14-EY5/2			Pyrite; minor	−0.11	0.50	0.55
270.5 m			Chalcopyrite	0.36	0.73	0.54
				0.41	1.09	0.88
				0.30	0.23	0.18
SSD14-EY5/3			Pyrite; minor	1.64	1.24	0.40
270.5 m			Chalcopyrite	0.94	0.54	0.05
				0.69	0.56	0.21
				0.37	0.50	0.31
				0.81	0.61	0.19
				0.58	0.73	0.43
SSD14-EY5/4			Chalcopyrite;	1.20	0.91	0.30
270.5 m			minor pyrite	0.19	0.31	0.21
				0.24	0.09	−0.04

(continued)

Table 3 (continued)

Sample	Host lithology	Deposit	Sulfur species	$\delta^{34}S$‰	$\delta^{33}S$‰	$\Delta^{33}S$‰
SSD51-81/1	Fe-carbonate	Sulphur	Pyrite	1.99	1.75	0.73
439.90 m	altered siltstone	Springs		2.33	1.85	0.65
		(HW)		2.34	1.68	0.48
SSD51-81/2			Pyrite	2.44	1.76	0.51
439.90 m				1.66	1.79	0.94
				2.08	1.45	0.38
				2.89	2.01	0.52
				1.81	1.31	0.37
				1.91	1.72	0.74
SSD51-102	Fe-carbonate	Sulphur	Pyrite	−2.71	−0.17	1.23
679.15 m	altered siltstone	Springs		−2.09	0.35	1.42
		(HW)		−2.70	−0.02	1.37
				−1.60	0.63	1.46
				−2.29	0.18	1.36
				−1.54	0.67	1.46
RMD2-01	Interbedded chert	Road	Pyrite	1.82	1.92	0.98
110.1 m	and	Master		1.38	1.49	0.78
	Fe-carbonates	(HW)		1.50	1.76	0.98
RMD2-05	Fe-carbonate	Road Master	Pyrite	0.80	1.76	1.35
119.4 m	altered siltstone	(HW)		0.47	1.28	1.04
RMD2-10	Chert	Road Master	Pyrite	−0.20	−0.08	0.03
144.6 m		(MS)		0.50	0.16	−0.10
				0.06	−0.05	−0.08
RMD2-EY1/1	Semi-massive	Road Master	Pyrite	0.38	0.46	0.26
149.4 m	sulfide	(MS)		0.77	0.91	0.51
				0.60	0.40	0.10
				1.49	1.03	0.27
				0.95	0.76	0.28
RMD2-EY1/2			Pyrite	0.26	0.52	0.39
149.4 m				0.69	0.38	0.03
				0.65	0.50	0.16
RMD2-EY1/3			Pyrite; minor	1.40	0.53	−0.19
149.4 m			sphalerite	1.21	0.79	0.17
				1.26	0.82	0.17
RMD2-12/1	Chert	Road Master	Vein pyrite	−0.48	−0.16	0.09
151.6 m		(FW)		−0.27	−0.13	0.00
				−0.62	−0.40	−0.08
RMD2-12/1			Vein pyrite	−0.04	0.12	0.14
151.6 m				−1.18	−0.39	0.22
				−0.99	−0.18	0.33
RMD2-17/1	Chert	Road Master	Vein pyrite	0.13	0.30	0.23
168.8 m		(FW)		0.60	0.24	−0.07
				0.35	0.27	0.09
RMD2-17/1			Concordant	0.19	0.21	0.11
168.8 m			pyrite	−0.48	−0.27	−0.03
				−0.88	−0.88	−0.43

(continued)

Table 3 (continued)

Sample	Host lithology	Deposit	Sulfur species	δ^{34}S‰	δ^{33}S‰	Δ^{33}S‰
KCD23-01 537.2 m	Silicified volcanic	Kangaroo Caves (FW)	Pyrite	2.25	1.19	0.03
				2.86	1.38	–0.10
				2.49	1.37	0.09
KCD23-02 546.8 m	Quartz-carbonate-pyrite-chlorite-sericite alteration	Kangaroo Caves (FW)	Pyrite	2.04	0.92	–0.13
				1.27	0.05	–0.60
				1.67	0.79	0.03
				1.31	0.71	0.04
				2.07	1.13	0.06
KCD23-05/1 556.25m	Quartz-carbonate-barite-pyrite-sericite alteration	Kangaroo Caves (FW)	Barite	6.28	3.02	–0.20
				6.09	3.05	–0.08
KCD23-05/2 556.25 m			Vein pyrite	0.38	0.23	0.04
				0.04	0.65	0.63
				0.32	0.41	0.25
KCD23-05/3 556.25 m			Disseminated pyrite	2.38	0.52	–0.70
				0.48	–0.07	–0.31

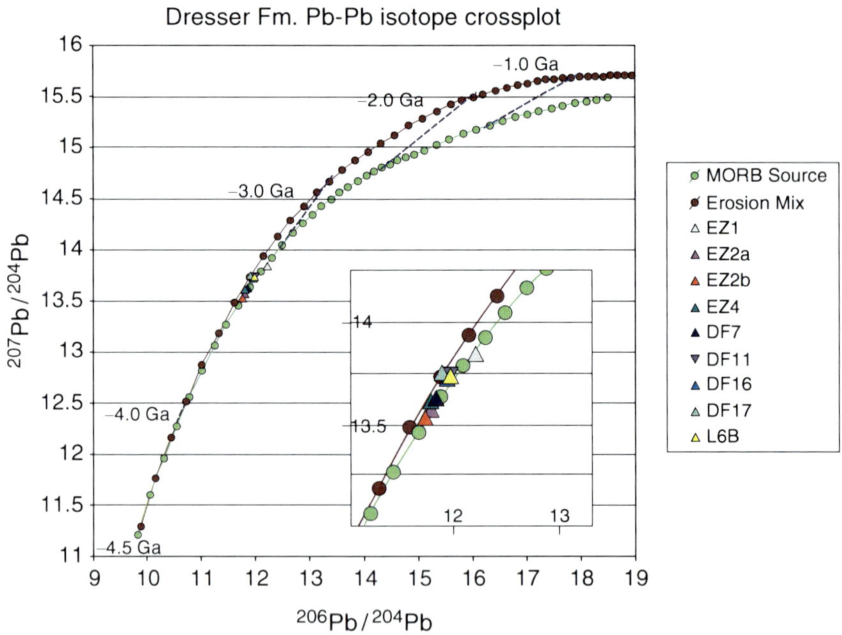

Fig. 4 Pyrite ^{207}Pb/^{204}Pb vs. ^{206}Pb/^{204}Pb of Dresser Formation pyrite plotted relative to MORB-source mantle and Erosion Mix Pb-isotope growth curves of Kramers and Tolstikhin (1997)

pyrite Pb with the appropriate model curve. The combined average of these two values per sample provided both an estimated age and error of 1σ (Table 1).

With the exception of sample EZ1 (see further below for discussion), the calculated Pb–Pb model ages (Table 1) of the sulfides extracted from the Dresser

Formation samples yield a tight cluster with an average $^{207}Pb/^{206}Pb$ age of 3,517 ± 23 Ma. Recent U–Pb SHRIMP dating of zircons isolated from a felsic tuff layer within the formation has provided a maximum depositional age of 3,481 ± 3.6 Ma respectively (Van Kranendonk et al. 2008), which is broadly consistent with the Pb–Pb model age of the stratiform barite-chert units. On the other hand, the monzogranite exposed in the core of the dome has a conventional U–Pb zircon age of 3,459 ± 18 Ma and a single porphyritic felsic intrusion has an overlapping SHRIMP-RG U–Pb zircon age of 3,438 ± 12 Ma (Harris et al. 2009). Based on these ages and local cross cutting relations between felsic porphyry dykes and quartz veins with colloform and comb textures, Harris et al. (2009) concluded that the barite-rich hydrothermal system at North Pole was related to this 3.45 Ga magmatism and contemporaneous with eruption of the overlying 3,458 ± 2 Ma Panorama Formation (Thorpe et al. 1992). Our geochronology results are not consistent with this interpretation of the timing of emplacement of the barite mineralisation in the deeper parts of the Dresser stratigraphy but do not preclude a subsequent quartz vein mineralisation overprint related to the 3.45 Ga magmatism.

The Pb–Pb model age of sample EZ1 of 3,312 ± 42 Ma is significantly younger than the rest of the sample set, and is interpreted to reflect late stage recrystallisation. Given the disparity between the calculated age of this and the other samples, EZ1 is classed as an outlier.

6 Stable Isotopes

Oxygen isotopes of vein and alteration minerals in hydrothermal systems are used to infer fluid sources and the temperature of formation of these minerals. However, processes such as fluid-rock interaction, boiling and mixing may modify the oxygen (and hydrogen) isotopic composition of the ore fluid and preclude unambiguous determination of fluid sources particularly in the absence of independent temperature estimates. The $\delta^{18}O$ values of whole rocks surrounding many ore deposits vary systematically, with low $\delta^{18}O$ values commonly associated with the most intense zones of alteration (Huston 1999). In the case of seafloor hydrothermal systems, $\delta^{18}O$ whole rock values commonly decrease with depth in the volcanic pile and in feeder zones beneath deposits, which is interpreted to reflect increasing temperature with depth and in the feeder system (Huston 1999 and references therein). However, other factors such as water/rock ratio and the $\delta^{18}O$ of the hydrothermal fluid also affect the final whole rock $\delta^{18}O$ values. Nevertheless, modelling of oxygen (and hydrogen) isotope systematics may allow distinction among these effects and provide constraints on district scale hydrology.

Sulfur isotopes are also used to determine the sources of sulfur present as sulfides and sulfates and may provide constraints on the extent of oxidation-reduction reactions in hydrothermal systems. Moreover, as discussed in Section 2, the Archean surface environment has multiple sulfur sources that have different $\delta^{34}S$ and $\Delta^{33}S$, which allows determination of whether sulfur in a hydrothermal system left the atmosphere in an oxidized or reduced form or was of magmatic origin (cf. Jamieson et al. 2006).

6.1 Dresser Formation

Oxygen isotope values were obtained for nine Dresser Formation black chert drill core samples (Table 2) to provide constraints on fluid sources and the likely depositional temperature of the chert-barite units. Originally interpreted as diagenetic replacement of evaporitic gypsum, these units are considered to be of hydrothermal origin by the majority of workers (e.g. Nijman et al. 1999; Hickman and Van Kranendonk 2004; Huston and Logan 2004; Van Kranendonk and Pirajno 2004; Van Kranendonk et al. 2008). Clean quartz chips were hand picked that allowed a distinction among whole rock and vein-sourced material for several of the samples. $\delta^{18}O$ values of the Dresser samples fall within a relatively narrow range from 13.5‰ to 16.1‰, with a mean of 14.8‰ ± 0.7‰ (n = 13). Vein quartz and black chert generally have similar $\delta^{18}O$ values (Table 2), and the results overlap the $\delta^{18}O$ values of quartz vein surface samples with epithermal textures from the North Pole Dome that range from 13.7‰ to 16.0‰ (n = 9) (Harris et al. 2009).

Two independent studies provide an indication of the possible temperature regime during deposition of the chert-barite units. Glikson et al. (2008) reported reflectance values for the various CM types identified in the black chert drill cores and obtained temperatures of 170–200°C (Ro = 1.2–1.8) and 230–240°C (Ro = 2.5–2.8) for the lowest Ro populations using the conversion of Aizawa (1990). Harris et al. (2009) undertook fluid inclusion microthermometry of quartz-chalcedony veins at various levels in the Dresser stratigraphy. These authors provided little information on the location and nature of the material analysed and appear to have focused on colloform and comb quartz including material interlayered with and overgrowing barite. The relationship between this material and the bedded cherts, barite and carbonates intersected in the Pilbara Drilling Project (Philippot et al. 2007; Van Kranendonk et al. 2008) is unclear; however, fluid inclusions in quartz have homogenization temperatures from 110–260°C, with the majority less than 210°C (Harris et al. 2009). On the other hand fluid inclusions in barite exhibited a wider range of homogenisation temperatures from 158–365°C (Harris et al. 2009), which in part may reflect post-entrapment modification. Using the quartz-water fractionation equation of Matsuhisa et al. (1979), the water in equilibrium with cherts with a mean $\delta^{18}O$ of 14.8‰ ± 0.7‰ at 150–200°C has a $\delta^{18}O$ value of −0.6‰ to 3.2‰ inconsistent with magmatic or metamorphic fluid sources (Taylor 1992).

The use of $\delta^{18}O$ values in the interpretation of fluid composition and/or palaeotemperatures from Precambrian seafloor sediments is recognised as uncertain due to the possibility that the $\delta^{18}O$ values may have been modified by isotopic reequilibration during diagenesis at elevated temperature (i.e. burial metamorphism) and postdepositional processes, such as regional metamorphism (e.g. de Ronde et al. 1997; Faure and Mensing 2005). However, the lack of a strong metamorphic overprint, as expressed by the Pb isotope results and the inferred thermal history indicators (Glikson et al. 2008) suggests these samples have not been significantly altered by any of these processes. The 2.6‰ variation in the $\delta^{18}O$ values reported in this study is most likely the result of somewhat variable hydrothermal fluid temperatures, resulting in isotopic fractionation within the system (de Ronde et al. 1997). This interpretation allows for calculation of palaeotemperatures from the chert $\delta^{18}O$ results using a model fluid composition and a rearrangement of the Matsuhisa et al. (1979) quartz-water fractionation equation. Modelling undertaken in previous studies (e.g. de Ronde et al. 1997; Robert and Chaussidon 2006) has shown that the $\delta^{18}O$ of seawater is buffered by mid-ocean ridge hydrothermal circulation, which implies that seawater has probably had a constant $\delta^{18}0$ value of between −1.0‰ and + 1.0‰ since Archean times. Assuming a likely value of −1‰ for the Archean ocean and a seawater-dominated hydrothermal system (de Ronde et al. 1997; Muehlenbachs 1998), infers a possible depositional temperature for the Dresser Formation charts of 132–160°C. If an assumed value of −2‰ is adopted, as has been suggested (Land and Lynch 1996; Lécuyer and Allemand 1999), then temperatures would be even lower for the same samples from 123–148°C. Conversely, a +2‰ $\delta^{18}O$, equivalent to the $\delta^{18}O$ value of modern seafloor systems (Shanks et al. 1995) would give a temperature range of 166–203°C. These temperatures and the previously calculated fluid compositions are consistent with a hydrothermal origin for the chert-barite units where sea water or low latitude meteoric water was the major component of the hydrothermal system.

Multiple sulfur isotope values were obtained for fine grained sulfides separated from drill core samples of black chert, and sulfide and sulfate concentrates from outcrop samples of stratiform barite (Table 2). Our data show significant variations in $\delta^{34}S$ and $\Delta^{33}S$ among the different sulfide types. Sulfides from the black cherts exhibit a wide range of $\delta^{34}S$ values from −8.93‰ to 4.82‰ ($\delta^{34}S$ = 1.24‰ ± 2.88‰; n = 25) and positive $\Delta^{33}S$ anomalies up to 1.91‰ ($\Delta^{33}S$ = 0.45‰ ± 0.6‰; n = 25). Samples with the most positive $\Delta^{33}S$ also have positive $\delta^{34}S$ values; however, there is only a modest positive correlation between $\delta^{34}S$ and $\Delta^{33}S$ (Fig. 5). Runnegar et al. (2001) and Ueno et al. (2008) also found that disseminated pyrite in the bedded black cherts and black chert veins have positive $\Delta^{33}S$ anomalies, which they attributed to an atmospheric elemental sulfur source. The similar $\Delta^{33}S$-$\delta^{34}S$ systematics of the bedded cherts and chert veins supports a common hydrothermal origin for these units in line with field relationships (Ueno et al. 2008) although it is unclear whether the cherts are truly exhalative or formed by silicification of precursor lithologies. In this context, Philippot et al. (2007) reported that sedimentary sulfides from bedded carbonates and volcanoclastic sandstones overlying the main chert-barite unit in drill holes PDP2B and PDB2C have a limited range of $\delta^{34}S$ values (−1.4‰ ± 2.7‰) and positive $\Delta^{33}S$ anomalies of 1.0‰ ± 0.6‰. The $\delta^{34}S$ values of the black chert pyrites overlap those of the bedded carbonates rather than the volcanoclastic sandstones (Fig. 6a), which provides support for the interpretation of Van Kranendonk et al. (2008) that the barite-chert units formed at least in part by hydrothermal replacement of the bedded carbonate rocks.

The positive $\Delta^{33}S$ of sulfides analysed in the current and previous studies potentially implicates an elemental sulfur source of UV-photolysis origin, with sulfide formation occurring as a result of reduction or disproportionation of elemental sulfur (S^o) to sulfide (H_2S). The modest range of $\delta^{34}S$ values is consistent with inorganic reduction or disproportionation of elemental sulfur to sulfide or microbial

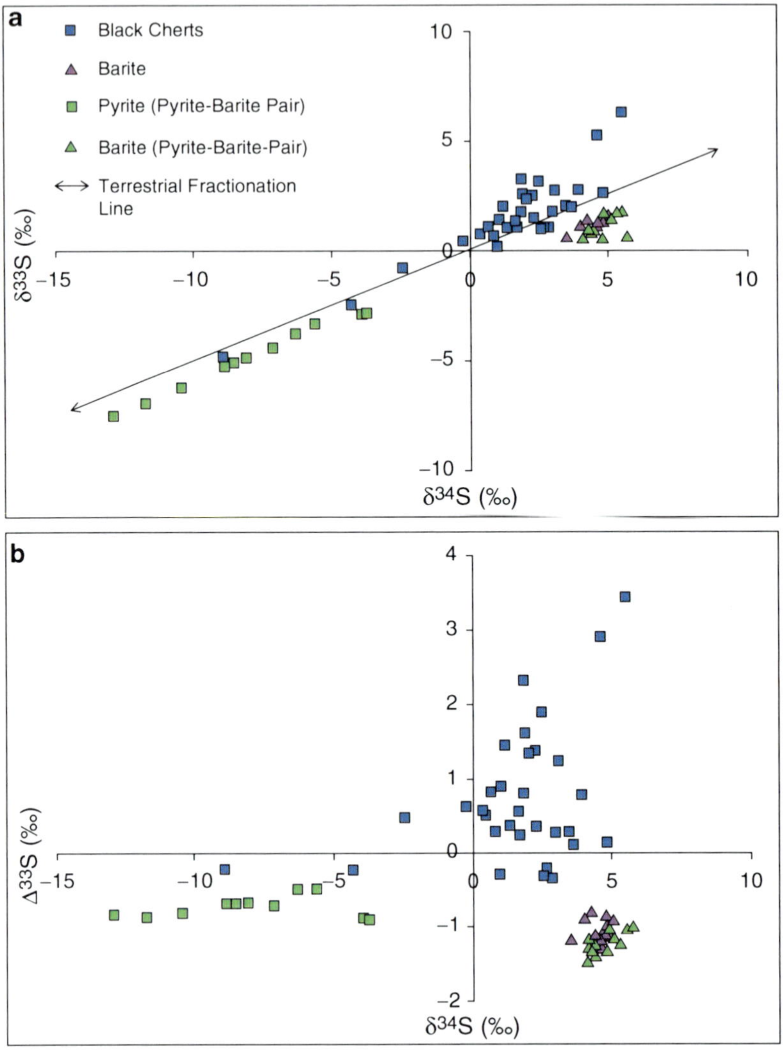

Fig. 5 (**a**, **b**) Multiple sulfur isotope compositions of sulfides and sulfates from the Dresser Formation black chert and barite units plotted with comparable data from previous work (Ueno et al. 2008; Shen et al. 2009). In this plot we have not distinguished between vein and bedded samples that have similar multiple sulfur isotope systematics

reduction of elemental sulfur to sulfide that produce ^{34}S isotope fractionations of only a few per mil (Ohmoto and Goldhaber 1997; Canfield et al. 1998; Smith 2000). The majority of black chert pyrites have positively correlated $\delta^{34}S$-$\Delta^{33}S$ values (Fig. 6a) at odds with compositions expected to result from atmospheric photolysis of volcanic gases with 193-nm UV radiation, which may reflect different atmospheric conditions (Farquhar and Wing 2003; Ono et al. 2003; Bao et al. 2007). Reactions between organic matter in the sediments and mass dependently

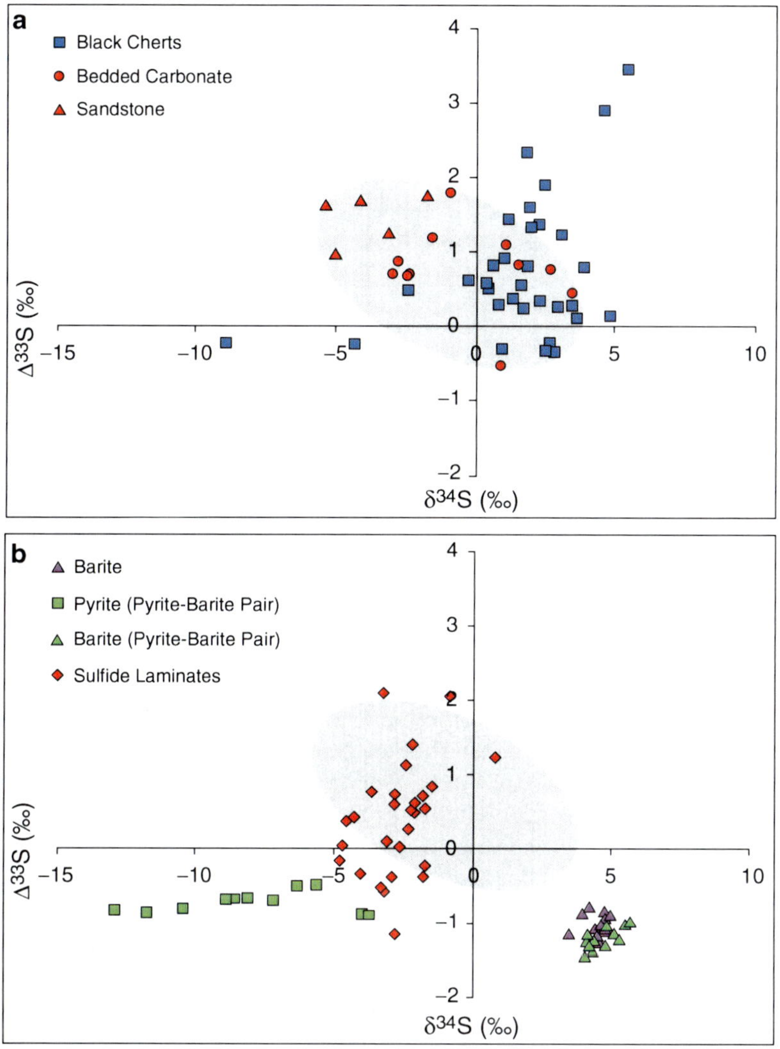

Fig. 6 (**a**) Plot of $\Delta^{33}S$ versus $\delta^{34}S$ shows that Dresser Formation black chert $\Delta^{33}S$ and $\delta^{34}S$ are positively correlated and overlap those of the bedded carbonates from drillholes PDP2B and PDP2C. Several pyrite analyses from DF2 and DF11 plot off the main trend and appear to have been mass dependently fractionated. (**b**) Plot of $\Delta^{33}S$ versus $\delta^{34}S$ shows that the macroscopic sulfide laminates from drillholes PDP2B and PDP2C have compositions intermediate between the volcanoclastic sandstones and bedded carbonates with positive $\Delta^{33}S$ anomalies (grey elliptical field) and the hydrothermal sulfides with negative $\Delta^{33}S$ anomalies. Data for drillholes PDP2B and PDP2C from Philippot et al. (2007)

fractionated sulfate-bearing hydrothermal fluids could also produce positive $\Delta^{33}S$ anomalies; however, the predominantly positive $\delta^{34}S$ values in the bedded black cherts and black chert veins militate against thermochemical sulfate reduction (TSR) that typically produces ^{34}S-depleted sulfide even in the case of the amino acid experiments (Watanabe et al. 2009). On the other hand, several black chert pyrites have more depleted $\delta^{34}S$ values and near zero $\Delta^{33}S$ (DF 2 and DF 11; Table 2), which may suggest they were produced by thermochemical or microbial reduction of sulfate of magmatic origin.

Pyrite occurring with barite in sample L6-B has negative $\delta^{34}S$ (−3.93‰ to −3.76‰) and $\Delta^{33}S$ (−0.89‰ to −0.88‰). The L6-B barite is enriched in ^{34}S relative to the associated sulfide by some 8‰, but has a similar, albeit somewhat more negative $\Delta^{33}S$ (−1.45‰ to −1.25‰) consistent with a common sulfur source for sulfide and sulfate (Table 2). More generally the bedded barites analysed in the current study have positive $\delta^{34}S$ values of 4.46‰ ± 0.40‰ (n = 15) and negative $\Delta^{33}S$ value of −1.08‰ ± 0.20‰ (n = 15) that are similar to those found previously by Farquhar et al. (2000), Philippot et al. (2007), Ueno et al. (2008) and Shen et al. (2009) (Fig. 5). The negative $\Delta^{33}S$ anomalies in sulfates and some sulfides of the Dresser Formation are strong evidence for the incorporation of sulfate of UV-photolysis origin and at odds with the interpretation of Van Kranendonk et al. (2008) that sulfate in the Dresser Formation resulted largely from disproportionation of magmatic SO_2. Moreover, the consistency of the negative $\Delta^{33}S$ anomalies at North Pole across vein and stratiform barite indicate a common sulfur source with a significant component of MIF-sulfur. While some of the barite in the Dresser Formation may have formed by diagenetic replacement of gypsum (Shen et al. 2001, 2009), the majority of the barite shows no evidence of gypsum precursors and precipitated from solution (Runnegar et al. 2001; Van Kranendonk et al. 2008; Harris et al. 2009). Because of its highly insoluble nature, barite forms most readily when a barium-bearing reduced fluid mixes with an oxidized sulfate-bearing fluid. This is a common occurrence in Phanerozoic VHMS systems where barium-bearing hydrothermal fluids mix with sulfate-bearing seawater at and beneath the seafloor (Huston and Logan 2004). The contrasting positive and negative $\Delta^{33}S$ anomalies of the sulfide-barite and sulfide-silica veins and stratiform deposits and their cyclic formation (Van Kranendonk et al. 2008; Ueno et al 2008) suggest the Dresser barites were formed in a similar way. The interpreted occurrence of evaporites in the Dresser stratigraphy indicates that contemporaneous seawater contained sufficient sulfate at least locally to precipitate gypsum. The variations in $\Delta^{33}S$ of the pyrite laminates of Philippot et al. (2007) may also have resulted from mixing of positive and negative MIF sulfide (Fig. 6b).

Ueno et al. (2008) and Shen et al. (2009) have shown that pyrite coexisting with barite has similar or slightly higher $\Delta^{33}S$ and lower $\Delta^{36}S$ than the barite (Figs. 5b and 6b), which they interpreted in terms of microbial sulfate reduction. On the other hand, Philippot et al. (2007) identified one sulfide-barite sample with positive $\Delta^{33}S$ and negative $\delta^{34}S$ that they suggested was due to microbial sulfur disproportionation. These interpretations are consistent with the quadruple sulfur isotope systematics of the sulfide-barite pairs but take little account of the nature of the samples and the likely environment of formation. Specifically, the sulfide-barite veins both feed and crosscut the stratiform sulfide-barite units that likely formed at temperatures significantly in excess of 100°C based on fluid inclusion studies (Harris et al. 2009) and modelling of the oxygen isotopic composition of the barites (Runnegar et al. 2002). We obtained similar temperatures from modelling of the oxygen isotopic composition of the black cherts that are spatially associated with the sulfide-barite units. These inferred temperatures preclude direct microbial involvement in the formation of the predominant sulfide in barite with negative $\Delta^{33}S$ and $\delta^{34}S$ values, which most likely formed through TSR using particulate or soluble organic matter sourced from the sediments or the hydrothermal system. Observed differences in $\Delta^{33}S$ and $\Delta^{36}S$ of the sulfide relative to coexisting sulfate can be explained by microbial sulfate reduction (Ueno et al. 2008; Shen et al. 2009) but are equally compatible with thermochemical reduction of negative MIF sulfate utilizing kerogen and/or abiogenic organic compounds (cf. Watanabe et al. 2009).

6.2 *Sulphur Springs*

VHMS mineralisation at Sulphur Springs and elsewhere in the Panorama Zn–Cu field occur mainly at the contact between the Sulphur Springs Group and the overlying Gorge Creek Group. The Kangaroo Caves Formation comprises basic to felsic volcanics and is intruded by the synvolcanic Strelley Granite that is interpreted to have been the heat source driving hydrothermal convection in the district (Brauhart et al. 1998, 2000). Resources have been defined at the Sulphur Springs, Kangaroo Caves and Bernts deposits (Brauhart et al. 2000) and base mineralisation interested at a number of other localities including the Roadmaster prospect. Brauhart et al. (2000) used oxygen isotopes of whole rocks to investigate fluid flow patterns and temperatures of water-rock interaction in the Panorama field. They showed there was a steady decrease in whole rock $\delta^{18}O$ values from the top to the bottom of the volcanic pile and zones of low $\delta^{18}O$ values underlying many of the VHMS deposits, which they interpreted to reflect increasing temperature with depth and in feeder zones consistent with a convective hydrothermal model. The systematically zoned alteration zones underlying Sulphur Springs and other VHMS deposits in the field record seawater-rock interaction at depth and the return of this hydrothermal fluid to the seafloor (Brauhart et al. 2000; Huston et al. 2001). A zone of sulfur depletion at the base of the volcanic pile characterised by $\delta^{34}S$ whole rock values up to 10‰ is interpreted to result from inorganic seawater sulfate reduction coupled with dissolution of volcanic sulfides (Huston et al. 2001). On the other hand sulfides from the Kangaroo Caves deposit exhibit a narrow range of $\delta^{34}S$ values around 0‰, which suggests volcanic and/or primary magmatic sulfur predominates in the VHMS deposits (Huston et al. 2001).

In the current study, multiple sulfur isotope values were obtained for sulfides microdrilled from drill core samples of hanging wall (HW) and massive sulfide mineralisation from the Sulphur Springs deposit, HW, massive sulfide and footwall (FW) mineralisation from the Roadmaster prospect and FW mineralisation from the Kangaroo Caves deposit (Table 3). Sulfides from the HW of the Sulphur Springs deposit exhibit the widest range of $\delta^{34}S$ values ($\delta^{34}S$ = 0.27‰ ± 1.49‰; n = 38) and positive $\Delta^{33}S$ anomalies up to 2.15‰ ($\Delta^{33}S$ = 1.16‰ ± 0.52‰; n = 38) (Fig. 7). Sulfides from the HW of the Roadmaster prospect exhibit a somewhat smaller range of $\delta^{34}S$ values ($\delta^{34}S$ = 1.19‰ ± 0.55‰; n = 5) and positive $\Delta^{33}S$

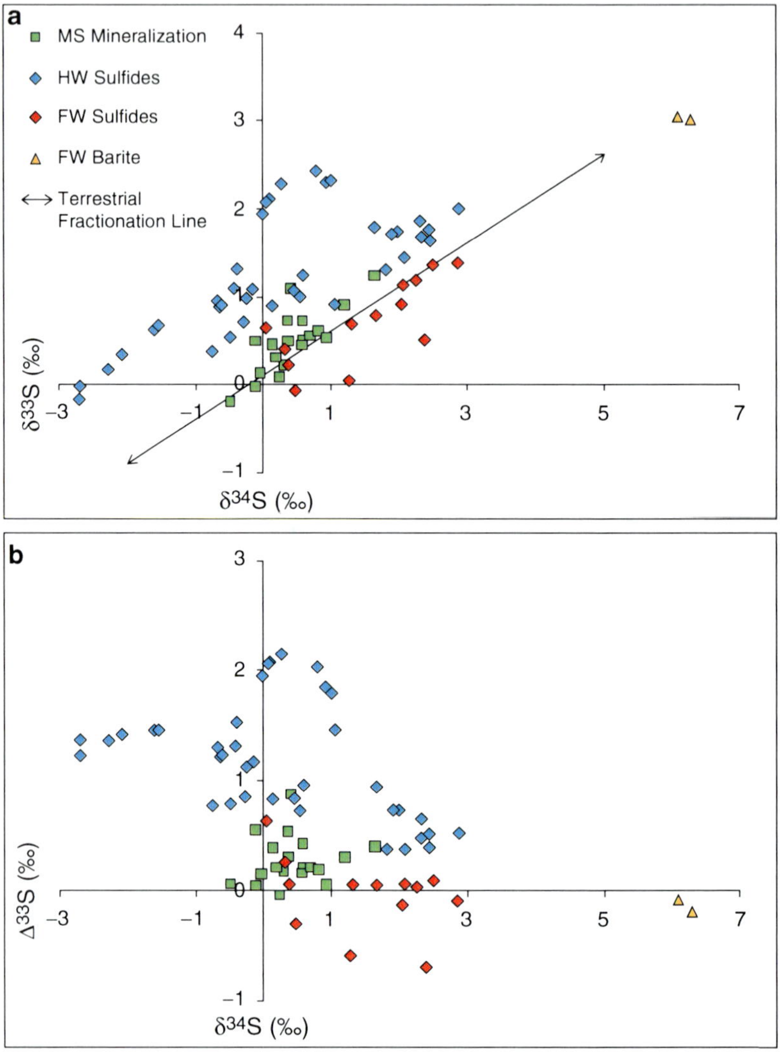

Fig. 7 (**a**, **b**) Multiple sulfur isotope compositions of sulfides and sulfates from the Sulphur Springs and Kangaroo Caves VHMS deposits separated on the basis of position within the individual ore systems that show $\Delta^{33}S$ increases with stratigraphic height (HW = hanging wall; MS = massive sulfide; FW = footwall)

anomalies up to 1.35‰ ($\Delta^{33}S$ = 1.03‰ ± 0.21‰; n = 5) (Fig. 8). HW sulfides comprise pyrite with lesser sphalerite and chalcopyrite and are preferentially hosted in Fe-carbonate altered siltstones. Siderite and ferroan carbonate is attributed to hydrothermal alteration of fine silts and organic matter rich muds of pelagic origin (Vearncombe et al. 1998), which may have provided a local elemental sulfur source of UV-photolysis origin with positive MIF.

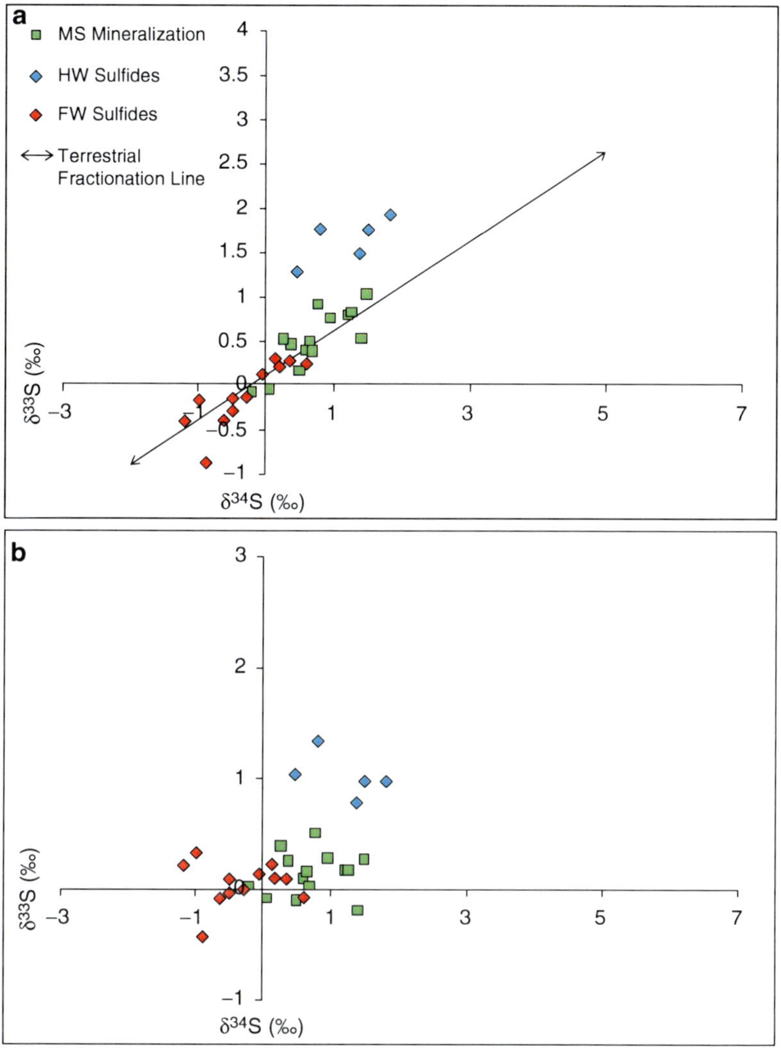

Fig. 8 (**a**, **b**) Multiple sulfur isotope compositions of sulfides from the Roadmaster prospect separated on the basis of position within the ore system that shows $\Delta^{33}S$ increases with stratigraphic height (HW = hangingwall; MS = massive sulfide; FW = footwall)

Our $\delta^{34}S$ data show a more limited range of values for massive sulfide mineralisation at Sulphur Springs ($\delta^{34}S = 0.43‰ \pm 0.50‰$; n = 19) and Roadmaster ($\delta^{34}S = 0.72‰ \pm 0.51‰$; n = 14) (Figs. 7 and 8). The $\Delta^{33}S$ data for massive sulfide mineralisation at Sulphur Springs ($\Delta^{33}S = 0.27‰ \pm 0.22‰$; n = 19) and Roadmaster ($\Delta^{33}S = 0.14‰ \pm 0.19‰$; n = 14) show mass dependent fractionation and positive $\Delta^{33}S$ anomalies in different sulfide generations of the same sample (Table 3), that indicates these sulfides had different sulfur sources and were not in isotopic equilibrium (cf. Jamieson et al. 2006). Mineral precipitation forming massive sulfide mineralisation on the seafloor occurs primarily through mixing between hydrothermal fluid and ocean bottom water (Ohmoto et al. 1983). Sulfides in the HW volcanics likely also have isotopic compositions that reflect mixing between late stage hydrothermal fluids and connate pore waters of marine origin. In this context, the lack of variation in sulfide $\delta^{34}S$ and variably positive $\Delta^{33}S$ values of the massive sulfide mineralisation are attributed to mixing between a hydrothermal fluid dominated by MDF igneous sulfur leached from the volcanic pile and connate waters of marine origin with reduced sulfur derived from aqueous disproportionation of atmospheric elemental sulfur. HW mineralisation exhibits a wider range of sulfide $\delta^{34}S$ and more positive $\Delta^{33}S$ values that likely indicates a higher contribution of MIF sulfur from connate fluids and/or the sedimentary host rocks of the Gorge Creek Group. The lack of negative MIF sulfur in the massive sulfide and HW mineralisation is noteworthy as is the general paucity of sulfate minerals in the deposits of the Panorama field.

Barite is a minor phase in the massive sulfide mineralisation at Sulphur Springs and Kangaroo Caves and is variably overprinted by hydrothermal silica in FW stockworks. Pyrite occurring with barite in sample KCD23-05 from the FW of the Kangaroo Caves deposit has near zero $\delta^{34}S$ (0.04–2.38‰) and both small positive and negative $\Delta^{33}S$ anomalies (−0.70‰ to 0.63‰) (Table 3). The KCD23-05 barite is enriched in ^{34}S relative to pyrite by some 4–6‰ and exhibits mass dependent fractionation ($\Delta^{33}S$ = −0.20‰ to −0.08‰). The small difference between the $\delta^{34}S$ values of the pyrite and the barite and the different $\Delta^{33}S$ values indicate that isotopic equilibrium was not attained between sulfate and sulfide during mixing of fluids and mineral precipitation. More generally our pyrite data show a modest range of $\delta^{34}S$ values for FW mineralisation at Kangaroo Caves ($\delta^{34}S$ = 1.50‰ ± 0.95‰; n = 13) and Roadmaster ($\delta^{34}S$ = −0.31‰ ± 0.56‰; n = 12) (Figs. 7 and 8). The $\delta^{34}S$ values overlap with previous data from the Kangaroo Caves deposit for pyrite ($\delta^{34}S$ = 0.10‰ ± 0.625‰; n = 8) and barite ($\delta^{34}S$ = 5.19‰ ± 0.36‰; n = 4) (Huston et al. 2001). On the other hand, FW mineralisation at the Kangaroo Caves deposit ($\Delta^{33}S$ = 0.27‰ ± 0.22‰; n = 19) and Roadmaster prospect ($\Delta^{33}S$ = 0.27‰ ± 0.22‰; n = 19) shows mass dependent fractionation and small positive and negative $\Delta^{33}S$ anomalies in different sulfide generations of the same sample (Table 3), which provides further evidence for fluid mixing and a possible component of seawater sulfate with negative sulfur MIF within the FW feeder zone. Thus there is a general trend of increasing $\Delta^{33}S$ with stratigraphic height in an individual ore system (Figs. 7 and 8); however, this zonation does not reflect absolute stratigraphic height in the district as the Sulphur Springs deposit at the top of the Kangaroo Caves Formation and the Roadmaster prospect deeper in the stratigraphy exhibit similar trends. This may suggest the hydrothermal fluid was dominated by dissolved rock sulfur of magmatic origin ($\Delta^{33}S \approx 0$‰) but included a small component of sulfur from inorganic reduction of seawater sulfate with negative $\Delta^{33}S$ (cf. Huston et al. 2001). The composition of this fluid seems to have evolved through time as disseminated pyrite with negative $\Delta^{33}S$ is overprinted by veinlet pyrite with MDF sulfur in the FW mineralisation at Roadmaster (RMD2-17) and Kangaroo Caves (KCD23-05).

7 Summary and Conclusions

Geochronology and stable isotope studies of the Dresser Formation show that the stratiform chert-barite units formed in a seafloor hydrothermal system with characteristics intermediate between a conventional seawater-dominated white smoker system and epithermal style vein mineralisation. The $\delta^{18}O$ values of the Dresser black cherts and quartz veins with epithermal textures are very similar and consistent with an intermediate temperature hydrothermal system dominated by sea water and/or low latitude meteoric water. Recent studies of modern shallow submarine hydrothermal systems have shown that active venting in water depths less than 500 m can result in epithermal style mineralisation or mineralisation transitional in character with classic massive sulfide deposits that form under deeper water conditions (e.g. Petersen et al. 2002; Sillitoe and Hedenquist 2003; Gemmell et al. 2004). In such systems black silica veins and quartz veins with epithermal textures and mineralogy indicative of near neutral hydrothermal fluids can occur with or overprint acid-sulphate style alteration as has been documented for the Dresser hydrothermal system (Van Kranendonk and Pirajno 2004; Van Kranendonk et al. 2008; Harris et al. 2009). While bedded silica-barite deposits can form directly on the seafloor at the site of shallow hot springs, solubility and kinetic considerations make it more likely that the Dresser silica-barite units were formed within unconsolidated sediments where hydrothermal fluids mixed with seawater as has been suggested for the Kuroko tetsusekiei ores of Japan (Ohmoto et al. 1983). The contrasting positive and negative $\Delta^{33}S$ anomalies of the sulfide-barite and sulfide-silica veins and stratiform deposits and their cyclic formation support this hypothesis.

Black cherts have positive $\Delta^{33}S$ values up to 1.91‰ that overlap those of a previous study (Ueno et al. 2008) and implicate an elemental sulfur source of UV-photolysis origin. The modest range of $\delta^{34}S$ values for the black cherts is consistent with inorganic reduction or disproportionation of elemental sulfur to sulfide or microbial reduction of elemental sulfur to sulfide. Reduction or disproportionation of elemental sulfur must have occurred in the hydrothermal convection cell as barite and coexisting pyrite have similar negative $\Delta^{33}S$ values suggesting a common seawater source different from that of the black chert pyrite. Atypically for Archean sulfide-sulfate pairs, the $\delta^{34}S$ values of pyrite are up to 22‰ less than that of coexisting barite and also have distinctive $\delta^{33}S/\delta^{34}S$ slopes consistent with microbial sulfate reduction (Shen et al. 2001, 2009; Ueno et al. 2008). These authors concluded on this basis that there was a strong case for the activity of sulfate reducing microbes at ca. 3.5 Ga. We note, however, that the reported quadruple sulfur isotope systematics of pyrite-barite pairs are equally permissive of a thermochemical origin for the pyrite based on the experimental work of Watanabe et al. (2009). Microbial processes were active in low temperature parts of the Dresser hydrothermal system based on the work of Glikson et al. (2008); however, the inferred temperatures of formation of the barite and associated pyrite based on fluid inclusion studies (Harris et al. 2009) and oxygen isotope modelling (Runnegar et al. 2002) are too high to permit microbial sulfate reduction and more compatible with thermochemical reduction of seawater sulfate. Moreover, the majority of sedimentary sulfides older than 2.8 Ga have positive $\Delta^{33}S$ implying an elemental sulfur source of UV-photolysis origin and/or thermochemical reduction of mass dependently fractionated sulfate (Fig. 1), which is inconsistent with an early Archean origin for microbial sulfate reduction. Multiple sulfur isotopes are unable to distinguish microbial and inorganic sulfur reduction such that these data do not preclude an early origin for sulfur reducers.

In contrast to the controversy surrounding the Dresser Formation, Sulphur Springs and other VHMS deposits of the 3.24 Ga Panorama Zn–Cu field are similar in morphology and mineralogy to modern seafloor massive sulfide deposits albeit with somewhat different sulfur isotope systematics. Our sulfur isotope data for Sulphur Springs, Kangaroo Caves and Roadmaster are similar to other Paleoarchean VHMS deposits in that they exhibit a narrow range of $\delta^{34}S$ values in sulfides from −3‰ to 3‰, whereas Phanerozoic massive sulfide deposits range from −5‰ to 20‰ (Huston 1999; Huston et al. 2001). Barite from the Kangaroo Caves deposit is enriched in ^{34}S relative to pyrite by some 4–6‰ that is typical for Paleoarchean deposits and contrasts with Phanerozoic deposits where the difference is typically 16–22‰ (Huston 1999). The variably positive $\Delta^{33}S$ anomalies of HW, massive sulfide and FW mineralisation, disequilibrium relations among sulfides and sulfates in the same sample and general trend of increasing sulfide $\Delta^{33}S$ with stratigraphic height in individual ore systems most likely reflects temperature evolution and fluid mixing at and beneath the seafloor through the life of the hydrothermal system. In this context recent modelling of hydrothermal fluid flow in the Panorama field shows that individual deposits could have formed in some 5,000 years, whereas district scale hydrothermal activity may have lasted for >200,000 years (Schardt and Large 2009). The absence of sulfides with significant negative $\Delta^{33}S$ anomalies contrasts with late Archean VHMS deposits (Jamieson et al. 2006) and suggests that volcanic sulfur not seawater sulfate was the dominant sulfur source for the hydrothermal system.

Paleoarchean seafloor mineral systems from the Pilbara Craton of Western Australia such as the Dresser Formation and Panorama Zn–Cu field are unusual in that they contain barite, whereas barite and other sulfates are generally absent from younger Archean VHMS mineralisation (Huston and Logan 2004). Previous workers have suggested these differences may reflect temporal and/or spatial variations in seawater sulfur content and redox conditions (Eastoe et al. 1990; Huston 1999; Huston and Logan 2004). In this context the negative $\Delta^{33}S$ anomalies of barite and associated pyrite of the Dresser Formation indicate the Paleoarchean seas were at least locally sulfate bearing. The presence of negative $\Delta^{33}S$ anomalies in the shallow water Dresser Formation and their general absence in the deeper water VHMS deposits of the Panorama field may suggest the occurrence of sulfate-bearing surface waters and more anoxic bottom waters.

References

Allwood AC, Walter MR, Kamber BS et al (2006) Stromatolite reef from the Early Archaean era of Australia. Nature 441(7094):714–718

Allwood AC, Walter MR, Burch IW et al (2007) 3.43 billion-year-old stromatolite reef from the Pilbara Craton of Western Australia: ecosystem-scale insights to early life on Earth. Precambr Res 158(3–4):198–227

Aizawa J (1990) Paleotemperatures from fluid inclusions and coal rank of carbonaceous material of the Tertiary formations in northwestern Kyushu, Japan. J Jap Assoc Min Petrol Econ Geol 85:145–154

Bao HM, Rumble D, Lowe DR (2007) The five stable isotope compositions of Fig Tree barites: Implications on sulfur cycle in ca. 3.2 Ga oceans. Geochim Cosmochim Acta 71: 4868–4879

Barley ME (1993) Volcanic, sedimentary and tectonostratigraphic environments of the 3.46 Ga Warrawoona Megasequence: a review. Precambr Res 60:47–67

Barley ME (1998) Archaean volcanic-hosted massive sulphides. AGSO J Aust Geol Geophys 17:69–73

Baublys KA, Golding SD, Young E et al (2004) Simultaneous determination of $\delta^{33}S_{V\text{-}CDT}$ and $\delta^{34}S_{V\text{-}CDT}$ using masses 48, 49 and 50 on a continuous flow isotope ratio mass spectrometer. Rapid Communications in Mass Spectrometry 18:2765–2769

Bekker A, Holland HD, Wang PL et al (2004) Dating the rise of atmospheric oxygen. Nature 427:117–120

Brasier MD, Green OR, Jephcoat AP et al (2002) Questioning the evidence for Earth's oldest fossils. Nature 416:76–81

Brasier M, Green O, Lindsay J et al (2004) Earth's oldest (3.5 Ga) fossils and the 'Early Eden hypothesis': questioning the evidence. Orig Life Evol Biosph 34:257–269

Brasier MD, Green OR, Lindsay JF et al (2005) Critical testing of Earth's oldest putative fossil assemblage from the 3.5 Ga Apex chert, Chinaman Creek, Western Australia. Precambr Res 140:55–102

Brauhart CW (1999) Regional alteration systems associated with Archean volcanogenic massive sulfide deposits at Panorama, Pilbara, Western Australia. Unpublished Ph.D., Department of Geology and Geophysics, University of Western Australia, Perth

Brauhart CW, Groves DI, Morant P (1998) Regional alteration systems associated with volcanogenic massive sulfide mineralization at Panorama, Pilbara, Western Australia. Econ Geol 93:292–302

Brauhart CW, Huston DL, Andrew AS (2000) Oxygen isotope mapping in the Panorama VMS district, Pilbara Craton, Western Australia: applications to estimating temperatures of alteration and to exploration. Miner Deposita 35:727–740

Buick R, Dunlop JSR (1990) Evaporitic sediments of Early Archaean age from the Warrawoona Group, North Pole, Western Australia. Sedimentology 37:247–277

Buick R, Brauhart CW, Morant P et al (2002) Geochronology and stratigraphic relationships of the Sulphur Springs Group and Strelley granite: a temporally distinct igneous province in the Archaean Pilbara Craton, Australia. Precambr Res 114:87–120

Canfield DE (2005) The early history of atmospheric oxygen: homage to Robert M. Garrels. Annu Rev Earth Planet Sci 33:1–36

Canfield DE, Raiswell R (1999) The evolution of the sulfur cycle. Am J Sci 299:697–723

Canfield DE, Thamdrup B, Fleischer S (1998) Isotope fractionation and sulfur metabolism by pure and enrichment cultures of elemental sulfur-disproportionating bacteria. Limnol Oceanogr 43:253–264

Cates NL, Mojzsis SJ (2006) Chemical and isotopic evidence for widespread Eoarchean metasedimentary enclaves in southern West Greenland. Geochim Cosmochim Acta 70:4229–4257

Clayton RN, Mayeda TK (1963) The use of bromine pentafluoride in the extraction of oxygen from oxides and silicates for isotopic analysis. Geochim Cosmochim Acta 27:43–52

Corliss JB, Baross JA, Hofmann SE (1981) An hypothesis concerning the relationship between submarine hot springs and the origin of life on Earth. Oceanologica Acta Suppl 4:59–69

de Ronde CEJ, Channer DM, Faure K et al (1997) Fluid chemistry of Archean seafloor hydrothermal vents: implications for the composition of circa 3.2 Ga seawater. Geochim Cosmochim Acta 61:4025–4042

Doolittle WF (1999) Phylogenetic classification and the universal tree. Science 284:2124–2128

Duck LJ, Glikson M, Golding SD et al (2004) Characteristics of organic matter in sediments from the 3.24 Ga Sulphur Springs VHMS deposit, Western Australia. Geochim Cosmochim Acta Suppl 68(11S):A798

Duck LJ, Glikson M, Golding SD et al (2005) Early Archaean carbonaceous material from the Pilbara, Western Australia: its nature, characteristics and possible sources. Geochim Cosmochim Acta Suppl 69:A861

Duck LJ, Glikson M, Golding SD et al (2007) Microbial remains and other carbonaceous forms from the 3.24 Ga Sulphur Springs black smoker deposit, Western Australia. Precambr Res 154:205–220

Eastoe CJ, Compston W, Williams IS et al (1990) Sulfur isotopes in Early Proterozoic volcanogenic massive sulfide deposits: new data from Arizona and implications for ocean chemistry. Precambr Res 46:353–364

Ertem G (2004) Montmorillonite, oligonucleotides, RNA and origin of life. Orig Life Evol Biosph 34:549–570

Farquhar J, Wing BA (2003) Multiple sulfur isotopes and the evolution of the atmosphere. Earth Planet Sci Lett 213:1–13

Farquhar J, Bao H, Thiemens M (2000) Atmospheric influence of Earth's earliest sulfur cycle. Science 289:756–758

Farquhar J, Savarini J, MH AS et al (2001) Observation of wavelength-sensitive mass-independent effects during SO2 photolysis: implications for the early atmosphere. J Geophys Res 106:32829–32839

Farquhar J, Peters M, Johnston DT et al (2007) Isotopic evidence for Mesoarchaean anoxia and changing atmospheric sulphur chemistry. Nature 449:706–709

Faure G, Mensing TM (2005) Isotopes: principles and applications, 3rd edn. Wiley, New Jersey

Fedo CM, Whitehouse MJ (2002) Metasomatic origin of quartz-pyroxene rock, Akilia, Greenland, and implications for Earth's earliest life. Science 296:448–1452

Fiebig J, Woodland AB, Spangenberg J et al (2007) Natural evidence for rapid abiogenic hydrothermal generation of CH_4. Geochim Cosmochim Acta 71:3028–3039

Früh-Green GL, Kelley DD, Bernascono SM (2003) 30,000 years of hydrothermal activity at the Lost City vent field. Science 301:495–498

Galtier N, Tourasse N, Gouy M (1999) A nonhyperthermophilic common ancestor to extant life forms. Science 283:220–221

Gemmell JB, Sharpe R, Jonasson IR et al (2004) Sulfur isotope evidence for magmatic contributions to submarine and subaerial gold mineralization: conical Seamount and Ladolam Gold Deposit, Papua New Guinea. Econ Geol 99:1711–1726

Glikson M, Duck LJ, Golding SD et al (2008) Microbial remains in some Earth oldest rocks: comparison with Methanocaldococcus jannaschii, a potential analogue. Precambr Res 164:87–200

Glikson M, Hickman A, Duck LJ et al (2010) Integration of observational and analytical methodologies to characterize organic matter in early Archaean rocks: distinguishing biological from abiotically synthesized CM structures. Chapter 8, 88 to 2010

Golding, SD, Young E (2005) Multiple sulfur isotope evidence for dual sulfur sources in the 3.24 Ga Sulphur Springs deposit. Geochim Cosmochim Acta Suppl 69(10):A449

Golding SD, Young E, Duck LJ et al (2006) Multiple sulfur isotope constraints on microbial processes in Archaean seafloor environments. Geochim Cosmochim Acta Suppl 70(18):A208

Green MG, Sylvester PJ, Buick R (2000) Growth and recycling of early Archaean continental crust: geochemical evidence from the Coonterunah and Warrawoona Groups, Pilbara Craton, Australia. Tectonophysics 322:69–88

Habicht KS, Gade M, Thamdrup B et al (2002) Calibration of sulfate levels in the Archean ocean. Science 298:2372–2374

Harris AC, White NC, McPhie J et al (2009) Early Archean hot springs above epithermal veins, North Pole, Western Australia: new insights from fluid inclusion microanalysis. Econ Geol 104:793–814

Henley RW (1996) Chemical and physical context for life in terrestrial hydrothermal systems: chemical reactors for the early development of life and hydrothermal ecosystems. In: Bock GR, Goode JA (eds) Evolution of hydrothermal ecosystems on Earth (and Mars?). Wiley, New York

Hickman AH, Van Kranendonk MJ (2004) Diapiric processes in the formation of Archaean continental crust, East Pilbara granite-greenstone terrane, Australia. In: Eriksson PG,

Altermann W, Nelson DR et al (eds) The Precambrian Earth: tempos and events. Elsevier, Amsterdam/Boston; Dev Precambr Geol 12:118–139

Hofmann A, Bolhar R (2007) Carbonaceous cherts in the Barberton Greenstone Belt and their significance for the study of early life in the Archean record. Astrobiology 7:355–388

Holm NG, Charlou JL (2001) Initial indications of abiotic formation of hydrocarbons in the rainbow ultramafic hydrothermal system, Mid-Atlantic Ridge. Earth Planet Sci Lett 191:1–8

Hu G, Rumble D, Wang PL (2003) An ultraviolet laser microprobe for the in situ analysis of multisulfur isotopes and its use in measuring Archean sulfur isotope mass-independent anomalies. Geochim Cosmochim Acta 67:3101–3117

Hulston JR, Thode HG (1965) Variations in the S33, S34 and S36 contents of meteorites and their relation to chemical and nuclear effects. J Geophys Res 70:3475–3484

Huston DL (1999) Stable isotopes and their significance for understanding the genesis of volcanic-hosted massive sulfide deposits: a review. Rev Econ Geol 8:157–179

Huston DL, Logan GA (2004) Barite, BIFs and bugs: evidence for the evolution of the Earth's early hydrosphere. Earth Planet Sci Lett 220:41–55

Huston DL, Brauhart CW, Drieberg SL et al (2001) Metal leaching and inorganic sulfate reduction in volcanic-hosted massive sulfide mineral systems: evidence from the paleo-Archean panorama district, Western Australia. Geology 29:687–690

Jamieson JW, Wing BA, Hannington MD et al (2006) Evaluating isotopic equilibrium among sulfide mineral pairs in Archean ore deposits: case study from the Kidd Creek VMS deposit, Ontario, Canada. Econ Geol 101:1055–1061

Kakegawa T, Ohmoto H (1999) Sulfur isotope evidence for the origin of 3.4 to 3.1 Ga pyrite at the Princeton gold mine, Barberton Greenstone Belt, South Africa. Precambr Res 96:209–224

Kakegawa T, Kawai H, Ohmoto H (1999) Origins of pyrites in the 2.5 Ga Mt McRae Shale, the Hamersley District, Western Australia. Geochim Cosmochim Acta 62:3205–3220

Kamber BS, Whitehouse MJ (2007) Micro-scale sulphur isotope evidence for sulphur cycling in the late Archean shallow ocean. Geobiology 5:5–17

Kamber BS, Collerson KD, Moorbath S et al (2003) Inheritance of early Archaean Pb-isotope variability from long-lived Hadean protocrust. Contr Miner Petrol 145:25–46

Kasting JF (2001) The rise of atmospheric oxygen. Science 293:819–820

Kaufman AJ, Johnston DT, Farquhar J et al (2007) Late Archean biospheric oxygenation and atmospheric evolution. Science 317:1900–1903

Kelley DS, Karson JA, Früh-Green GL et al (2005) A serpentinite-hosted ecosystem: the Lost City hydrothermal field. Science 307:1428–1434

Kitajima K, Maruyama S, Utsunomiya S et al (2001) Seafloor hydrothermal alteration at an Archaean mid-ocean ridge. J Metamorph Geol 19:583–599

Kramers JD, Tolstikhin IN (1997) Two terrestrial lead isotope paradoxes, forward transport modelling, core formation and the history of the continental crust. Chem Geol 139:75–110

Krapez B (1993) Sequence stratigraphy of the Archaean supracrustal belts of the Pilbara Block, Western Australia. Precambr Res 60:1–45

Land LS, Lynch FL (1996) $\delta^{18}O$ values of mudrocks: more evidence for an ^{18}O-buffered ocean. Geochim Cosmochim Acta 60:3347–3352

Lécuyer C, Allemand P (1999) Modelling of the oxygen isotope evolution of seawater: implications for the climate interpretation of the $\delta^{18}O$ of marine sediments. Geochim Cosmochim Acta 63:351–361

Lindsay JF, Brasier MD, McLoughlin N et al (2005) The problem of deep carbon – an Archean paradox. Precambr Res 143:1–22

Macleod G, McKeown C, Hall AJ et al (1994) Hydrothermal and oceanic pH conditions at 4 Ga relevant to the origin of life. Orig Life Evol Biosph 24:19–41

Martin W, Russell MJ (2007) On the origin of biochemistry at an alkaline hydrothermal vent. Phil Trans R Soc B 362:1887–1925

Martin W, Baross J, Kelley D, Russell MJ (2008) Hydrothermal vents and the origin of life. Nat Rev Microbiol 6:805–814

Matsuhisa Y, Goldsmith JR, Clayton RN (1979) Oxygen isotope fractionation in the system quartz-albite-anorthite-water. Geochim Cosmochim Acta 43:1131–1140

McCollom TM, Seewald JS (2007) Abiotic synthesis of organic compounds in deep-sea hydrothermal environments. Chem Rev 107:382–401

Mojzsis SJ, Coath CD, Greenwood JP et al (2003) Mass-independent isotope effect in Archaean (2.5 to 3.8 Ga) sedimentary sulfides determined by ion microprobe analysis. Geochim Cosmochim Acta 67:1635–1658

Morant P (1995) The Panorama Zn–Cu VMS deposits, Western Australia. AIG Bull 16:75–84

Morant P (1998) Panorama zinc-copper deposits. In: Berkman DA, Mackenzie DH (eds) Geology of Australian and Papua New Guinean mineral deposits. The Australasian Institute of Mining and Metallurgy, Melbourne

Muehlenbachs K (1998) The oxygen isotopic composition of the oceans, sediments and the seafloor. Chem Geol 145:263–273

Nijman W, de Bruijne KCH, Valkering ME (1998) Growth fault control of Early Archaean cherts, barite mounds and chert-barite veins, North Pole dome, Eastern Pilbara, Western Australia. Precambr Res 88:25–52

Nijman W, de Bruijne KCH, Valkering ME (1999) Growth fault control of Early Archaean cherts, barite mounds and chert-barite veins, North Pole Dome, Eastern Pilbara, Western Australia (Part 2). Precambr Res 95:247–274

Ohmoto H, Goldhaber MB (1997) Sulfur and carbon isotopes. In: Barnes HL (ed) Geochemistry of hydrothermal ore deposits, 3rd edn. Wiley, New York

Ohmoto H, Mizukami M, Drummond SE et al (1983) Chemical processes of Kuroko formation. Econ Geol Mongr 5:570–604

Ohmoto H, Watanabe Y, Ikemi H et al (2006) Sulphur isotope evidence for an oxic Archaean atmosphere. Nature 442:908–911

Ono S, Eigenbrode JL, Pavlov AA et al (2003) New insights into Archaean sulfur cycle from mass-independent sulfur isotope records from the Hamersley Basin, Australia. Earth Planet Sci Lett 213:15–30

Ono S, Shanks WC III, Rouxel OJ et al (2007) S-33 constraints on the seawater sulfate contribution in modern seafloor hydrothermal vent sulfides. Geochim Cosmochim Acta 71:1170–1182

Ono S, Kaufman AJ, Farquhar J et al (2009) Lithofacies control on multiple-sulfur isotope records and Neoarchean sulfur cycles. Precambr Res 169:58–67

Orberger B, Rouchon V, Westall F et al (2006) Microfacies and origin of some Archean cherts (Pilbara, Australia). In: Reimold WU, Gibson RL (eds) Processes on the early Earth. Geol Soc Am Spec Pap 405:133–156

Papineau D, Mojzsis SJ (2006) Mass-independent fractionation of sulfur isotopes in sulfides from the pre-3770 Ma Isua Supracrustal Belt, West Greenland. Geobiology 4:227–238

Partridge MA, Golding SD, Baublys KA et al (2008) Pyrite paragenesis and multiple sulfur isotope distribution in late Archean and early Paleoproterozoic Hamersley Basin sediments. Earth Planet Sci Lett 272:41–49

Pavlov AA, Kasting JF (2002) Mass-independent fractionation of sulfur isotopes in Archaean sediments: strong evidence for an anoxic Archaean atmosphere. Astrobiology 2:27–41

Pavlov AA, Kasting JF, Eigenbrode JL et al (2001) Organic haze in Earth's early atmosphere: source of low-13C Late Archean kerogens. Geology 29:1003–1006

Petersen S, Herzig PH, Hannington MD et al (2002) Submarine gold mineralization near Lihir Island, New Ireland Fore-Arc, Papua New Guinea. Econ Geol 97:1795–1814

Philippot P, Van Zuilen M, Lepot K et al (2007) Early Archaean microorganisms preferred elemental sulfur, not sulfate. Science 317:1534–1537

Pinti DL, Mineau R, Clement V (2009) Hydrothermal alteration and microfossil artefacts of the 3,465-million-year-old Apex chert. Nat Geosci 2:640–643

Poole AM, Jeffares DC, Penny D (1999) Early evolution: prokaryotes, the new kids on the block. BioEssays 21(10):880–889

Proskurowksi G, Lilley MD, Seewald JS et al (2008) Abiogenic hydrocarbon production at Lost City hydrothermal field. Science 319:604–607

Robert F, Chaussidon M (2006) A palaeotemperature curve for the Precambrian oceans based on silicon isotopes in cherts. Nature 443:969–972
Romero AB, Thiemens MH (2003) Mass-independent sulfur isotopic composition in present-day sulfate aerosols. J Geophys Res Atmosph 108(D16):4524
Runnegar B, Dollase WA, Ketcham RA et al (2001) Early Archaean sulfates from Western Australia first formed as hydrothermal barites not gypsum evaporates. Geol Soc Am Abstr 33:A404
Runnegar B, Coath DM, Lyons JR et al. (2002) Mass-independent and mass-dependent sulfur processing through the Archean. Geochim Cosmochim Acta Suppl 66(15A):A655
Russell MJ (2003) Origin and evolution of life: clues from ore deposits. Inst Mining Metallurgy Trans Sec B Appl Earth Sci 112:177–178
Russell MJ (2007) The alkaline solution to the emergence of life: energy, entropy and early evolution. Acta Biotheor 55:133–179
Russell MJ, Hall AJ (1997) The emergence of life from iron monosulphide bubbles at a submarine hydrothermal redox and pH front. J Geol Soc Lond 154:377–402
Russell MJ, Hall AJ (2006) The onset and early evolution of life. In: Kesler SE, Ohmoto H (eds) Evolution of early Earth's atmosphere, hydrosphere, and biosphere – constraints from ore deposits. Geol Soc Am Memoir 198:1–32
Russell MJ, Hall AJ, Boyce AJ et al (2005) 100th Anniversary special paper: on hydrothermal convection systems and the emergence of life. Econ Geol 100:419–438
Savarino J, Romero A, Cole-Dai J et al (2003) UV induced mass-independent sulfur isotope fractionation in stratospheric volcanic sulfate. Geophys Res Lett 30:2131
Schardt C, Large RR (2009) New insights into the genesis of volcanic-hosted massive sulfide deposits on the seafloor from numerical modeling studies. Ore Geol Rev 35:333–351
Schopf JW (1993) Microfossils of the Early Archean apex chert: new evidence of the antiquity of life. Science 260:640–646
Schopf JW (2006) Fossil evidence of Archaean life. Phil Trans R Soc B 361:869–885
Shanks WC, Böhlke JK, Seal RR (1995) Stable isotopes in Mid-Ocean Ridge hydrothermal systems: interactions between fluids, minerals, and organisms. In: Humphris SE, Zierenberg RA, Mullineaux LS et al (eds) Seafloor hydrothermal systems: physical, chemical, biological, and geological interactions. Geophys Monogr 91:194–221
Shen Y, Buick R, Canfield DE (2001) Isotopic evidence for microbial sulfate reduction in the early Archaean era. Nature 410:77–81
Shen Y, Farquhar J, Masterson A et al (2009) Evaluating the role of microbial sulfate reduction in the early Archean using quadruple isotope systematics. Earth Planet Sci Lett 279:383–391
Shock EL (1990) Geochemical constraints on the origin of organic compounds in hydrothermal systems. Orig Life Evol Biosph 20:331–367
Shock EL (1996) Hydrothermal systems as environments for the emergence of life. In: Bock GR, Goode JA (eds) Evolution of hydrothermal ecosystems on Earth (and Mars?). Wiley, New York
Shock EL, Schulte MD (1998) Organic synthesis during fluid mixing in hydrothermal systems. J Geophys Res 103(E12):28513–28527
Sillitoe RH, Hedenquist JW (2003) Volcanotectonic settings, ore-fluid compositions and epithermal precious metal deposits. In: Simmons SF, Graham I (eds) Volcanic, geothermal, and ore-forming fluids: rulers and witnesses of processes within the Earth. Soc Econ Geol Spec Publ 10:315–343
Simoneit BRT (1995) Evidence for organic synthesis in high temperature aqueous media – facts and prognosis. Orig Life Evol Biosph 25(1–3):119–140
Smith JW (2000) Isotopic fractionation accompanying sulfur hydrolysis. Geochem J 34:95–99
Stetter KO (1996) Hyperthermophiles in the history of life. In: Bock GR, Goode JA (eds) Evolution of hydrothermal ecosystems on Earth (and Mars?). Wiley, New York
Strauss H (2003) Sulphur isotopes and the early Archaean sulfur cycle. Precambr Res 126:349–361
Summons RE, Jahnke LL, Simoneit BRT (1996) Lipid biomarkers for bacterial ecosystems: studies of cultured organisms, hydrothermal environments and ancient systems. In: Bock GR, Goode JA (eds) Evolution of hydrothermal ecosystems on Earth (and Mars?). Wiley, New York

Taylor BE (1992) Degassing of H_2O from rhyolite magma during eruption and shallow intrusion, and the isotopic composition of magmatic water in hydrothermal systems. Geol Surv Jpn Mem 279:190–195

Terabayashi M, Masadab Y, Ozawa H (2003) Archean ocean-floor metamorphism in the North Pole area, Pilbara Craton, Western Australia. Precambr Res 127:167–180

Thiemens MH (2006) History and applications of mass-independent isotope effects. Annu Rev Earth Planet Sci 34:217–262

Thorpe RI, Hickman AH, Davis DW et al (1992) Constraints to models for Archaean lead evolution from precise zircon U–Pb geochronology for the Marble Bar region, Pilbara Craton, Western Australia. In: Glover JE, Ho SE (eds) The Archaean: terrains, processes and metallogeny. Geology Department and University Extension, The University of Western Australia, Western Australia

Tice MM, Lowe DR (2004) Photosynthetic microbial mats in the 3,616-Myr-old ocean. Nature 431:549–552

Towe KM (2000) The Archaean atmosphere and sedimentary sulfides. Science 289:1297

Ueno Y, Isozaki Y, Yurimoto H et al (2001) Carbon isotopic signatures of individual Archean microfossils (?) from Western Australia. Int Geol Rev 43:196–212

Ueno Y, Yoshioka H, Maruyama SY et al (2004) Carbon isotopes and petrography of kerogens in 3.5-Ga hydrothermal dikes in the North Pole area, Western Australia. Geochim Cosmochim Acta 68:573–589

Ueno Y, Yamada K, Yoshida N et al (2006) Evidence from fluid inclusions for microbial methanogenesis in the early Archaean. Nature 440:516–519

Ueno Y, Ono S, Rumble D et al (2008) Quadruple sulfur isotope analysis of ca. 3.5 Ga Dresser formation: new evidence for microbial sulfate reduction in the early Archean. Geochim Cosmochim Acta 72:5675–5691

Van Kranendonk MJ (1998) Litho-tectonic and structural components of the North Shaw 1:100000 sheet, Archaean Pilbara Craton. In: 1997–98 Annual Review, Geological Survey of Western Australia

Van Kranendonk MJ (2006) Volcanic degassing, hydrothermal circulation and the flourishing of early life on Earth: a review of the evidence from c. 3490–3240 Ma rocks of the Pilbara Supergroup, Pilbara Craton, Western Australia. Earth Sci Rev 74:197–240

Van Kranendonk MJ, Pirajno F (2004) Geochemistry of metabasalts and hydrothermal alteration zones associated with c. 3.45 Ga chert and barite deposits: implications for the geological setting of the Warrawoona Group, Pilbara Craton, Australia. Geochem Explor Environ Anal 4:253–278

Van Kranendonk MJ, Hickman AH, Smithies RH et al (2002) Geology and tectonic evolution of the Archean North Pilbara Terrain, Pilbara Craton, Western Australia. Econ Geol 97:695–732

Van Kranendonk MJ, Smithies RH, Hickman AH et al (2004) Event stratigraphy applied to 700 million years of Archaean crustal evolution, Pilbara Craton, Western Australia. In: 2003–04 Annual Review. Geological Survey of Western Australia

Van Kranendonk MJ, Hickman AH, Huston DL (2006a) Geology and mineralization of the east Pilbara – a field guide. Western Australia Geological Survey, Record 2006/16

Van Kranendonk MJ, Hickman AH, Smithies RH et al (2006b) Revised lithostratigraphy of Archean supracrustal and intrusive rocks in the northern Pilbara Craton, Western Australia. Western Australia Geological Survey, Record 2006/15

Van Kranendonk MJ, Philippot P, Lepot K et al (2008) Geological setting of Earth's oldest fossils in the ca. 3.5 Ga Dresser Formation, Pilbara Craton, Western Australia. Precambr Res 167:93–124

van Zuilen MA, Lepland A, Arrhenius G (2002) Reassessing the evidence for the earliest traces of life. Nature 418:627–630

Vearncombe S (1995) Volcanogenic massive sulphide-sulphate mineralisation at Strelley, Pilbara Craton, Western Australia. Unpublished PhD, Department of Geology and Geophysics, University of Western Australia, Perth

Vearncombe S, Barley ME, Groves DI et al (1995) 3.26 Ga black smoker-type mineralization in the Strelley Belt, Pilbara Craton, Western Australia. J Geol Soc Lond 152:587–590

Vearncombe S, Vearncombe JR, Barley ME (1998) Fault and stratigraphic controls on volcanogenic massive sulphide deposits in the Strelley Belt, Pilbara Craton, Western Australia. Precambr Res 88:67–82
Walter MR (1996) Ancient hydrothermal ecosystems on Earth: a new palaeobiological frontier. In: Bock GR, Goode JA (eds) Evolution of hydrothermal ecosystems on Earth (and Mars?). Wiley, New York
Watanabe Y, Farquhar J, Ohmoto H (2009) Anomalous fractionations of sulfur isotopes during thermochemical sulfate reduction. Science 324:370–373
Westall F, Southam G (2006) The early record of life. In: Benn K, Marescha J-C, Condie KC (eds) Archean geodynamics and environments. American Geophysical Union, Washington, DC; Geophys Monogr Ser 164:283–304
Whitehouse MJ, Kamber BS, Fedo CM et al (2005) Integrated Pb- and S-isotope investigation of sulphide minerals from the early Archaean of southwest Greenland. Chem Geol 222:112–131
Williams LB, Canfield B, Voglesonger KM et al (2005) Organic molecules formed in a "primordial womb". Geology 33:913–916

Archaean Hydrothermal Systems in the Barberton Greenstone Belt and Their Significance as a Habitat for Early Life

Axel Hofmann

Abstract Hydrothermal systems have played an important role in shaping the 3.54–3.23 Ga volcano-sedimentary succession of the Barberton greenstone belt. Evidence for relatively low-temperature (≤150°C) seafloor hydrothermal activity is widely recorded in extensive silicification of volcanic and sedimentary rocks, leaching of elements commonly mobile during water-rock interaction, and extensive hydraulic fracturing. Evidence for the presence of high-temperature hydrothermal vents is scarce and restricted to a massive sulphide deposit near the top of the succession. Many of the zones affected by seafloor alteration are spatially associated with traces of early life, such as carbonaceous matter and bioalteration features. Diffuse venting of low-temperature hydrothermal fluids was a widespread phenomenon on the Palaeoarchaean seafloor, making it an ideal habitat for hyperthermophiles and the possible birthplace of life during earlier times.

Keywords Barberton greenstone belt • Archaean • Hydrothermal activity • Seafloor alteration • Silicification

1 Introduction

The Barberton greenstone belt contains a well preserved Palaeoarchaean volcano-sedimentary succession that provides abundant evidence for the interaction of submarine volcanic rocks and seafloor sediments with the hydrosphere and biosphere. Petrological and geochemical studies have provided evidence for the existence, as early as 3.5 Ga, of a rich microbial ecosystem. This is preserved as filamentous and spherical structures of carbonaceous matter and microbial mats in

A. Hofmann (✉)
Palaeoproterozoic Mineralization Research Group, Department of Geology, University of Johannesburg, Johannesburg 2006, South Africa
e-mail: hofmannaxel@hotmail.com

S.D. Golding and M. Glikson (eds.), *Earliest Life on Earth: Habitats, Environments and Methods of Detection*, DOI 10.1007/978-90-481-8794-2_3,

cherts (Walsh 1992; Westall et al. 2001, 2006; Tice and Lowe 2006; Glikson et al. 2008) as well as possible bioalteration features in pillow lavas and hyaloclastites (Furnes et al. 2004; Banerjee et al. 2006). Many of these traces of life are associated with rocks that show evidence for hydrothermal activity on the Archaean ocean floor. Submarine hydrothermal systems, driven by circulation of seawater through volcanic rocks, constitute a possible site where life emerged and evolved. This paper provides a summary of features preserved in the Barberton greenstone belt that may be directly related to the presence of hydrothermal systems. The importance of these systems as a habitat of ancient life is briefly evaluated.

2 Geological Setting

The Barberton greenstone belt (Fig. 1) is one of the key belts for greenstone studies and represents a type locality of Palaeoarchaean supracrustal sequences. The Barberton greenstone belt consists of a NE–SW striking succession of supracrustal rocks, termed the Swaziland Supergroup, which ranges in age from 3.54 to 3.22 Ga (Viljoen and Viljoen 1969; de Ronde and de Wit 1994; Lowe and Byerly 2007a). The belt has a strike length of ca. 130 km, width of 10–35 km, and an approximate depth of 4–5 km, and is surrounded by granitoid domes and intrusive sheets, ranging in age from 3.51 to 3.10 Ga. The Swaziland Supergroup is subdivided into three stratigraphic units, the Onverwacht, Fig Tree and Moodies Groups (Fig. 2). These units have been tightly folded into a number of synclines that are separated by anticlines or shear zones. The metamorphic grade is greenschist facies, but locally reaches amphibolite facies close to the contact with the surrounding granitoid domes that intruded during several episodes of magmatism at ca. 3.45, 3.22, and 3.1 Ga (Kamo and Davis 1994).

2.1 Onverwacht Group

The Onverwacht Group formed between 3.54 and 3.30 Ga (Kröner et al. 1996; Byerly et al. 1996) and consists of komatiites, komatiitic basalts, basalts and minor dacites. Sedimentary rocks make up less than 10% of the succession and formed in a deep to shallow marine environment. South of the Inyoka Fault, the Onverwacht Group has been subdivided into six formations, the Sandspruit, Theespruit, Komati, Hooggenoeg, Kromberg and Mendon Formations (Viljoen and Viljoen 1969; Lowe and Byerly 2007a). The Sandspruit and Theespruit Formations are separated from the Komati Formation by a fault zone, but the younger units are stratigraphically conformable. The Onverwacht Group is best developed and least deformed in the southwestern part of the belt northeast of Tjakastad (Fig. 1). Metamorphic grade is mainly greenschist facies, but locally reaches amphibolite facies, in particular in the Sandspruit and Theespruit Formations. Onverwacht Group rocks north of the Inyoka Fault have been grouped together as the Weltevreden Formation (Lowe and Byerly 2007a).

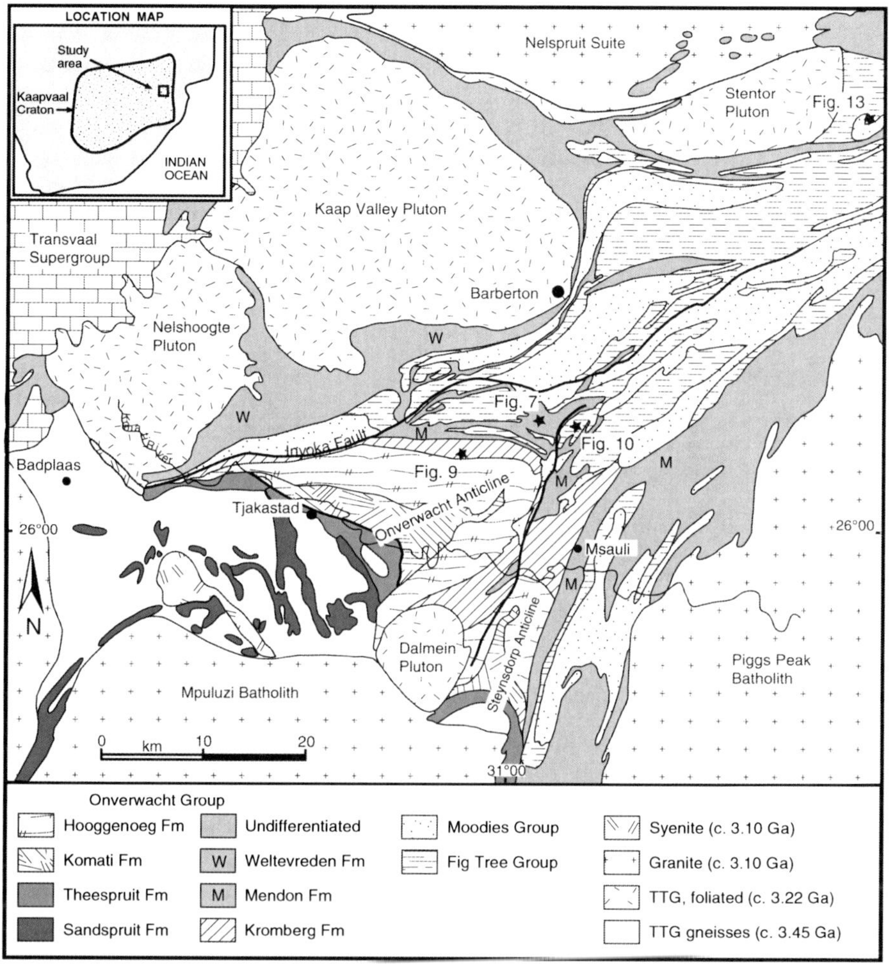

Fig. 1 Geological map of the southwestern part of the Barberton greenstone belt (Modified from Hofmann and Bolhar 2007). TTG, tonalite-trondhjemite-granodiorite

2.2 *Fig Tree Group*

The 3.26–3.23 Ga Fig Tree Group consists of a several kilometres thick siliciclastic-volcaniclastic sedimentary sequence that is capped by felsic volcanic rocks (Condie et al. 1970; Heinrichs 1980; Hofmann 2005; Lowe and Byerly 2007a). South of the Inyoka Fault, a variety of lithofacies are present that formed in deep- to shallow-water to alluvial environments (southern facies). The Loenen Formation is a sequence of sandstones and shales that contain abundant felsic detritus. It is overlain by ferruginous shales and jaspilitic banded iron formations of the Ngwenya Formation. The Mapepe Formation consists of shale, tuffaceous shale, laminated felsic tuff, chert-clast conglomerates and, locally, baryte beds.

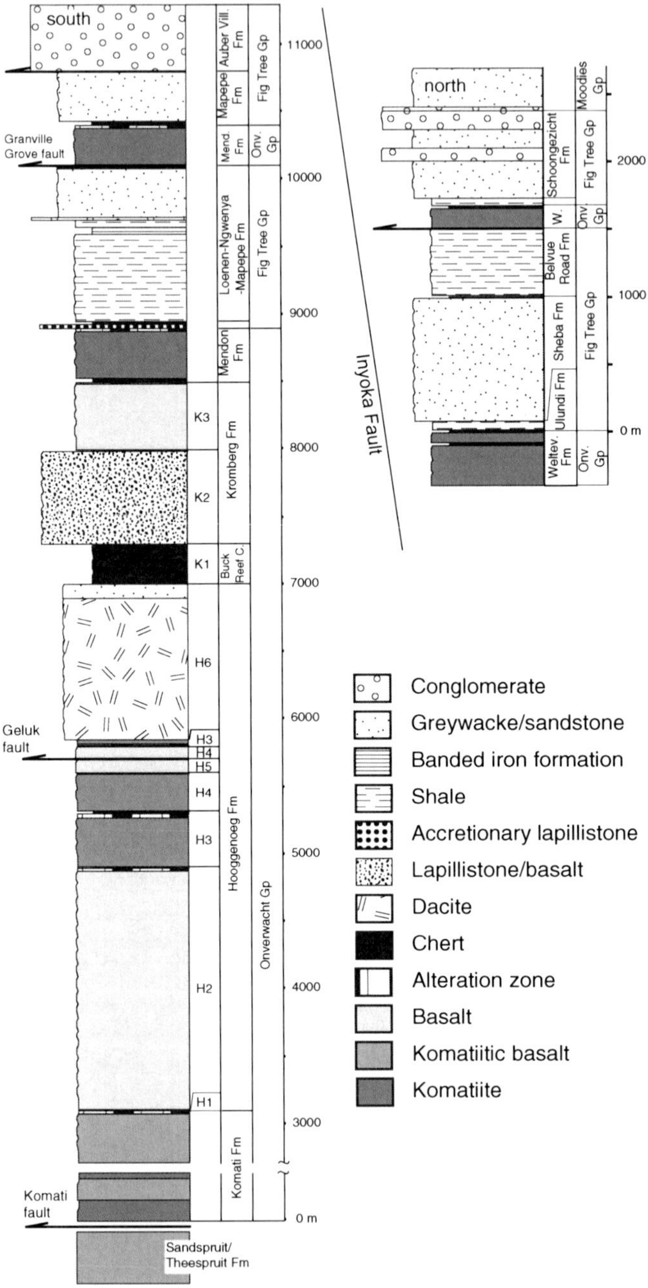

Fig. 2 Stratigraphic sections of the Swaziland Supergroup from the western limb of the Onverwacht Anticline (Fig. 1) and from areas north of the Inyoka Fault, as exemplified by the stratigraphy of the Stolzburg Syncline (Modified from Hofmann 2005)

Zircon dates from the Mapepe Formation vary from 3.26 to 3.23 Ma (Kröner et al. 1991; Byerly et al. 1996). The ca. 3.25 Ga Auber Villiers Formation is approximately 1.5 km thick and consists of dacitic, plagioclase-phyric volcaniclastic rocks and terrigeneous chert-clast conglomerates and sandstones.

In the northern part of the belt, the Fig Tree Group is mainly characterized by turbiditic sandstones and shales of the Ulundi, Sheba and Belvue Road Formations that formed in a relatively deep-water environment (northern facies). Kohler and Anhaeusser (2002) defined the Bien Venue Formation, which consists of quartz-muscovite schist derived from dacitic to rhyodacitic volcaniclastic protoliths dated at ca. 3.25 Ma. Subordinate rock types include banded chert, phyllite, and biotite-plagioclase and chlorite schists, derived, in turn, from dacitic and basaltic precursors. Massive sulphide and baryte deposits are also present. The Schoongezicht Formation overlies the Belvue Road Formation in the central and northwestern part of the greenstone belt. It consists of plagioclase-rich turbidites intercalated with shale at the base and cross-bedded volcaniclastic sandstones and dacite clast conglomerates at the top. Intercalated felsic volcanic rocks have been dated at 3,226 ± 1 Ma (Kamo and Davis 1994).

2.3 *Moodies Group*

The Moodies Group was deposited close to 3.227 Ga, but not later than 3.11 Ga (Kamo and Davis 1994). It consists of shallow-marine to fluvial sandstone and conglomerate with minor shale and banded iron-formation. The Moodies Group is approximately 3 km thick and consists of several fining-upward sequences ranging from conglomerate or pebbly quartzose sandstone at the base, to a thick sandstone unit, to capping siltstone and shale (Anhaeusser 1976).

3 Hydrothermal Systems in the Onverwacht Group

The Onverwacht Group is a thick pile of submarine lava flows that erupted intermittently over a time period of ca. 250 Ma. Breaks in volcanic activity resulted in the deposition of thin interflow sedimentary units characterized by predominantly tuffaceous volcaniclastic material. At the same time, hydrothermal processes were operating on, and immediately below, the seafloor, leaving behind extensive zones of alteration that are mainly characterized by the presence of silicified volcanic rocks, bedded cherts, and chert veins.

3.1 *Silica Alteration Zones*

Silicification of volcanic rocks immediately below sedimentary chert horizons is a common phenomenon of Palaeoarchaean volcano-sedimentary successions in general and the Barberton greenstone belt in particular (Kitajima et al. 2001; Terabayashi et al. 2003; Van Kranendonk and Pirajno 2004; Van Kranendonk 2006; Hofmann and Wilson 2007; Hofmann and Harris 2008). Because of the high silica content of the alteration zones, early workers in the Barberton greenstone belt regarded some of these zones to consist of felsic volcanic rocks, which were later shown to be silicified mafic and ultramafic rocks (Paris et al. 1985; Lowe and Byerly 1986; Duchac and Hanor 1987). The origin of the silicification has been the subject of debate, with interpretations ranging from weathering of volcanic flow tops (Lowe and Byerly 1986) to various processes of hydrothermal alteration (de Wit et al. 1982; Paris et al. 1985; Duchac and Hanor 1987; Hanor and Duchac 1990; de Wit and Hart 1993).

In order to test competing models for the origin of the alteration zones, Hofmann and Harris (2008) investigated several sections of silicified volcanic rocks underlying bedded carbonaceous cherts that included most stratigraphic units of the Onverwacht Group, ranging from 3.54 to 3.30 Ga in age. These authors observed that each and every chert horizon is underlain by rocks that are invariably silicified, and locally carbonatised, in zones 10–>50 m thick, irrespective of their composition. Silicification affected lithologies that range from ultramafic to felsic in composition and include volcanic, volcaniclastic, and epiclastic rocks. Carbonate alteration is common in mafic to ultramafic rocks, but also in coarse-grained sedimentary rocks (Rouchon et al. 2009). An example of such an alteration zone developed in pillow basalt and capped by a thin chert horizon from the top of the Hooggenoeg Formation is shown in Figs. 3 and 4. In all alteration zones, the intensity of silicification increases stratigraphically upward towards the contact with the chert, showing a gradual increase in SiO_2 content from the original igneous value to ~90%. Enrichment of SiO_2 is associated with a depletion of most other major elements, as monitored by element/Al_2O_3 or element/Hf ratios, including Fe_2O_3, MnO, MgO, and especially CaO and Na_2O. K_2O, together with Rb, and Ba, are enriched in these zones. In terms of trace elements,

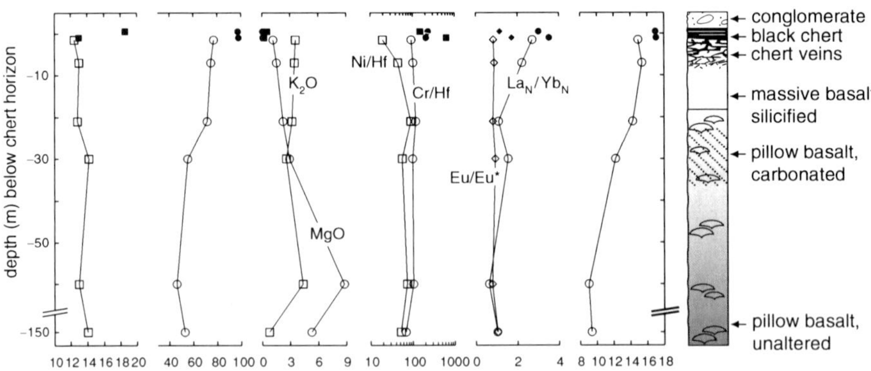

Fig. 3 Variations of selected geochemical data of silicified basalts (*open symbols*) with depth below chert H5c of the Hooggenoeg Formation. Analytical data of a bedded chert and a chert vein (*filled symbols*) are also shown (From Hofmann and Harris 2008)

Fig. 4 Features of the upper Hooggenoeg Formation (H5v). (**a**) Pavement of unaltered ocelli-bearing pillow basalt ca. 120 m below the capping chert bed. (**b**) Silicified and carbonatised pillow basalt ca. 20 m below the chert bed. The interpillow space is filled by secondary carbonate. (**c**) Network of carbonaceous chert veins transecting silicified basalt 2 m below the chert bed. (**d**) Photomicrograph (plane polarised light; scale bar is 1 mm) of highly silicified basalt shown in c. Large irregular patches consist of microcrystalline quartz and minor carbonate. Former plagioclase laths have been completely replaced by an intergrowth of quartz and phyllosilicates

depletion was observed for some metals (Ni, Co, Cu, Zn), Sr, the heavy rare earth elements (REE) and Y, while Eu became enriched. Several alteration zones were observed to show zoning, with zones of carbonate alteration present 20–30 m below the top.

Hofmann and Harris (2008) observed that $\delta^{18}O$ values of silicified volcanic rocks show a positive linear relationship with silica content and plot between the mantle value (ca. 5.7‰) and a $\delta^{18}O$ value of ca. 18‰ extrapolated for a composition of 100% SiO_2. A Si isotope study was reported by Abraham et al. (2007) from the same alteration zone shown in Fig. 3. Unsilicified basalts display negative $\delta^{30}Si$ values (−0.15‰ to −0.40‰), similar to average mafic magmatic rocks, whereas $\delta^{30}Si$ for silicified basalts shifts towards more positive values (up to 0.8‰) with increasing grade of silicification, with overlying chert displaying the highest Si isotope values (up to 1.1‰).

Hofmann and Harris (2008) concluded that the element depletion–enrichment patterns and oxygen isotope data indicate low-temperature (ca. 100–150°C) hydrothermal processes for the origin of the alteration zones. Diffuse venting over large tracts of ocean floor was envisaged to have given rise to silicification of the volcanic rocks and of seafloor sediments to form cherts. The Si isotope variation may be attributed to simple mixing of a mafic volcanic and seawater endmember composition, both of which had values similar to the modern day. Apparent zonations of alteration, with silicification at the top and carbonate alteration at depth may be a result of differing fluid chemistries with depth. In the deeper part of the alteration zones, hydrothermal fluids may have been more alkaline, becoming more acidic towards the contact with the likely acidic Archaean ocean due to mixing with seawater.

3.2 Bedded Cherts

Bedded chert horizons are typically 1–20 m thick and were deposited on the seafloor in between phases of extrusive submarine volcanic activity. With the exception of the Komati Formation, bedded cherts are common throughout the Onverwacht Group and the lower part of the Fig Tree Group, but are rare to absent in the upper Fig Tree and Moodies Groups. They typically consist of microcrystalline quartz with minor amounts of phyllosilicates, Fe-Mg-Ca-carbonates, Fe-Mn and Ti-oxide minerals, disseminated sulphides and carbonaceous matter. Cherts are highly siliceous rocks (SiO_2 commonly >95 wt%) that contain clastic or tuffaceous material of ultramafic to felsic composition replaced by microquartz. Despite the fact that addition of silica has resulted in dilution of most other elements, they have relatively high alkali contents and are enriched in many metals, particularly in Co, Ni, and Cu. The $\delta^{18}O$ values of cherts are somewhat higher than for underlying silicified volcanic rocks and are generally in the range of 15–20‰. The highest $\delta^{18}O$ value yet obtained for cherts of the Barberton greenstone belt is 22.1‰ (Knauth and Lowe 2003).

Bedded chert horizons consist of a variety of silicified sediments (Lowe 1999; Lowe and Fisher Worrel 1999; Hofmann 2005; Hofmann et al. 2006; Tice and Lowe 2006; Hofmann and Bolhar 2007). Silicified volcaniclastic sediments are common and include silicified beds of komatiitic ash and accretionary lapilli that are typically represented by light green to greenish grey cherts, the colouration due to finely dispersed chlorite and Cr-rich sericitic mica (Fig. 5a, b). Laminated cherts of various shades of grey to black represent mixtures of volcaniclastic material and carbonaceous matter. Carbonaceous cherts include thinly bedded heterolithic rocks (black and white banded cherts), and laminated and massive black chert varieties. Black and white banded cherts are made up of layers of carbonaceous chert and white-weathering translucent chert (Fig. 5c, d). Silicified orthochemical deposits include rare silicified evaporites (Fig. 5c), possibly primary sea-floor silica deposits in the form of translucent cherts, and banded iron formation and associated ferruginous rocks. Deposition took place predominantly in a low-energy, sub-wave base setting with episodic, high-energy current events during which coarse volcaniclastic material and, in rare cases, meteorite impact ejecta were deposited.

Fig. 5 Features of bedded chert. (**a**) Planar and wavy bedded green chert, consisting predominantly of komatiitic ash (Kromberg Formation). (**b**) Photomicrograph (plane polarised light; scale bar is 1 mm) of silicified volcaniclastic sandstone with accretionary lapilli and glass shards (Middle Marker). (**c**) Thinly bedded black-and-white chert transected by "evaporite" pseudomorphs (Buck Reef Chert). (**d**) Photomicrograph (plane polarised light; scale bar is 1 mm) of black-and-white chert showing lamina rich in carbonaceous grains (Buck Reef Chert)

It is generally agreed that silicification took place shortly after deposition of the precursor sediment on the seafloor (Lowe 1999), but the mechanism is a matter of debate. Many researchers regard the silicification as a direct result of hydrothermal processes on the seafloor (de Wit et al. 1982; Paris et al. 1985; Duchac and Hanor 1987). Hofmann and Harris (2008) suggested silicification being the result of diffuse flow of hydrothermal fluids that were also responsible for the silicification of the underlying rocks. Others refute the evidence for hydrothermal activity, and ascribe the formation of these rocks to marine processes that acted in a hot Archean ocean (e.g. Knauth and Lowe 2003; Tice and Lowe 2006). Cherts of the Barberton greenstone belt have much lower $\delta^{18}O$ values relative to Phanerozoic cherts (difference of $\geq$10‰ on average; Perry 1967; Knauth and Lowe 2003). These signatures have been interpreted in various ways, but mainly being the result of one or more of the following: (1) lower $\delta^{18}O$ values of Archaean seawater (Perry 1967; Kasting et al. 2006; Jaffres et al. 2007), (2) higher temperatures of the Archaean ocean (ca. 70°C; Perry 1967; Knauth and Lowe 2003), (3) temperatures of hydrothermal fluid–seawater mixtures and not the temperature of the Archaean ocean (Hofmann 2005; Hofmann and Harris 2008).

If seawater had $\delta^{18}O$ values as low as −13‰ at 3.4 Ga (Jaffres et al. 2007), this would imply that the alteration took place at temperatures of around 20°C. Such low temperatures are inconsistent with the geochemical and mineralogical changes in the alteration zones that need elevated temperatures to form. The model of a hot Archaean ocean, potentially resulting in downward infiltration of hot fluids from the seafloor and leading to top down alteration of volcanic rocks, is inconsistent with a number of observations that indicate the presence of upwelling hydrothermal fluids, such as depletion of metals (Co, Ni and Cu) in subsurface volcanic rocks and their relative enrichment in overlying chert (Hofmann 2005; Hofmann and Harris 2008), and the development of chert veins, regarded as hydraulic fractures beneath chert beds necessitating high fluid pressures at times (Hofmann and Bolhar 2007).

3.3 Chert Veins

Veins of carbonaceous chert occur immediately beneath sedimentary chert horizons, transecting underlying silicified volcanic rocks. They are 0.1–3 m wide and cross-cut the host rock both perpendicular and sub-parallel to stratification (Fig. 6). A detailed field study of chert veins revealed the importance of syndepositional hydrothermal activity for their origin (Hofmann and Bolhar 2007). The geometry of the chert-filled veins (e.g. the presence of stratiform veins and the branching of vertical veins into stratiform ones), multiple vein fillings, in situ brecciation of earlier generations of vein fillings, and brecciation of the host rock suggest that they represent hydraulic fractures that were initiated by the forceful intrusion of overpressured fluids.

Vein and bedded cherts of a single stratigraphic section are petrographically, isotopically as well as geochemically very similar. They are all characterized by sand-sized carbonaceous grains of apparent detrital origin. The differences between several pairs of vein chert and directly overlying bedded chert is <2‰ for C isotopes and <1.1‰ for O isotopes. The same applies to REE patterns and trace element ratios (Hofmann and Bolhar 2007; Hofmann and Harris 2008). All these observations suggest a genetic link between the veins and the sedimentary horizons.

While some workers regard chert veins as feeder channels for silicified sedimentary deposits in the Pilbara craton (e.g. Brasier et al. 2002; Van Kranendonk 2006), in the Barberton belt they have been regarded to represent fractures that were filled with sediment from above (Lowe and Byerly 1986; Hofmann and Bolhar 2007). In the model presented by Hofmann and Bolhar (2007), diffuse upflow of hydrothermal fluids in a shallow, but laterally very extensive hydrothermal system that was initially open to the ocean is envisaged during deposition of the sedimentary chert precursor. This system became temporarily closed during progressive sedimentation and hydrothermal silicification of the seafloor sediments that resulted in the bedded chert horizons. The cherts acted as a poorly permeable barrier for ascending fluids, resulting in the build-up of fluid overpressure and hydraulic fracturing of the volcanic rocks and the cap rocks at times. The hydraulic fractures were filled with

Fig. 6 Features of chert veins. (**a**) Two generations of chert veins cutting across highly silicified komatiite of the Mendon Formation. The vein filled with botryoidal chert is sub-parallel to stratification in the host rock. (**b**) Anastomosing chert veins in highly silicified sandstone below the Buck Reef Chert. The veins are subparallel to bedding. (**c**) Chert vein (margins not shown) containing fragments of silicified sedimentary host rock in a black chert matrix. Note strong brecciation of the fragments due to hydraulic fracturing. (**d**) Chert vein (note margin to silicified sediment in lower left) consisting of several generations of chert forming layers subparallel to the vein margins

sedimentary material from as yet unsilicified sediments overlying the chert for some time before the hydraulic system was sealed again and the next cycle of pressure build-up and fracturing took place. This eventually resulted in multiple injections of fluidized sedimentary material and hydrothermal fluids in cross-cutting veins, partly brecciating the host rock and incorporating host rock fragments into the veins.

3.4 Ironstone Pods: Archaean Hydrothermal Systems or Products of Recent Weathering

De Wit et al. (1982) and de Ronde et al. (1994) described discontinuous occurrences of lens-shaped "pods" of massive, specular haematite and goethite, up to 200 m in length, from the contact between Onverwacht komatiites and overlying sedimentary rocks at the base of the Fig Tree Group. One of these pods, which mainly consists of haematite (Fig. 7), was interpreted as a Fe-oxide-rich mound and

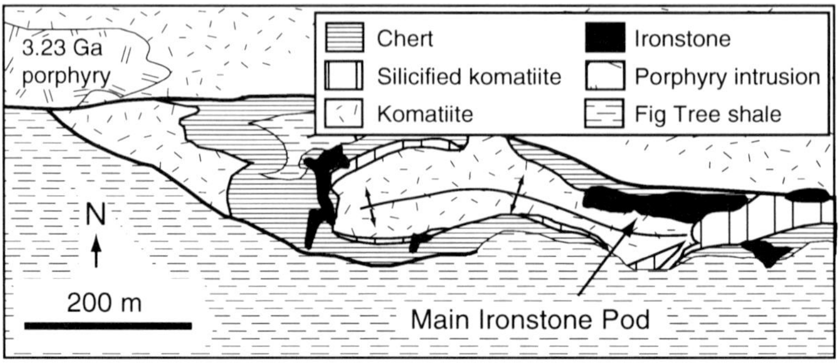

Fig. 7 Geological map of the ironstone pod locality on Mendon Farm (Modified from Lowe and Byerly 2007b). The dip of the strata is subvertical. See Fig. 1 for locality

discharge vent of 90–150°C hot hydrothermal fluids of Archaean age (de Ronde et al. 1994). Subsequent studies of the pod included mass spectroscopy that revealed a large variety of organic compounds (de Ronde and Ebbesen 1996), suggesting abundant organic productivity associated with the processes that led to ironstone pod formation. Fluid inclusion leachates obtained from quartz samples associated with the pod (Channer et al. 1997) yielded Br^-/Cl^- and I^-/Cl^- ratios higher than seawater, which was interpreted as a result of a secular change in ocean halide ratios through time. Additional fluid inclusion data on seawater and hydrothermal fluid endmember composition were reported by de Ronde et al. (1997).

Lowe and Byerly (2003) studied Fe-oxide-rich deposits at different localities in the Barberton greenstone belt, including the ironstone pod described by de Ronde et al. (1994). They concluded that the ironstones were deposited on and directly below the modern ground surface by active groundwater and spring systems during relatively recent (Cenozoic) times. This reinterpretation resulted in some debate in the literature (de Ronde et al. 2004; Lowe and Byerly 2004), which has since cooled down with additional work published by Lowe and co-workers (Hren et al. 2006; Lowe and Byerly 2007b).

Debate has centred on the pod originally described by de Ronde et al. (1994) to be of Archaean age (Fig. 7). De Ronde et al. (2004) cited the following main arguments for the ancient age of this pod: (1) The pod is regarded to be part of the Archaean stratigraphy; it is cut by a porphyry intrusion and clasts derived from ironstone pods are present in overlying sedimentary units. (2) It contains rounded chert boulders interpreted as talus derived from scarps in the vicinity of the mound. (3) Quartz-haematite-goethite stockwork veins are present in the pod. (4) Its haematitic composition is unlike low-T spring deposits. Lowe and Byerly (2004, 2007b) argued for a recent origin of the ironstone pod, through partial replacement of Archaean chert host rocks by Fe-oxide-rich material. These authors based their interpretation on the following main observations: (1) No evidence was found that the pods are part of the stratigraphy, as they are not cut by Archaean porphyry or provided detritus upon erosion during Archaean times as claimed by de Ronde et al. (1994). Instead, they locally observed layering in the pod that dips parallel to the present slope.

(2) Internal chert blocks are not rotated and represent in situ remnants not affected by replacement of chert by haematite. (3) The pods do not show evidence for deformation that characterizes the surrounding rocks and contain common open cavities, which would not have been preserved up to the present. Lowe and Byerly (2004, 2007b) further argue that the quartz stockworks represent remnants of Archaean quartz veins that were only partially replaced by ironstone.

Our own observations of ironstone pods in the Barberton greenstone belt tend to agree with a relatively recent origin of these features. Nevertheless, a large part of the work by de Ronde and co-workers was done on fluid inclusions of quartz veins that are of undoubted Archaean age. These quartz veins are commonly associated with chert veins interpreted to have formed during seafloor hydrothermal activity (Hofmann and Bolhar 2007) and may thus still provide evidence for hydrothermal fluid composition and temperature of an Archaean hydrothermal system. This is in contrast to the claim by Lowe and Byerly (2007b) that these features are a result of late-stage hydrothermal alteration of the belt (as late as 2.7 Ga). If the haematite-rich ironstone pod is a result of relatively recent spring activity, two questions still remain. One issue is the predominance of haematite in the pod, which generally only precipitates from relatively high-temperature fluids. This led Lowe and Byerly (2007b) to suggest a hot spring origin for the ironstone pod. No hot spring has so far been reported from the greenstone succession of the Barberton belt. Another major issue is the source of the iron. Lowe and Byerly (2007b) suggest that the iron was likely derived from the weathering of siderite-rich sedimentary units at depth, such as banded iron-formation. This is far from proven, and will require further investigations.

3.5 *Heat Source for Onverwacht Hydrothermal Activity*

Hydrothermal systems consist of two essential components, a heat source and a fluid phase (Pirajno 1992). While the fluid phase is most readily attributed to seawater circulating through the volcanic pile of the Onverwacht Group, the heat source is less easy to define. Most hydrothermal systems operating on the seafloor today are driven by heat generated from shallow-level magma chambers, especially those developed close to the mid-ocean ridges (Alt 1995). Other heat sources may include shallow-level magma chambers, intrusive igneous bodies or volcanic complexes associated with volcanic islands, oceanic plateaus and island arcs. Crustal sections close to these heat sources may experience extensive alteration related to the interaction with actively circulating >350°C hot hydrothermal fluids, which may locally discharge on the seafloor through black smokers. Further away from these zones fluid temperatures will be much lower, and hydrothermal fluid discharge is much more diffuse (Alt 1995; German and von Damm 2003).

Although the Onverwacht Group was deposited in a submarine setting and formed mainly as a result of subaquatic deposition of mafic-ultramafic lavas, there is no evidence that the volcanic rocks formed close to an Archaean mid-oceanic ridge. The lavas accumulated during an extensive period of time far longer than the average life span of modern oceanic crust. The alteration zones are laterally extensive and tabular, bearing no resemblance to the heavily faulted spreading centres along modern mid-ocean ridges. Instead, the Onverwacht Group is characterized by laterally continuous submarine lava flows more akin to large submarine shield volcanoes or oceanic plateaus (Cloete 1999; Hofmann and Harris 2008). Focused hydrothermal upflow zones in the form of epidosites and black smoker-type massive sulphides are lacking throughout the Onverwacht Group, suggesting diffuse upflow of hydrothermal fluids through the volcanic rocks.

Although there are several sub-volcanic intrusions in the Barberton sequence, ranging from ultramafic to felsic in composition, that could have been the heat source for the hydrothermal activity, there are many more examples of stratiform alteration zones in the Barberton greenstone belt that cannot be directly linked to such intrusions. Each volcanic sequence capped by chert is altered at the top, indicating that the alteration took place during a time interval of relative volcanic quiescence, when there was enough time for the deposition of finely laminated sediments before the onset of the next volcanic episode. Hofmann and Harris (2008) attributed the hydrothermal activity to heat derived from cooling of volcanic rocks and a high regional heat flow in an oceanic plateau-like setting in the vicinity of a long-lived source of hot, upwelling mantle that resulted in the establishment of shallow subseafloor convection cells. Shallow-level igneous intrusions may have provided additional heat, but they were not the main factor for elevated heat flow. Extensive silicification was the result of diffuse venting of hydrothermal fluids over broad areas of the ancient ocean floor, likely aided by seawater chemistry close to silica saturation and thus more conducive for widespread silicification (Hofmann and Harris 2008).

3.6 *Buck Reef Chert: Deposition During Hydrothermal or Normal Marine Conditions?*

While most of the Onverwacht cherts can be readily associated with active hydrothermal activity during their formation, this may not necessarily be the case for the Buck Reef Chert (BRC) for which a hydrothermal influence has been most strongly debated. The BRC is an unusually thick (up to 350 m) sequence of predominantly black-and-white banded cherts at the contact between the Hooggenoeg and Kromberg Formations along the western limb of the Onverwacht Anticline. It overlies a shallow intrusive to extrusive sequence (member H6 of Lowe and Byerly 2007a) of dacitic volcanic rocks, up to 2 km thick, dated at 3.45 Ga (de Wit et al. 1987; Armstrong et al. 1990; De Vries et al. 2006) and an epiclastic sedimentary unit of dacite-derived conglomerates and sandstones. The BRC has been subdivided into three facies (Fig. 8; Lowe and Fisher Worrel 1999; Tice and Lowe 2004, 2006): a basal evaporitic facies, containing silicified sandstones, chert and pseudomorphs after nahcolite; a lower division of platform facies black-and-white banded chert; a basin facies of banded ferruginous chert; and an upper division of black-and-white banded chert. A variety of fossil-like microstructures have been reported from the

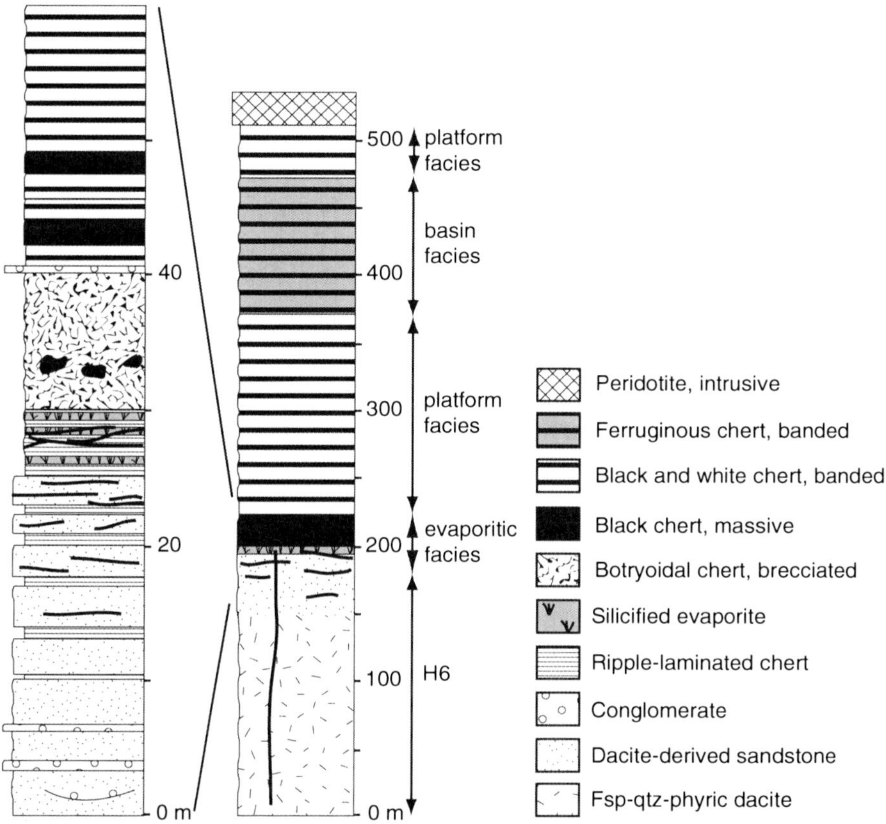

Fig. 8 Simplified section of the Buck Reef Chert (Modified from Hofmann and Bolhar 2007)

BRC (Walsh 1992; Tice and Lowe 2006). A felsic tuff at the base of the BRC has been dated at 3,416 ± 5 Ma (Kröner et al. 1991). Along the east limb of the Onverwacht Anticline, H6 is represented by a fining-upward sedimentary sequence of dacite-clast conglomerate, turbiditic sandstone and shale (Rouchon et al. 2009). These rocks are overlain by a sequence of ultramafic lapillistone and pillow basalt that are intercalated with several chert horizons that have been correlated with the BRC (Lowe and Byerly 2007a).

Debate has centred on the interpretation of the palaeoenvironment. De Vries (2004) and de Vries et al. (2006) consider the BRC to be the uppermost silicified part of the Buck Ridge volcano-sedimentary complex (BR-vsc), a series of arc-like basaltic and felsic volcanic rocks, capped by pervasively silicified sedimentary rocks. This complex is believed to be characterized by syndepositional normal faulting, which widely controlled felsic magma upwelling and sedimentation. De Vries et al. (2006) advocate hydrothermal venting as a common growth-fault related phenomenon directly related to the depositional history of the unit, giving rise to features such as chert veins, hydraulic breccia bodies and quartz-filled cavities.

In contrast, Lowe and Byerly (2007a) regard the BRC to have been deposited on a subsiding volcanic platform representing the eroded felsic volcanic edifice of H6 that was emplaced well before deposition of the BRC. Syn-depositional normal faulting was observed to have taken place during deposition of the lower half of the BRC (Lowe and Fisher Worrel 1999). Tice and Lowe (2004, 2006) suggested that the BRC consists of normal marine sediments that were silicified through interaction with hot marine waters supersaturated with amorphous silica and related abundant carbonaceous matter to microbial mats in a non-hydrothermal environment.

Hofmann and Bolhar (2007) and Hofmann and Harris (2008) suggested hydrothermal activity during deposition of at least the lower part of the BRC (the evaporitic facies). Dacite and dacite-derived clastic sedimentary rocks below and interbedded with the evaporitic cherts are highly silicified and show the same chemical variations with depth as observed in all the other alteration zone below chert horizons. Multiple generations of cross-cutting veins of chert are common in the alteration zone. These are both stratiform as well as at a high angle to bedding, the latter extending several 100 m below the BRC.

If the evaporitic facies was indeed affected by syn-depositional hydrothermal activity, then the chert clast breccias commonly observed in this zone may be attributed to subsurface dissolution of evaporites by hydrothermal fluids rather than to evaporite dissolution in the phreatic zone (Lowe and Fisher Worrel 1999). A study of fluid inclusions in quartz veins and quartz-filled cavities by de Vries and Touret (2007) from the BRC gave conflicting results, although it pointed towards relatively low pressure of ~100 bars for quartz precipitation, in support of a shallow water environment and synsedimentary hydrothermal activity.

In addition, the evaporites may not be precipitates from purely marine waters. Elongate crystal pseudomorphs interpreted to represent pseudomorphs after nahcolite ($NaHCO_3$; Lowe and Fisher Worrel 1999) are associated with finely laminated and rippled cherts and grew concurrently with deposition of the host sediment. The nahcolite crystals have been interpreted as primary evaporitic minerals that grew in a hypersaline marine coastal environment with high bicarbonate and sodium contents. If the pseudomorphs are indeed former nahcolite and if the BRC was deposited during hydrothermal activity, there is a possibility that elevated concentrations of bicarbonate and/or sodium are derived from hydrothermal fluids. Alteration in the underlying dacite and dacite-derived sediments has resulted in the almost complete removal of Na from at least the uppermost 100 m of rock succession (<<0.1 wt% Na_2O, Hofmann and Harris 2008), as the Na content of the unaltered magmatic rocks was likely similar in composition to the genetically related Theespruit Pluton (ca. 7 wt% Na_2O, Kleinhanns et al. 2003). It is thus possible that Na was derived from syn-depositional hydrothermal fluids. It needs to be emphasized that nahcolite has only been observed at the base of the BRC where it overlies altered felsic igneous rocks.

Chert veins were not observed cutting across the basal evaporitic facies. This may indicate that hydrothermal activity ceased before deposition of the overlying cherts or that the evaporitic facies acted as a barrier for ascending hydrothermal fluids. Cessation of hydrothermal activity can be attributed to reduced convection of seawater as the basal chert beds would have prevented continuous recharge of the hydrothermal system. If this was the case, deposition of cherts higher up in the BRC may have been less affected by hydrothermal activity. This may explain the abundance of soft-sediment deformation features in black-and-white banded cherts above the evaporitic facies (Tice and Lowe 2004), suggesting silicification acted more slowly. $\delta^{18}O$ values for the BRC are amongst the highest values obtained by Knauth and Lowe (2003) from the Barberton greenstone belt and range from 15.7‰ to 21.4‰, which is in line with subdued hydrothermal activity.

If BRC deposition was indeed influenced by hydrothermal activity, then several potential heat sources need to be considered. The sill-like dacite intrusion underlying the BRC could have been the heat source, but this is unlikely for several reasons. (1) The dacite is erosively overlain by dacite-derived sandstones that are silicified at the top and may have had sufficiently cooled down prior to subaerial exposure and erosion not to have been responsible for the hydrothermal alteration as seen at the top. (2) Available age data suggest that the dacite formed long before the BRC was deposited (ca. 3.45 versus 3.42 Ga, see above). The heat source could thus have been an elevated regional geotherm or renewed volcanic activity that gave rise to the overlying volcanic rocks. The BRC is overlain by undated mafic/ultramafic lapillistones and tuffs of the Kromberg Formation, it is cut by abundant ultramafic dykes and sills that may be genetically related to the overlying unit, it contains rare layers of ultramafic accretionary lapilli, and has been regarded as interleaved laterally with mafic and ultramafic rocks (Ransom et al. 1999; Lowe and Byerly 2007a). All these observations suggest that mafic/ultramafic magmatism was ongoing during deposition of the middle to upper part of the BRC. It is very well possible that extensional deformation of the sequence was related to this magmatic event.

With the evidence for syndepositional volcanism and potential extensional deformation, it is hard to evoke the absence of elevated heatflow and hydrothermal activity during BRC deposition. This leads to the evaluation of a possible hydrothermal control of the deposition of ferruginous cherts. Ferruginous cherts in the BRC are composed mainly of microquartz and goethite, although siderite is regarded to be the main iron-bearing mineral at depth (Tice and Lowe 2004, 2006). Deposition was regarded to have taken place in a low-energy environment, far below storm wave base, from a stratified Archaean ocean consisting of a shallow, CO_2-dominated layer and a deep, iron-rich layer. No hydrothermal influence was invoked, partly based on the lateral continuity of the different sedimentary facies, including the ferruginous basin facies (Tice and Lowe 2004, 2006). Recent mapping has revealed that the ferruginous facies may not be as continuous as previously thought (Fig. 9). In the map area, the unit of ferruginous cherts is discontinuous and shows changes in thickness along strike, with the maximum thickness being present in a lenticular area where rocks are strongly impregnated by goethite. The goethite forms layers that locally cross-cut bedding in the cherts and strongly resemble ironstone pod-like features described by Lowe and Byerly (2007b) from the BRC. Layered goethite was interpreted by these authors to represent precipitates from upward-migrating groundwater, with the Fe derived from dissolution of siderite bands in the ferruginous chert at depth. While the goethite is clearly secondary and

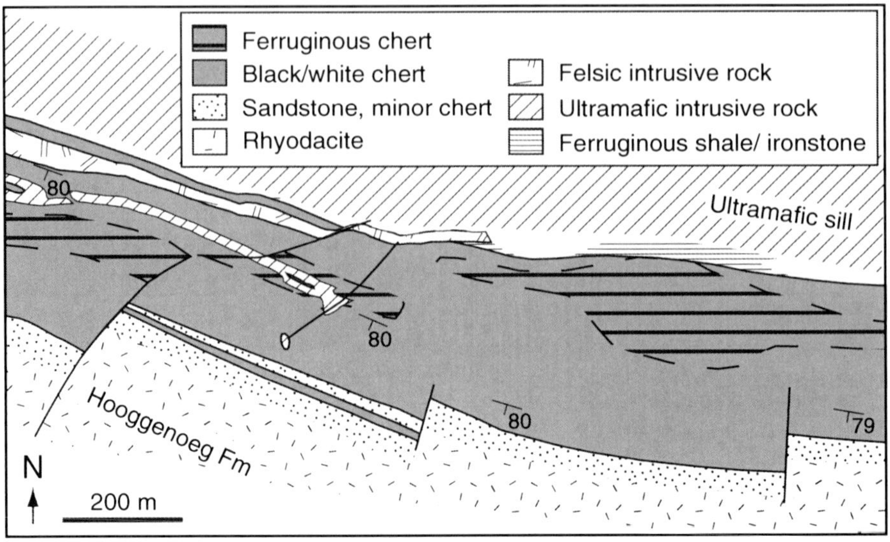

Fig. 9 Geological map of a portion of the Buck Reef Chert. The strata are younging to the NE. See Fig. 1 for locality

likely relatively young, this does not mean that it could not have been derived from weathering of a lenticular body of unusual ferruginous rock at depth, possibly including sulphide minerals and derived from seafloor hydrothermal activity. These speculations will require subsurface investigations by drilling.

4 Hydrothermal Systems of the Fig Tree Group

The 3.26–3.23 Ga Fig Tree Group is a volcano-sedimentary sequence that is dominated by relatively fine-grained siliciclastic sediments and felsic volcanic and volcaniclastic rocks. Carbonaceous cherts are uncommon, but many of the shale horizons are rich in carbonaceous matter (Reimer 1975), although reports of microfossil finds have been met with scepticism (Schopf and Walter 1983). Evidence for hydrothermal activity during Fig Tree deposition is mainly preserved in the presence of baryte horizons, a massive sulphide deposit and localized hydrothermal alteration of the sedimentary rocks.

4.1 Baryte

Layers of baryte are present at a particular stratigraphic level within the lower Mapepe Formation in the central part of the greenstone belt and are well developed in the

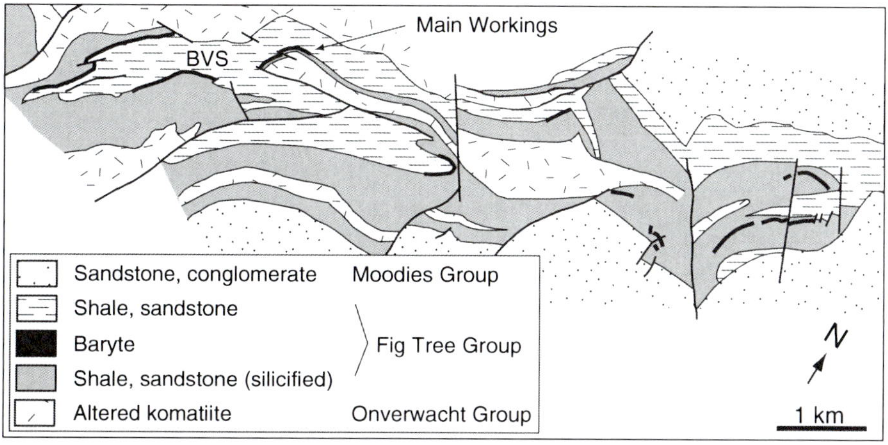

Fig. 10 Geological map of the Baryte Valley (After Pearton 1986). BVS: Baryte Valley Syncline. See Fig. 1 for locality

Fig. 11 (**a**) Outcrop of bedded detrital barite containing abundant blades of secondary barite oriented perpendicular to bedding. (**b**) Photomicrograph (plane polarised light; scale bar is 1 mm) of detrital barite admixed with minor detrital quartz and chromite

Baryte Valley Syncline area (Fig. 10) where they can be traced for several kilometres. The baryte is considered to be sedimentary in origin and has been studied by Heinrichs and Reimer (1977) and Reimer (1980). Baryte occurs in layers 13 cm thick on average (Fig. 11a) that are intercalated with green chert, dolomitic chert, chert pebble conglomerate and highly silicified sandstone and grit, all of which were deposited under a range of shallow-marine conditions (Lowe and Nocita 1999). Baryte occurs in three different varieties (Heinrichs and Reimer 1977; Reimer 1980). (1) Detrital baryte beds, consisting of slightly rounded baryte grains admixed with minor amounts of volcanic quartz, muscovite, pyrite, zircon and chromite, and showing cross-lamination (Fig. 11b). (2) Authigenic baryte crystals that form cauliflower-like structures locally and that grew at the sediment-water interface. (3) Authigenic, coarse baryte blades that replaced detrital baryte and grew as single crystals or bundles after deposition. Bao et al. (2007) reported multiple S and O isotope data from the baryte. No ^{17}O anomaly was observed, while the average $\delta^{18}O$ value is 10.6‰, close to that of the modern seawater sulphate value. Values of $\delta^{34}S$ range narrowly from +3.7 to +6.2‰ while $\Delta^{33}S$ values are negative, ranging from −0.69‰ to −0.31‰. Similar values were reported by Farquhar et al. (2000) from ca. 3.5 Ga baryte of the Warrawoona Group.

Detrital baryte was interpreted by Heinrichs and Reimer (1977) and Reimer (1980) to represent reworked baryte deposits of hydrothermal-exhalative origin. Multiple sulphur isotope values, in particular negative $\Delta^{33}S$ values, suggest that the sulphate was derived from seawater, and ultimately from photolysis reactions in the atmosphere (Bao et al. 2007). Hydrothermal activity may have been roughly contemporaneous with baryte reworking and deposition. On the western limb of the Baryte Valley Syncline (Fig. 10), baryte beds directly overlie a succession of highly silicified shales, now chert, that are transected by numerous chert veins. Baryte values are high (>1,000 ppm) in several samples of these rocks (Hofmann 2005; Hofmann and Bolhar 2007). At a disused baryte mine (Main Workings, Fig. 10), chert and baryte veins and baryte-cemented breccias have been observed in cherts underlying the baryte beds (Pearton 1986). This suggests a genetic link between chert veins and reworked baryte, i.e. syndepositional hydrothermal activity. Horizons consisting of interbedded chert and baryte also occur in the Pilbara craton, where they are associated with swarms of zoned veins of chert and baryte, indicative of a hydrothermal origin (Nijman et al. 1999; Van Kranendonk and Pirajno 2004; Van Kranendonk 2006). The timing of baryte deposition in the Fig Tree Group is unclear, as dating of the Mapepe Formation in different structural domains has yielded conflicting results (Kröner et al. 1991; Byerly et al. 1996).

4.2 Hydrothermally Altered Shales and Sandstones

Shales and greywackes of the southern facies of the Fig Tree Group in the central part of the Barberton greenstone belt have been affected by strong alteration, in contrast to greywackes of the northern facies (Hofmann 2005). This alteration led to similar compositional changes as observed in the silica alteration zones of the Onverwacht Group: K, Rb and Ba were added, whereas Mg, Ca, Na and Sr were lost (Fig. 12). Palaeoweathering was excluded as a possible source for the alteration due to the restricted occurrence of altered rocks in the central part of the belt and low Al_2O_3 contents of Fig Tree greywackes and shales, suggestive of a palaeoclimate dominated by mechanical, rather than chemical, weathering processes.

Because sedimentary baryte deposits are restricted to the central part of the greenstone belt where Fig Tree Group sedimentary rocks are highly altered, it is possible that this area was a geothermally active zone during Fig Tree times. Ba-rich hydrothermal fluids may have ascended mainly during lower Mapepe times and resulted, firstly, in local deposits of hydrothermal baryte that were reworked to

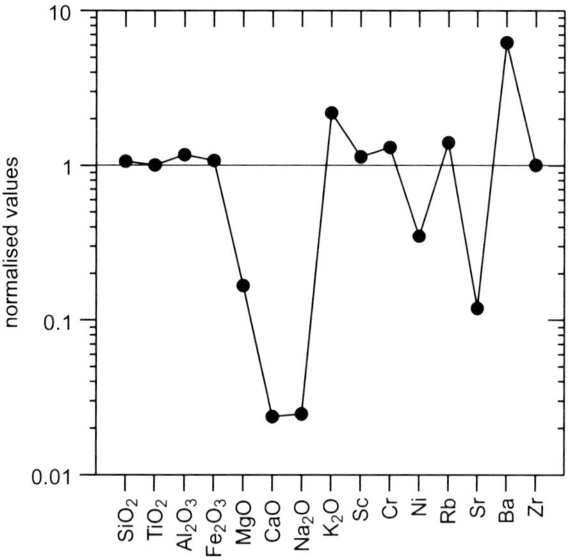

Fig. 12 Distribution of selected major and trace elements of average Fig Tree Group sedimentary rock (greywacke and shale) of the southern facies normalised to average greywacke of the northern facies. Note the addition of K_2O, Rb, and Ba, and the depletion of MgO, CaO, Na_2O, Ni and Sr in rocks of the southern facies, which is interpreted to be a result of hydrothermal alteration

form detrital baryte beds and, secondly, in the metasomatic alteration of Fig Tree strata in hydrothermal upflow zones. Syndepositional alteration may be related to the vicinity of the "baryte belt" to a tectonically active zone underlying, and possibly controlling parts of, the Fig Tree sedimentary basin (e.g. Proto-Inyoka zone of Heinrichs and Reimer 1977).

4.3 Bien Venue Massive Sulphide Deposit

The Bien Venue volcanogenic massive sulphide deposit is situated in felsic volcanic rocks of the Bien Venue Formation of the northern Fig Tree Group (Fig. 13; Murphy 1990; Ward 1999; Kohler and Anhaeusser 2002). Host rocks include a variety of tightly folded rhyolitic to rhyodacitic pyroclastic rocks, commonly containing quartz phenocrysts. The host rocks have been affected by hydrothermal alteration (silicification, sericitisation and chloritisation), followed by greenschist facies metamorphism and complex deformation, resulting in the formation of quartz-sericite schists. The ore body consists of 50–70 m wide, stratabound lenses of massive and disseminated sulphides, mainly pyrite, sphalerite and chalcopyrite, with minor galena, tennantite and native silver. The footwall ore is dominated by pyrite and chalcopyrite. Towards the hangingwall, sphalerite, galena and baryte become common. This zonation of the ore body is similar to that in the much less deformed VMS deposits from the Strelley belt

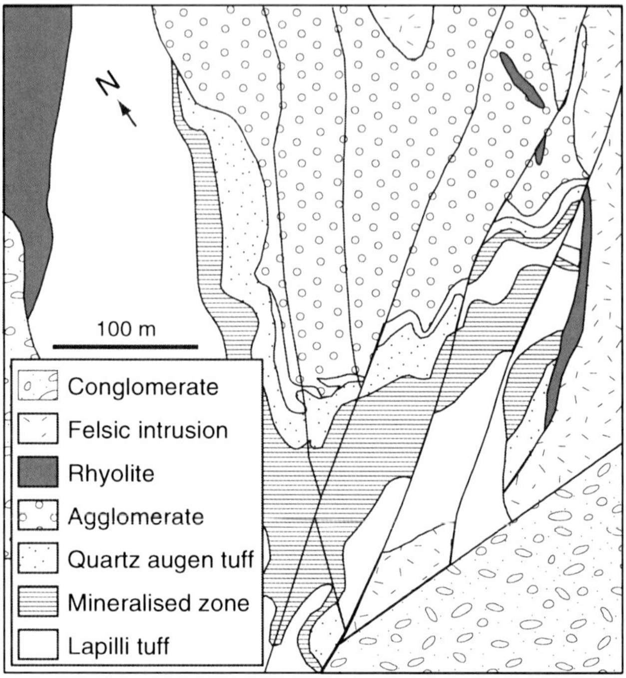

Fig. 13 Geological map of the Bien Venue massive sulphide deposit (After Murphy 1990). See Fig. 1 for locality

of the Pilbara craton that formed at almost the same time (e.g. Vearncombe et al. 1995). Murphy (1990) suggested that the sulphide mineralisation formed from Kuroko-type sea-floor exhalative activity, based on characteristic features such as stratabound mineralisation, ore zonations and spatial association with felsic volcanic rocks. While filamentous microfossils have been reported from cherts associated with the VMS deposit of the Strelley belt (Rasmussen 2000), no geobiological studies have so far been reported from the Bien Venue deposit.

4.4 Heat Source for Fig Tree Hydrothermal Activity

Felsic volcanic rocks of the Bien Venue Formation have been dated at 3,256 ± 1 Ma (Kohler 2003), which is the likely age of the Bien Venue VMS deposit. This is the same age as reported for the felsic volcanic Auber Villiers Formation (3,256 ± 4 Ma, Kröner et al. 1991; 3,253 ± 2 Ma, Byerly et al. 1996), suggesting a belt-wide event of felsic volcanic activity. A similar age has also been reported from dacitic tuffs at the base of the Mapepe Fm at one locality (3,258 ± 3 Ma, Byerly et al. 1996), which could suggest that at least some of the Mapepe baryte deposits formed at that time. The presence of baryte associated with the Bien Venue deposit would fit the interpretation that the Fig Tree Group both north and south of the Inyoka Fault was affected by a major hydrothermal event at ca. 3.255 Ga associated with felsic volcanic activity. No TTG plutons of this age have so far been identified in the Barberton Greenstone belt, although many of the highly sheared felsic rocks in the Bien Venue Formation may have originally formed as sub-volcanic sills. Instead there are a number of plutons with an age of 3.23 Ga that formed at the same time as the felsic volcanic rocks of the Schoongezicht Formation that forms the uppermost unit of the Fig Tree Group. Most authors (e.g. de Ronde and Kamo 2000; Kohler and Anhaeusser 2002; Lowe and Byerly 2007a) have interpreted the felsic volcanic activity in the Fig Tree Group to be associated with subduction-related magmatism in a continental arc or back-arc setting, environments conducive for hydrothermal activity.

5 Implications of Hydrothermal Activity for Early Life in the Barberton Greenstone Belt

Evidence for hydrothermal activity is widespread in the Barberton greenstone belt. This is not surprising, as the Swaziland Supergroup, with the exception of the Moodies Group, consists largely of volcanic and pyroclastic rocks that were intruded by TTG plutons and ultramafic layered complexes at several stages in the history of the succession. What remains to be answered more fully is how hydrothermal activity did affect the life forms that were present at that time and to what extent hydrothermal systems and sub-seafloor environments were occupied by microbial life. Furthermore, it is important to determine how hydrothermal alteration affected the morphology and composition of remnants of life as preserved in Barberton rocks. Research to answer these questions is still in its infancy.

Carbonaceous matter of likely organic origin (Walsh 1992; Westall et al. 2001; Tice and Lowe 2006; Glikson et al. 2008) is present in bedded cherts and in chert veins of the Onverwacht Group. In which environment this material formed is unresolved. If the carbonaceous matter formed in shallow parts of the ocean and represents background suspension sediments that settled out of the water column, its origin would be unrelated to seafloor hydrothermal activity. It is also possible that organic matter was produced by photo-/chemosynthetic bacteria close to the seafloor. Mat-like accumulations of carbonaceous matter have been described from bedded cherts (Westall et al. 2001; Tice and Lowe 2006). If biogenic and of chemosynthetic origin, low-temperature hydrothermal activity may thus have been conducive to life on the Archaean seafloor.

Alternatively, carbonaceous matter may have formed by subsurface biota that inhabited the hydrothermal systems and were deposited on the seafloor during hydrothermal effusions. Supporting the idea of a subsurface habitat are tubular features described from the rims of pillow basalts from the 3.3 Ga Kromberg Formation and interpreted to represent corrosion features by chemolithoautotrophs that colonized the originally glassy pillow rims (Furnes et al. 2004). If these structures are truly biogenic in origin, the microbes must have thrived while hydrothermal fluids were circulating the basaltic subseafloor environment.

Although thermophilic organisms may well have occupied the low-temperature hydrothermal systems at that time (e.g. Ueno et al. 2004), most of the carbonaceous matter present in the chert veins likely originated from the downward flow of sediment-laden seawater from the seafloor and was not produced at depth. This is indicated by its ubiquitous presence in subvertical veins extending to the palaeoseafloor, but its absence in stratiform chert veins that are commonly filled with botryoidal chert precipitate. The latter is interpreted to have formed when the hydrothermal system was closed off from the seafloor due to the impermeable chert cap (Hofmann and Bolhar 2007). A possible abiological carbon source includes the reduction of CO_2 during serpentinisation of ultramafic rocks at depth (e.g. Horita and Berndt 1999) and redistribution of carbonaceous matter during hydrothermal venting, although these processes have been discounted to have played a major role for the origin of carbonaceous matter in the Barberton greenstone belt (Hofmann and Bolhar 2007). However, remobilisation of carbonaceous matter did take place, as indicated by its presence as secondary material in pore space and in veins (van Zuilen et al. 2007), indicating that morphological interpretations of carbonaceous structures need to be conducted with care.

Shales of the Fig Tree Group are rich in carbonaceous material (0.16–1.34 wt% TOC; Reimer 1975), which occurs as compacted carbonaceous grains. This material has very similar isotopic signatures compared to Onverwacht carbonaceous cherts ($\delta^{13}C = -29‰$ on average; Hofmann and Bolhar 2007), suggesting that most of the carbon must represent material that either formed and was reworked on the seafloor or settled out of the water column. A biogenic origin appears most plausible in the absence of a tenable model for the abiogenic origin of carbonaceous matter in thick and spatially widespread black shale sequences with light carbon isotope signatures comparable to those in younger settings. Carbonaceous matter is both present in hydrothermally altered and unaltered shales and throughout the Fig Tree Group, suggesting no direct link between life and hydrothermal activity, at least during Fig Tree times.

Acknowledgments Research in the Barberton greenstone belt was supported by Deutsche Forschungsgemeinschaft (Ho 2507/1–1/2), University of the Witwatersrand Research Committee and National Research Foundation of South Africa (FA2005040400027). Numerous people provided access, support, and hospitality, including Johan Eksteen and Property Mokoena (Mpumalanga Parks Board), Colin Wille (Taurus Estate), Jan Maarten and Wilma van Rensburg (Sappi Forests) and Roelf le Roux and Chris Rippon (Barberton Mines).

References

Abraham K, Opfergelt S, Cardinal D, Hofmann A, Foley S, André L (2007) Si isotopes as a clue for understanding Eoarchaean silicification. Geochim Cosmochim Acta 71(Suppl 1):A3

Alt JC (1995) Subseafloor processes in mid-ocean ridge hydrothermal systems. In: Humphris SE, Zierenberg RA, Mullineaux LS, Thompson RE (eds) Seafloor hydrothermal systems: physical, chemical, biological, and geological interactions. Geophys Monogr 91:85–114

Anhaeusser CR (1976) The geology of the Sheba Hills area of the Barberton Mountain Land, South Africa, with particular reference to the Eureka syncline. Trans Geol Soc S Afr 79:253–280

Armstrong RA, Compston W, de Wit MJ, Williams IS (1990) The stratigraphy of the 3.5–3.2 Ga Barberton greenstone belt revisited: a single zircon ion microprobe study. Earth Planet Sci Lett 101:90–106

Banerjee NR, Furnes H, Muehlenbachs K, Staudigel H, de Wit MJ (2006) Preservation of microbial biosignatures in 3.5 Ga pillow lavas from the Barberton Greenstone Belt, South Africa. Earth Planet Sci Lett 241:707–722

Bao H, Rumble D, Lowe DR (2007) The five stable isotope compositions of Fig Tree barites: implications on sulfur cycle in ca. 3.2 Ga oceans. Geochim Cosmochim Acta 71:4868–4879

Brasier MD, Green OR, Jephcoat AP, Kleppe AK, van Kranendonk MJ, Lindsay JF, Steele A, Grassineau NV (2002) Questioning the evidence for Earth's oldest fossils. Nature 416:76–81

Byerly GR, Kröner A, Lowe DR, Todt W, Walsh MM (1996) Prolonged magmatism and time constraints for sediment deposition in the early Archean Barberton greenstone belt: evidence from the Upper Onverwacht and Fig Tree groups. Precambrian Res 78:125–138

Channer DMD, de Ronde CEJ, Spooner ETC (1997) The Cl^{-}-Br^{-}-I^{-} composition of ~3.23 Ga modified seawater: implications for the geological evolution of ocean halide chemistry. Earth Planet Sci Lett 150:325–335

Cloete M (1999) Aspects of volcanism and metamorphism of the Onverwacht Group lavas in the southwestern portion of the Barberton greenstone belt. Mem Geol Surv S Afr 84:232 pp

Condie KC, Macke JE, Reimer TO (1970) Petrology and geochemistry of early Precambrian graywackes from the Fig Tree Group, South Africa. Geol Soc Am Bull 81:2759–2776

de Ronde CEJ, de Wit MJ (1994) Tectonic history of the Barberton greenstone belt, South Africa: 490 million years of Archean crustal evolution. Tectonics 13:983–1005

de Ronde CEJ, Ebbesen TW (1996) 3.2 b.y. of organic compound formation near sea-floor hot springs. Geology 24:791–794

de Ronde CEJ, Kamo SL (2000) An Archaean arc-arc collisional event: a short-lived (ca. 3 Myr) episode, Weltevreden area, Barberton greenstone belt, South Africa. J Afr Earth Sci 30:219–248

de Ronde CEJ, de Wit MJ, Spooner ETC (1994) Early Archean (>3.2 Ga) Fe-oxide-rich, hydrothermal discharge vents in the Barberton greenstone belt, South Africa. Geol Soc Am Bull 106:86–104

de Ronde CEJ, Channer DMD, Faure K, Bray CJ, Spooner ETC (1997) Fluid chemistry of Archean seafloor hydrothermal vents: implications for the composition of circa 3.2 Ga seawater. Geochim Cosmochim Acta 61:4025–4042

de Ronde CEJ, de Wit MJ, Spooner ETC, Channer DMdeR, Christenson BW, Bray CJ, Faure K (2004) Ironstone pods in the Archean Barberton greenstone belt, South Africa: Earth's oldest seafloor hydrothermal vents reinterpreted as Quaternary subaerial springs. Geology, Online Forum, p e68

de Vries ST (2004) Early Archaean sedimentary basins: depositional environment and hydrothermal systems. Examples from the Barberton and Coppin Gap Greenstone Belts. Geologica Ultraiectina 244:160 pp

de Vries ST, Touret JLR (2007) Early Archaean hydrothermal fluids; a study of inclusions from the ~3.4 Ga Buck Ridge Chert, Barberton Greenstone Belt, South Africa. Chem Geol 237:289–302

de Vries ST, Nijman W, Armstrong RA (2006) Growth-fault structure and stratigraphic architecture of the Buck Ridge volcano-sedimentary complex, upper Hooggenoeg Formation, Barberton Greenstone Belt, South Africa. Precambr Res 149:77–98

de Wit MJ, Hart RA (1993) Earth's earliest continental lithosphere, hydrothermal flux and crustal recycling. Lithos 30:309–335

de Wit MJ, Hart RJ, Martin A, Abbott P (1982) Archaean abiogenic and probable biogenic structures associated with mineralized hydrothermal vent systems and regional metasomatism, with implications for greenstone belt studies. Econ Geol 77:1783–1802

de Wit MJ, Armstrong RA, Hart RJ, Wilson AH (1987) Felsic igneous rocks within the 3.3 to 3.5 Ga Barberton greenstone belt: high-level crustal equivalents of the surrounding tonalite-trondhjemite terrain, emplaced during thrusting. Tectonics 6:529–549

DMd C, de Ronde CEJ, Spooner ETC (1997) The Cl^{-}-Br^{-}-I^{-} composition of ~3.23 Ga modified seawater: implications for the geological evolution of ocean halide chemistry. Earth Planet Sci Lett 150:325–335

Duchac KC, Hanor JS (1987) Origin and timing of metasomatic silicification of an early Archean komatiite sequence, Barberton Mountain Land, South Africa. Precambr Res 37:125–146

Farquhar J, Bao HM, Thiemens M (2000) Atmospheric influence of Earth's earliest sulfur cycle. Science 289:756–758

Furnes H, Banerjee NR, Muehlenbachs K, Staudigel H, de Wit M (2004) Early life recorded in Archean pillow lavas. Science 304:578–581

German CR, von Damm KL (2003) Hydrothermal processes. Treatise Geochem 6:181–222

Glikson M, Duck LJ, Golding S, Hofmann A, Bolhar R, Webb R, Baiano J, Sly L (2008) Microbial remains in some earliest Earth rocks: comparison with a potential modern analogue. Precambr Res 164:187–200

Hanor JS, Duchac KC (1990) Isovolumetric silicification of early Archean komatiites: geochemical mass balances and constraints on origin. J Geol 98:863–877

Heinrichs T (1980) Lithostratigraphische Untersuchungen in der Fig Tree Gruppe des Barberton Greenstone Belt zwischen Umsoli und Lomati (Südafrika). Göttinger Arbeiten zur Geologie und Paläontologie 22:118 pp

Heinrichs T, Reimer TO (1977) A sedimentary barite deposit from the Archean Fig Tree Group of the Barberton Mountain Land (South Africa). Econ Geol 72:1426–1441

Hofmann A (2005) The geochemistry of sedimentary rocks from the Fig Tree Group, Barberton greenstone belt: implications for tectonic, hydrothermal and surface processes during mid-Archaean times. Precambr Res 143:23–49

Hofmann A, Bolhar R (2007) The origin of carbonaceous cherts in the Barberton greenstone belt and their significance for the study of early life in mid-Archaean rocks. Astrobiology 7:355–388

Hofmann A, Harris C (2008) Stratiform alteration zones in the Barberton greenstone belt: a window into subseafloor processes 3.5–3.3 Ga ago. Chem Geol 257:224–242

Hofmann A, Wilson AH (2007) Silicified basalts, bedded cherts and other sea floor alteration phenomena of the 3.4 Ga Nondweni greenstone belt, South Africa. In: Van Kranendonk MJ, Smithies RH, Bennett V (eds) Earth's oldest rocks. Dev Precambr Geol 15:571–605

Hofmann A, Reimold UW, Koeberl C (2006) Archean spherule layers in the Barberton Greenstone Belt, South Africa: a discussion of problems related to the impact interpretation. In: Reimold WU, Gibson R (eds) Processes on the early Earth. Geol Soc Am Spec Pap 405:33–56

Horita J, Berndt ME (1999) Abiogenic methane formation and isotopic fractionation under hydrothermal conditions. Science 285:1055–1057

Hren MT, Lowe DR, Tice MM, Byerly GR, Chamberlain CP (2006) Stable isotope and rare earth element evidence for recent ironstone pods within the Archean Barberton greenstone belt, South Africa. Geochim Cosmochim Acta 70:1457–1470

Jaffres JBD, Shields GA, Wallmann K (2007) The oxygen isotope evolution of seawater: a critical review of a long-standing controversy and an improved geological water cycle model for the past 3.4 billion years. Earth Sci Rev 83:83–122

Kamo SL, Davis DW (1994) Reassessment of Archean crustal development in the Barberton Mountain Land, South Africa, based on U-Pb dating. Tectonics 13:167–192

Kasting JF, Tazewell Howard M, Wallmann K, Veizer J, Shields G, Jaffres J (2006) Paleoclimates, ocean depth, and the oxygen isotopic composition of seawater. Earth Planet Sci Lett 252:82–93

Kitajima K, Maruyama S, Utsunomiya S, Liou JG (2001) Seafloor hydrothermal alteration at an Archaean mid-ocean ridge. J Metamorph Geol 19:581–597

Kleinhanns IC, Kramers JD, Kamber BS (2003) Importance of water for Archaean granitoid petrology: a comparative study of TTG and potassic granitoids from Barberton Mountain Land, South Africa. Contrib Mineral Petrol 145:377–389

Knauth LP, Lowe DR (2003) High Archean climatic temperatures inferred from oxygen isotope geochemistry of cherts in the 3.5 Ga Swaziland Supergroup, South Africa. Geol Soc Am Bull 115:566–580

Kohler EA (2003) The geology of the Archaean granitoid-greenstone terrain in the vicinity of Three Sisters, Barberton Greenstone Belt. Geol Surv S Afr Bull 133:150 pp

Kohler EA, Anhaeusser CR (2002) Geology and geodynamic setting of Archaean silicic metavolcaniclastic rocks of the Bien Venue Formation, Fig Tree Group, northeast Barberton greenstone belt, South Africa. Precambr Res 116:199–235

Kröner A, Byerly GR, Lowe DR (1991) Chronology of early Archaean granite-greenstone evolution in the Barberton Mountain Land, South Africa, based on precise dating by single zircon evaporation. Earth Planet Sci Lett 103:41–54

Kröner A, Hegner E, Wendt JI, Byerly GR (1996) The oldest part of the Barberton granitoid-greenstone terrain, South Africa: evidence for crust formation between 3.5 and 3.7 Ga. Precambr Res 78:105–124

Lowe DR (1999) Petrology and sedimentology of cherts and related silicified sedimentary rocks in the Swaziland Supergroup. In: Lowe DR, Byerly GR (eds) Geologic evolution of the Barberton Greenstone Belt, South Africa. Geol Soc Am Spec Pap 329:83–114

Lowe DR, Byerly GR (1986) Archean flow-top alteration zones formed initially in a low-temperature sulphate-rich environment. Nature 324:245–248

Lowe DR, Byerly GR (2003) Ironstone pods in the Archean Barberton greenstone belt, South Africa: Earth's oldest seafloor hydrothermal vents reinterpreted as Quaternary subaerial springs. Geology 31:909–912

Lowe DR, Byerly GR (2004) Ironstone pods in the Archean Barberton greenstone belt, South Africa: Earth's oldest seafloor hydrothermal vents reinterpreted as Quaternary subaerial springs. Reply, Geology, p e69

Lowe DR, Byerly GR (2007a) An overview of the geology of the Barberton greenstone belt and vicinity: implications for early crustal development. In: Van Kranendonk MJ, Smithies RH, Bennett V (eds) Earth's oldest rocks. Elsevier, Amsterdam. Dev Precambr Geol 15:481–526

Lowe DR, Byerly GR (2007b) Ironstone bodies of the Barberton greenstone belt, South Africa: products of a Cenozoic hydrological system, not Archean hydrothermal vents! Geol Soc Am Bull 119:65–87

Lowe DR, Fisher Worrel G (1999) Sedimentology, mineralogy, and implications of silicified evaporites in the Kromberg Formation, Barberton Greenstone Belt, South Africa. In: Lowe DR, Byerly GR (eds) Geologic evolution of the Barberton Greenstone Belt, South Africa. Geol Soc Am Spec Pap 329:167–188

Lowe DR, Nocita BW (1999) Foreland basin sedimentation in the Mapepe Formation, southern-facies Fig Tree Group. In: Lowe DR, Byerly GR (eds) Geologic evolution of the Barberton Greenstone Belt, South Africa. Geol Soc Am Spec Pap 329:233–258

Murphy PW (1990) The petrology, geochemistry and classification of the Bien Venue massive sulphide deposit, Baberton Mountainland. MSc Thesis, University of Natal, Durban

Nijman W, de Bruijne KCH, Valkering ME (1999) Growth fault control of Early Archaean cherts, barite mounds and chert-barite veins, North Pole Dome, Eastern Pilbara, Western Australia. Precambr Res 95:247–274

Paris I, Stanistreet IG, Hughes MJ (1985) Cherts of the Barberton greenstone belt interpreted as products of submarine exhalative activity. J Geol 93:111–129

Pearton TN (ed) (1986) Excursion guidebook for Barberton Mountain Land, Geocongress '86. Geol Soc S Afr, 82 pp

Perry EC (1967) The oxygen isotope geochemistry of ancient cherts. Earth Planet Sci Lett 3:62–66

Pirajno F (1992) Hydrothermal mineral deposits. Springer, Berlin, 709 pp

Ransom B, Byerly GR, Lowe DR (1999) Subaqueous to subaerial Archean ultramafic phreatomagmatic volcanism, Kromberg Formation, Barberton Greenstone Belt, South Africa. In: Lowe DR, Byerly GR (eds) Geologic evolution of the Barberton Greenstone Belt, South Africa. Geol Soc Am Spec Pap 329:151–166

Rasmussen B (2000) Filamentous microfossils in a 3235-million-year-old volcanogenic massive sulphide deposit. Nature 405:676–679

Reimer TO (1975) Untersuchungen über Abtragung, sedimentation und diagenese im frühen Präkambrium am Beispiel der Sheba Formation (Südafrika). Geol Jahrb B 17:108 pp

Reimer TO (1980) Archean sedimentary baryte deposits of the Swaziland Supergroup (Barberton Mountain Land, South Africa). Precambr Res 12:393–410

Rouchon V, Orberger B, Hofmann A, Pinti D (2009) Late diagenetic Fe-carbonates in Early Archean felsic sediments (Hooggenoeg Formation, Barberton Greenstone Belt, South Africa): implications for CO_2 sequestration and the chemical budget of seawater. Precambr Res. doi:10.1016/j.precamres.2009.04.010

Schopf JW, Walter MR (1983) Archean microfossils: new evidence of ancient microbes. In: Schopf JW (ed) Earth's earliest biosphere: its origin and evolution. Princeton University Press, Princeton, NJ, pp 214–239

Terabayashi M, Masada Y, Ozawa H (2003) Archean ocean-floor metamorphism in the North Pole area, Pilbara Craton, Western Australia. Precambr Res 127:167–180

Tice MM, Lowe DR (2004) Photosynthetic microbial mats in the 3,416-Myr-old ocean. Nature 431:549–552

Tice MM, Lowe DR (2006) The origin of carbonaceous matter in pre-3.0 Ga greenstone terrains: a review and new evidence from the 3.42 Ga Buck Reef chert. Earth Sci Rev 76:259–300

Ueno Y, Yoshioka H, Maruyama S, Isozaki Y (2004) Carbon isotopes and petrography of kerogens in 3.5-Ga hydrothermal silica dikes in the North Pole area, Western Australia. Geochim Cosmochim Acta 68:573–589

Van Kranendonk MJ (2006) Volcanic degassing, hydrothermal circulation and the flourishing of early life on Earth: a review of the evidence from c. 3490–3240 Ma rocks of the Pilbara Supergroup, Pilbara Craton, Western Australia. Earth Sci Rev 74:197–240

Van Kranendonk MJ, Pirajno F (2004) Geological setting and geochemistry of metabasalts and alteration zones associated with hydrothermal chert ± barite deposits in the ca. 3.45 Ga Warrawoona Group, Pilbara Craton, Australia. Geochem Explor Environ Anal 4:253–278

Van Zuilen MA, Chaussidon M, Rollion-Bard C, Marty B (2007) Carbonaceous cherts of the Barberton Greenstone Belt, South Africa: isotopic, chemical and structural characteristics of individual microstructures. Geochim Cosmochim Acta 71:655–669

Vearncombe S, Barley ME, Groves DI, McNaughton NJ, Mikucki EJ, Vearncombe JR (1995) 3.26 Ga black smoker-type mineralization in the Strelley Belt, Pilbara Craton, Western Australia. J Geol Soc Lond 152:587–590

Viljoen MJ, Viljoen RP (1969) An introduction to the geology of the Barberton granite-greenstone terrain. Geol Soc S Afr Spec Publ 2:9–28

Walsh MM (1992) Microfossils and possible microfossils from the early Archean Onverwacht Group, Barberton Mountain Land, South Africa. Precambr Res 54:271–293

Ward JHW (1999) The metallogeny of the Barberton greenstone belts, South Africa and Swaziland. Mem Geol Surv S Afr 86:108 pp

Westall F, de Wit MJ, Dann JC, van der Gaast S, de Ronde CEJ, Gerneke D (2001) Early Archean fossil bacteria and biofilms in hydrothermally influenced sediments from the Barberton greenstone belt, South Africa. Precambr Res 106:93–116

Westall F, de Ronde CEJ, Southam G, Grassineau N, Colas M, Cockell C, Lammer H (2006) Implications of a 3.472–3.333 Gyr-old subaerial microbial mat from the Barberton greenstone belt, South Africa for the UV environmental conditions on the early Earth. Phil Trans R Soc B 361:1857–1875

Birth of Biomolecules from the Warm Wet Sheets of Clays Near Spreading Centers

Lynda B. Williams, John R. Holloway, Brandon Canfield, Christopher R. Glein, Jeffrey M. Dick, Hilairy E. Hartnett, and Everett L. Shock

Abstract The role of clay minerals in the abiotic synthesis of organic molecules near seafloor spreading centers was simulated experimentally. Clays are common hydrothermal alteration products of volcanic glass and due to their nano-scale crystal size, provide extensive and variably charged surfaces that interact with aqueous organic species. Volcanic gases H_2 and CO_2 have been shown to react on magnetite surfaces to form methanol, a primary organic molecule, under hydrothermal conditions. Therefore, our experiments simulated the temperature and pressure conditions (300°C, 100 MPa) that exist beneath hydrothermal vents, in stockwork fractures through which hydrothermal fluids interact with fresh basalt. We examined the products of reactions between aqueous methanol and three common clay minerals found in those environments (montmorillonite, saponite, illite).

Montmorillonite reacted to ~60% illite over 6 weeks, while saponite and illite were mineralogically stable. Organic products extracted with dichloromethane from the two expandable smectite clays (montmorillonite, saponite) contained a variety of complex organic molecules including: alkanes, alkyl-benzenes, alkyl-naphthalenes, alkyl-phenols, alkyl-naphthols, alkyl-anthrols, methoxy and alkyl-methoxy-phenols, methoxy and alkyl-methoxy-naphthols, and long-chain methyl esters. Experiments with the non-expandable illite yielded only traces of alkanes and alkyl-benzene after 6 weeks. We infer that the interlayer surfaces of smectites provide crystallographic sites involved in the organic synthesis of polycyclic aromatic hydrocarbons. The largest variety and quantity of organic products was produced from montmorillonite as the layer charge increased during conversion to illite.

Keywords Abiotic organic synthesis • Biomolecules • Clay minerals • Smectite • Saponite • Hydrothermal experiments

---

L.B. Williams (✉), J.R. Holloway, C.R. Glein, H.E. Hartnett, and E.L. Shock
School of Earth and Space Exploration, Arizona State University, Tempe, Arizona 85287, USA
e-mail: lynda.williams@asu.edu

J.R. Holloway, B. Canfield, J.M. Dick, H.E. Hartnett, and E.L. Shock
Department of Chemistry and Biochemistry, Arizona State University, Tempe, Arizona 85287, USA

S.D. Golding and M. Glikson (eds.), *Earliest Life on Earth: Habitats, Environments and Methods of Detection*, DOI 10.1007/978-90-481-8794-2_4,

1 Introduction

Seafloor hydrothermal systems are environments on the Earth where life may have originated (e.g. Corliss et al 1981; Baross and Hoffmann 1985; Shock 1990, 1992; Holm 1992; Russell and Hall 1997). This idea is largely based on the phylogenetic analysis of thermophilic archaea and bacteria that inhabit seafloor hydrothermal systems. Based on 16s RNA sequencing, these microorganisms reside closest to the root on the tree of life, and therefore may be closely related to the last common ancestor of life on Earth (Woese 1987; Stetter 1996). An obvious question then is: If life did begin in a hydrothermal environment, what steps in chemical evolution eventually led to functional biomolecules?

Since the onset of seafloor spreading ~3.5 – 4 billion years ago (Burke and Kidd, 1978), reactants such as CO_2 and H_2 have been delivered to hydrothermal systems via magmatic/volcanic processes (Delaney et al 1998; Holloway and O'Day 2000). These inorganic compounds are able to participate in a number of reactions leading to simple organic compounds. The presence of catalytic mineral surfaces facilitates organic reactions (Ferris 2005). Previous experimental work has demonstrated the synthesis of methanol from CO_2 and H_2 in the presence of magnetite (Voglesonger et al 2001; Seewald et al 2006). Studies have also shown that numerous organic compounds, including acetic acid, can theoretically be created from mixing hydrothermal fluid and seawater (Shock 1996; Shock and Schulte 1998). The evidence is not limited to carbon compounds; ammonia, a potentially important inorganic reactant in prebiotic chemical evolution, has been experimentally synthesized from molecular nitrogen, nitrate and nitrite under seafloor hydrothermal conditions (Brandes et al 1998).

These simple compounds (acetic acid, methanol, ammonium) may comprise reactants for the synthesis of more complex organic molecules (e.g., amino acids, lipids, polycyclic aromatic compounds), given the right conditions for reaction progress and compound stability (Benetoli et al 2007). Williams et al., (2005) demonstrated that the surfaces of expandable clay minerals can play an important role in organic synthesis in nature. Amino acid synthesis under hydrothermal conditions has been demonstrated experimentally and theoretically (Fox and Windsor 1970; Hennet et al 1992; Yanagawa and Kobayashi 1992; Marshall 1994; Schulte and Shock 1995; Amend and Shock 1998; Islam et al 2001) and clay minerals have been shown to play a role in peptide synthesis (Balogh and Laszlo 1983). However, the role of changing oxidation state concurrent with mineralogical reactions has not been experimentally evaluated with respect to organic synthesis.

The chemical mechanisms for the formation of organic compounds in the presence of clays under hydrothermal conditions are not well known. Variables of importance to the quantity and distribution of organic compounds produced include: 1) the amounts and types of mineral surfaces that are able to enhance reactivity and protect molecules from thermal degradation and 2) the activity of dissolved hydrogen in the fluid phase (oxidation state). Through control of the oxidation state of the fluid phase, the activity of dissolved hydrogen determines the relative abundance of the different types of organic compounds, and under reducing conditions, may stabilize certain organic compounds from thermal degradation (Lewan 1997; Islam et al 2001; Seewald 2003).

Williams et al. (2005) showed that some varieties of smectite contribute to the abiotic synthesis of aromatic and polycyclic aromatic hydrocarbons (Table 1). Previous experiments (McCollom et al., 2001; Rushdi and Simoneit, 2001) showed that many organic compounds decompose at 300 °C in the absence of minerals. However some aqueous organic compounds may be stabilized by H_2 from water dissociation (Lewan, 1997). Canfield (2006) speculated that aqueous hydrogen activity in the montmorillonite interlayers is higher than in external fluid and assists in stabilizing polycyclic aromatic hydrocarbons. Other field (Simoneit 1993; Cruse et al 2006) and experimental (McCollom et al, 2001) evidence indicates that aromatic structures, once formed, are stable under hydrothermal conditions similar to our experiments.

2 The Primordial Womb Hypothesis

Important organic synthesis reactions occur in contact with clay minerals as a result of mineralogical changes that occur under the high T conditions of hydrothermal vents. Elemental substitutions occur during illitization of smectite, for example, that affect the charge distribution on the 001 surfaces of the silicate, coincident with organic complexation on those surfaces. The hypothesis is that the smectite – illite reaction at the elevated temperatures common in hydrothermal systems influences reactions among simple aqueous organic molecules and favors the production of more complex organic compounds that have the potential to be used as building blocks for biomolecules.

Williams et al (2005) conducted hydrothermal experiments in which common clay minerals (montmorillonite, saponite, and illite) were reacted with aqueous methanol at 300°C and 100MPa. These conditions were chosen to simulate the region of alteration beneath seafloor hydrothermal vents (black smokers), where hot waters circulate through glassy basalt and clay alteration is pervasive (Holloway and O'Day 2000; Giorgetti et al 2001). The experiments (see discussion below) demonstrated that organic synthesis reactions occur coincident with mineralogical changes in montmorillonite as this smectite reacts to form illite. The minerals that did not change mineralogically (saponite and illite) showed little evidence for organic synthesis. In the absence of minerals, methanol was converted to CO_2, CH_4, dimethyl ether, and H_2. This suggests that the expandable clay provides important reactive surfaces for the abiotic synthesis of poly-aromatic hydrocarbons at high temperatures.

Table 1 Summary of products from dilute methanol reacted with clay minerals. a) Major products observed in the quenched gas phase and, b) Minor products extracted from clays with dichloromethane (relative quantities normalized to hexamethylbenzene)

		Montmorillonite			Saponite				Illite			Blank
Total Reacted Carbon	Ref	Week 1	Week 3	Week 6	Ref	Week 1	Week 3	Week 6	Ref	Week 3	Week 6	Week 3
% carbon reacted		11.1	10.4	14.6			11.7	10.3		9.6	10.3	8.3
μmoles organic product (post-reaction gas phase)	5.6	220.2	203.9	291.3	2.7	n.a.	233.3	206.0	3.3	192.5	205.7	165.0
μmoles organic product (dichloromethane extracts)		0.8	4.3	8.9		0.3	0.4	0.6		0.03		0.0

A. Major Product (μmoles)		Montmorillonite				Saponite				Illite			
(quechead gas phase)		Control	Week 1	Week 3	Week 6	Control	Week 1	Week 3	Week 6	Control	Week 3	Week 6	Blank
hydrogen	H_2		2.7	0.2	0.04		n.a.	7.7	8.2		3.8	6.5	7.8
methane	CH_4	0.2	6.4	25.0	71.0	0.1	n.a.	3.8	5.7	0.1	1.5	3.2	8.6
carbon dioxide	CO_2	5.4	10.2	12.0	20.8	2.7	n.a.	15.0	14.6	3.2	7.6	9.4	5.9
dimethyl ether	C_2H_6O		203.0	164.3	192.2		n.a.	213.6	185.0		182.7	192.4	150.4
Other ^{13}C products	C_2-C_6		0.6	2.7	7.6		n.a.	0.9	0.8		0.5	0.8	
B. Minor Product (nmoles)													
Calibration standard	$C_{12}H_{18}$		453	2,703	5,128		35	157	281	28	25		
(dichloromethane extracts)													
n-alkyl-benzenes	C_7H_8 - $C_{14}H_{22}$		568	3,607	7,374		125	333	482	34	30		
n-alkyl-phenols	$C_9H_{12}O$ - $C_{11}H_{16}O$		32	49	72		54	34	36				
n-alkyl-benzenediols	$C_8H_{10}O_2$		34	39	49		18						
n-alkyl-methoxy-phenols	$C_7H_8O_2$ - $C_9H_{12}O_2$		68	73	102		63						
n-alkyl-naphthalenes	$C_{13}H_{14}$ - $C_{18}H_{24}$		58	357	949		32	28	45				
n-alkyl-naphthols	$C_{14}H_{16}O$ - $C_{16}H_{20}O$		21	43	84								
n-alkyl-methoxy-naphthols	$C_{12}H_{12}O_2$ - $C_{16}H_{20}O_2$		24	120	216								

n-alkyl-anthrols	$C_{16}H_{14}O$ - $C_{17}H_{16}O$	19	34	61				
SUM ALIPHATIC EXTRACT	C_{12} - C_{20}	18	19	49	32	5	4	21
SUM AROMATIC EXTRACT	C_7 - C_{18}	825	4,321	8,907	294	395	563	34

Control = clay + water (3 wks)

Blank = methanol + water (3 wks)

n.a.=not analyzed

3 Background

3.1 Organic Geochemistry

Experimental work on carbon chemistry has shown that organic molecules containing up to thirty-five carbon atoms (C_{35}), including alcohols, carboxylic acids, esters, aldehydes, ketones, alkanes and alkenes, can be produced by abiotic reactions under conditions that simulate seafloor hydrothermal systems (McCollom et al 1999, 2001; Rushdi and Simoneit 2001). Other researchers have shown that important biomolecules such as amino acids can be synthesized under P-T conditions similar to those found in hydrothermal vents but only using reactants at much higher concentrations than are generally found in nature (Yanagawa and Kobayashi 1992; Zamaraev et al 1997; Imai et al 1999; McCollom et al 1999; Ogasawara et al 2000; Ogata et al 2000; Washington 2000; Rushdi and Simoneit 2001; Alargov et al 2002; Tsukahara et al 2002; Yokoyama et al 2003). At the critical point for H_2O (374°C and 21.8 MPa) the dissociation constant of water (K_w) increases by 10^3 over standard conditions (25 °C, 0.1 MPa) giving increased concentrations of H^+ and OH^- that can participate in acid- and base-catalyzed reactions. In addition, at elevated T and P, the dielectric constant for water drops from roughly 80 to ~20, still high enough to dissolve ionic species, but low enough to dissolved many organic compounds, including hydrocarbons (Eckert et al 2004; Kruse and Dinjus 2007b). In summary, water is a much better solvent for organic compounds at high T and P and under these conditions becomes an effective medium for acid-base catalyzed reactions (Simoneit 1995; Savage 1999). Seafloor hydrothermal fluid temperatures approach and can exceed 374 °C; it is thus entirely likely that organic reactions, which cannot proceed at 25°C and 0.1MPa, may become quite favorable. A number of reactions have been demonstrated in hydrothermal experiments including dehydrogenation, hydration/ dehydration, carboxylation/ decarboxylation, and amination/ deamination (An et al 1997; Savage 1999; Seewald 2001; McCollom and Seewald 2001, 2003a, 2003b; Siskin and Kitritzky 2001; Katritzky et al 2001; Akiya and Savage 2002). Mineral surfaces found in seafloor hydrothermal systems have been shown to be effective catalysts that promote organic reactions as well (Voglesonger et al 2001; Cody et al 2000, 2004). Nonetheless, many of the organic molecules formed are not stable at high-temperatures and will decompose with time in hydrothermal fluids (Huber and Wachtershauser 1997; Andersson and Holm 2000; Rushdi and Simoneit 2001) unless protected (Williams et al 2005) or transported to more favorable conditions for stability such as cooler fluids, or more reducing environments (Islam et al 2001).

To date, the potential for abiotic organic synthesis has been evaluated from an energetic standpoint (Shock 1990a, 1992; Shock and Schulte 1998; Schoonen et al 1999), experimental work (Simoneit, 1992; McCollom et al., 1999, 2001) and field investigations (Sherwood-Lollar 2008; Proskurowski et al, 2008). Thermodynamic calculations have shown that reactions leading to compounds such as aldehydes, ketones, carboxylic acids, alcohols, and amino acids are energetically favorable at seafloor hydrothermal conditions. In mixing calculations, the amounts and types of organic compounds produced is dependent on the activity of dissolved hydrogen and therefore, oxidation state. Thermodynamic calculations for amino acid synthesis (Shock and Schulte 1990, 1998; Amend and Shock 1998) indicate that formation reactions for 11 out of the 20 essential amino acids are energetically favored under the hydrothermal conditions that support growth of hyper-thermophilic microbes. However, these calculations also suggest that a wide-range of physical-chemical conditions (T, P, pH, oxidation state) is required for stabilizing the various organic compounds needed for biosynthesis. An ideal geochemical environment for production of the full gamut of biomolecules needed for living organisms would be a chemical gradient that covers the entire range in various energy requirements (Russel and Hall 1997; Washington 2000). The seafloor hydrothermal environment provides the diverse conditions required for energetically favored reactions and protective environments that may produce and stabilize the compounds of life: proteins, carbohydrates, lipids and nucleic acids.

3.2 Clay Minerals

Smectite is a group of expandable, layered-silicates (Fig. 1) with fundamental units of two tetrahedral silicate layers sandwiching an octahedral layer. Smectites with Fe and Mg-rich compositions (Fe-smectite, saponite, beidellite, chlorite) are common alteration products of olivine, pyroxene and basaltic glass, all of which are found in seafloor hydrothermal systems (e.g., Koyama et al., 1973; Staudigal et al., 1981; Staudigal and Hart, 1983; McMurty et al., 1983; Alt and Honnorez, 1984; Alt et al., 1986; Haymon and Kastner, 1986; Deccarreau et al., 1987; Howard and Fisk, 1988; Desprairies et al., 1989; Munier, 1994; Pilcher et al., 1999; Munier et al., 2008; Dekov et al., 2008). Smectites are expandable because they have a low silicate layer charge (– 0.2 to – 0.6 electrostatic units (esu); Moore and Reynolds 1997), that allows the interlayer tetrahedral surface to expand, attracting H_2O, cations and polar organic compounds. Binding in the interlayer can occur via hydrogen bonds between organic compounds and the hydration spheres of interlayer cations, ion – dipole interactions between cations and polar functional groups, or through electrostatic (van der Waals) interactions. Uncharged polar compounds can also enter the interlayer without displacing cations (MacEwan 1944). Metal cations in the interlayer can enhance binding of organic anions because they can interact simultaneously with the organic compound and the clay surface (Ferris et al 1996).

When smectite interlayers expand, the interlayer cations can be exchanged, thus smectites have a higher cation exchange capacity (CEC) than clay minerals that do not expand (e.g. illite, chlorite). The non-expandable clay minerals have a higher layer charge than smectite (– 0.8 esu), and this charge is compensated, for example by K^+ (in illite) that substitutes in the interlayer. The particular mineral formed depends on the chemical substitutions in the silicate framework, the interlayer species that compensate the charge, and environmental conditions.

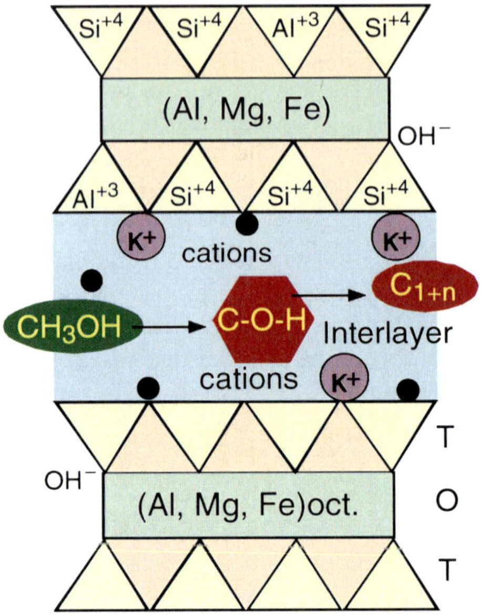

Fig. 1 Schematic cross section of a smectite showing the structural units (triangles = tetrahedral (T); rectangles = octahedral (O)). The interlayer region (blue) may contain water, cations and organic molecules

Smectites are a minor component of the minerals found in seafloor hydrothermal systems (formed by alteration of volcanic glass), but they have very high surface areas (~700 m^2/g) and they become highly reactive as the smectite recrystallizes to more stable phases such as illite or chlorite at elevated temperatures (Whitney and Northrup 1988). At temperatures and pressures typical of seafloor hydrothermal vents (ca. 300°C, 25 MPa), simple organic compounds produced by Fischer-Tropsch type (FTT) synthesis (Anderson 1984) may form, coincident with clay mineral recrystallization. Many small organic molecules are adsorbed in the interlayer of smectites and as the mineral recrystallizes, electrochemical variations over the extensive surface area of the clay (primarily the interlayer surfaces) may provide a template of variable surface energy that may promote organic synthesis (Williams et al 2005).

At elevated temperatures (300°C), smectite is unstable and dissolves layer by layer to form more stable non-expandable clays (Srodon 2000). Reaction progress leads to progressive ordering from randomly ordered (R0) unit cells of the unstable smectite, to 'nearest neighbor' ordering (R1) and eventually to long range ordering (R3) where the non-expandable mineral dominates (e.g., three illite layers surround each smectite layer). In terms of the mineral surfaces, the most dramatic electrochemical changes occur during R1-ordering as the magnitude of the charge imbalance is greatest during this highly reactive stage. Charge imbalance across the unit cell occurs as cations (Al^{3+}, $Fe^{2+, 3+}$ and Mg^{2+}) substitute in the tetrahedral or octahedral sites of the silicate framework. This gives rise to a negative charge on the basal tetrahedral surfaces of the clay (Fig. 1), usually compensated by adsorption of cations in the interlayer. Eventual collapse of the interlayer (due to high layer charge attraction) results in expulsion of electrochemically incompatible molecules previously held within the interlayer. Important reactive sites on the clay mineral surfaces include: (1) Neutral siloxane surface (acts like a weak Lewis base; largely hydrophobic); (2) Charged surface (due to cation substitutions in the tetrahedral or octahedral sheets); (3) Metal cation exchange sites; (4) Polarized waters of hydration; (5) Broken edge sites where positive charges may attract anions (Johnston, 1996).

The neutral interlayer surfaces of smectite are initially hydrophobic, due to the absence of Coulombic forces (Cowan and White 1958). Metal cations are attracted to the negatively charged surface, and can bond to specific sites (Fig. 2), which can modify the forces of attraction at the mineral surface. Vibrational spectroscopy indicates that polarized water molecules surround metal cations and function as sites of surface Brønsted acidity (Johnston, 1996). The acidity of the surface can be much greater (pH~2) than the solution pH (Mortland 1970; Pinnavaia and Mortland 1986); therefore protons can be donated to adjacent organic solutes or participate in surface reactions. Free water in the interlayer decreases as the total layer charge of the mineral increases, thus dehydration of the smectite interlayer is important to condensation reactions that promote organic synthesis (Thomas 1982). The charge density on the basal siloxane surfaces depends on whether substitutions occur in the tetrahedral or octahedral sheets of the silicate framework (Fig. 3). Tetrahedral substitutions increase the charge density of the basal siloxane surface more than

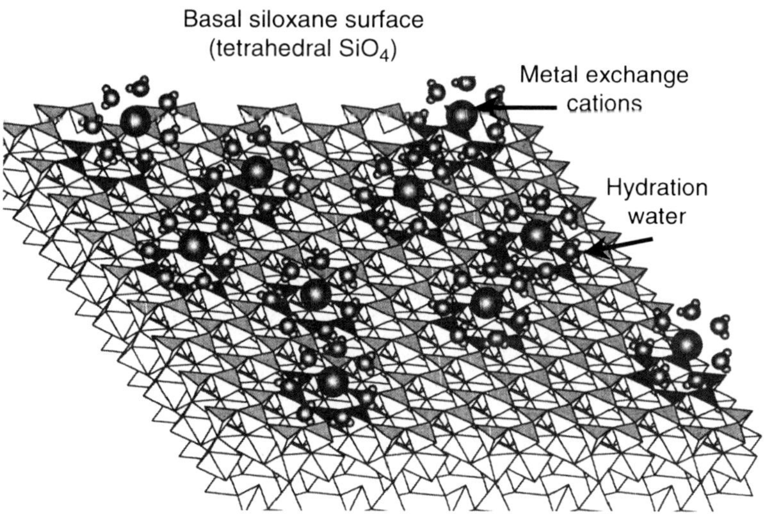

Fig. 2 Basal siloxane surface is neutral until isomorphic substitution in the silicate layers cause attraction of hydrated metal cations to Bronsted acid sites (modified from Johnston 1996)

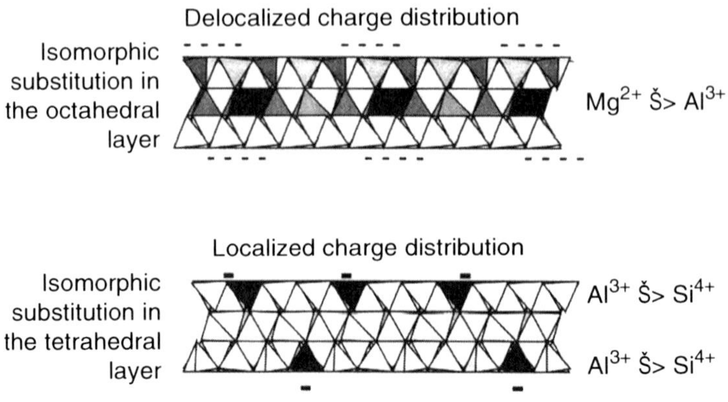

Fig. 3 Edge view of smectite showing the effect of isomorphic substitution in tetrahedral versus octahedral sites on charge distribution (modified from Johnston 1996)

octahedral substitutions. The magnitude of the total layer charge continually increases during illitization, limiting the expansion of the interlayer region and eventually restricting the size of organic molecules that can fit into the interlayer. The siloxane surface plays a critical role, increasing the surface Lewis acidity of the metal cations, promoting single electron transfer reactions, and stabilizing organic radical cations (Mortland and Raman 1968).

Bernal (1951) first proposed that clay minerals might act as substrates to adsorb and concentrate amino acids and nucleotides. Cairns-Smith further developed the hypothesis (2001, 2005) and reviewed the extensive research on clay minerals and the origin of life from 1980 through the turn of the 21st century. Industrial research primarily focuses on designing clay surfaces to promote a desired organic synthesis pathway (e.g., Ballantine et al 1981, Pinnavaia 1983; Ferris et al 1996; Bujdak and Rode 1999; Varma 2002), depending on the reactants and temperature. However, the mechanism(s) for these clay-organic complexation reactions are poorly understood.

From a geochemical viewpoint, the changes in charge density and magnitude across the clay surface is of fundamental importance in driving organic reactions. For example, Porter et al., (2000) reacted amino acids and activated nucleotides with Cu-saturated smectite subjected to cyclic wetting/drying, to produce short peptides and oligonucleotides. Because wetting and drying of smectite promotes illitization (Eberl et al 1986), the production of biomolecules reported by Porter et al (2000) may have been influenced by the silicate-promoted reaction (Williams et al 2005).

In summary, the basal siloxane surfaces of clay minerals represent multi-functional locations for organic synthesis reactions that are dictated by silicate structural changes resulting from recrystallization. Specific reactions include dehydration of alcohols, hydrolysis affected by Brønsted acidity and the hydration state of the clay, oxidation of various aromatic molecules to form radical cations, oligomerization and polymerization (Soma and Soma 1989; Balogh and Laszlo 1993; Varma 2002).

3.3 Natural Clays in Hydrothermal Systems

Smectite is observed in natural hydrothermal vents (Alt and Jiang 1991) and the discovery of a large (10 kg) deposit of saponite (trioctahedral Mg-smectite) in a black smoker chimney (Kelley et al 2002) indicates that clays are entrained by hydrothermal fluids, providing a mechanism for transport from high temperature, low pH, reducing conditions to lower temperature, higher pH, more oxidizing conditions. The latter state is conducive to end-stage oligomerization reactions (Ertem and Ferris 1998). Hydrothermal experiments that simulate seafloor hydrothermal smokers (Holloway 1997; Holloway and O'Day 2000) have shown that vent fluid compositions can be reproduced by reacting seawater with basaltic glass at 300°C, coincident with alteration of glass to smectite.

The oxidation states of fluids and clays will influence the particular chemical substitutions that occur in the silicate framework (e.g., Fe^{2+} for Fe^{3+} substitutions), and therefore influence the rate of illitization. While aqueous organic reactions may occur on time scales of minutes to days, clay mineral recrystallization occurs over timescales on the order of months, at 300°C. Therefore one would expect differences between the reactions of organics in solution and those adsorbed onto clay surfaces.

4 Experimental Design for Evaluating the Role of Clay in Biosynthesis Reactions

Three typical clay minerals found in seafloor hydrothermal environment (montmorillonite, saponite and illite) were reacted with aqueous methanol as the sole source of carbon, under conditions that mimic the hydrothermal environment temperature and pressure (300 °C and 100 MPa). Our approach to evaluating the role of clay minerals in the process of biosynthesis reactions initially requires a simplified experimental system. We did not attempt to simulate all of the chemical components of a seawater/hydrothermal fluid system because the complexity would be too great to evaluate the role of each component. As a starting point, we examined only three types of clay minerals (a di-and tri-octahedral smectite and illite) as representatives of the common clay structures found in hydrothermally altered volcanics. The fluid composition was limited to distilled-deionized water plus methanol (the effect of ionic strength on the process has not yet been evaluated). Furthermore, the concentration of methanol used in the experiments (2098 μmoles C) was greater than has been observed in natural hydrothermal fluids (Voglesonger et al., 2006) in order to assure that major products could be detected. The effect of more dilute C fluids on the reaction path has not been tested, but the role of the clay surfaces is likely to be consistent. Therefore our experiments were designed to answer the following questions:

1) Does methanol react to form more complex organic compounds in the presence of clay minerals?
2) Is methanol, a known product of a reaction between CO_2 and H_2 on magnetite (Voglesonger et al 2001), stable in the absence of the metal oxide?

3) Do organic synthesis reactions that take place in the smectite interlayer differ from reactions on the exterior or edge surfaces of clay minerals?
4) Do the organic products change with time in response to changes in the clay mineral structure and surface energy distribution?

Three sets of hydrothermal experiments were conducted over 6 weeks at 300°C, 100 MPa, in sealed gold capsules. Reactants were prepared by separation of <0.2μm size fraction of standard clay minerals and saturation with K^+ ions. This step removes the natural interlayer cations as a variable, and promotes illitization (Whitney and Northrup 1988). The presence of natural organic compounds already existing on the clay surfaces was examined by gas chromatography (GC) using dichloromethane (DCM) extracts from the clays prior to the experiments. No organic compounds were detected, however Fourier transform infrared (FT-IR) analysis of the clay showed a trace amount (<1 wt.%) of carbonate.

Two common smectite minerals (montmorillonite and saponite) that display different rates of recrystallization under the experimental conditions were studied (Fig 4.). Illite, which is non-expandable and stable under the experimental conditions, was used as a control to determine what organic products could be produced on a clay mineral without accessible interlayer surfaces and only exterior surfaces. Each clay mineral was reacted with equal amounts (by weight) of 25% aqueous ^{13}C-labeled methanol. The ^{13}C-label allowed us to distinguish products synthesized using the

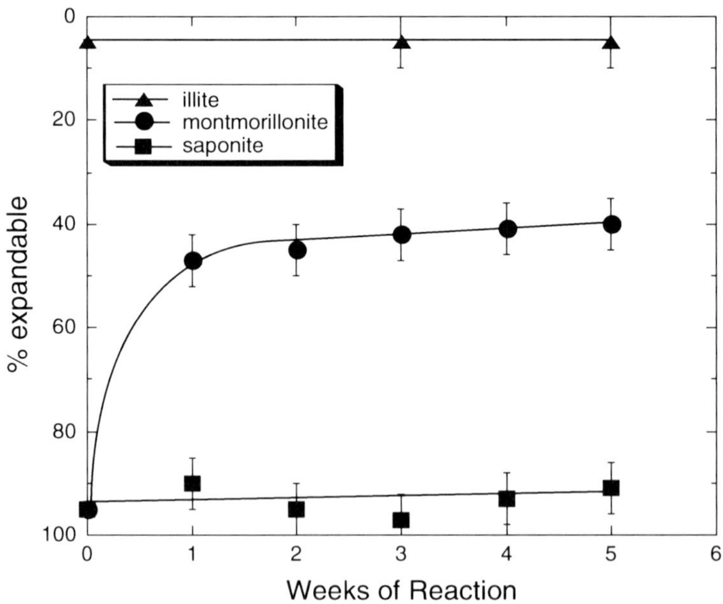

Fig. 4 Mineralogical reaction for clays used in our pilot study show no change in the expandability of saponite (>90% expandable) or illite (<5% expandable), however montmorillonite changes with time from >95% to only ~40% expandable

methanol as a carbon sourse from products derived from contaminant organic carbon or carbonate. The products were examined after 1, 3 and 6 weeks of reaction.

4.1 Experimental Methods

Reactants (clay, methanol, water) were sealed in sterilized and annealed Au capsules. Mineral reactants included a dioctahedral montmorillonite (SWy-1) and illite (IMt-1) from the Clay Minerals Repository (Purdue University, Indiana USA) and a triocthedral saponite from the Amargosa Desert (Eberl et al 1978). The smectites were K-saturated, dialyzed to remove Cl, and dried at 60°C following standard procedures for cation exchange of clays (Jackson 1979).

Gold tubing (5 mm outer diameter x 0.12 mm wall thickness) used for reaction capsules was pretreated in concentrated bleach, rinsed in deionized water, and heated at 500 °C for >12 hours in air to remove any organic contaminants. Tubing was then quenched, cut and one end of each tube arc-welded to form capsules ~30 mm in length. Duplicate capsules were loaded for reaction with 200 mg of dried, powdered clay and an equal mass of 10.4 M isotopically-labeled (^{13}C) methanol solution (2098 μmol carbon). Loaded capsules were chilled in liquid nitrogen to prevent loss of methanol, and arc-welded to seal the open end. They were then weighed, heated at 100°C for 10 min, and re-weighed to test for leakage. Additionally, blank capsules containing aqueous methanol without clay, as well as controls (clay/water mixtures without methanol) were run in parallel with the experimental samples.

Reaction capsules were pressurized in a hydrothermal pressure vessel consisting of a cold-sealed stainless steel tube (38mm outer diameter x 508 mm length) closed with a Teflon Bridgeman seal. Distilled H_2O was the pressurizing medium. The vessel was heated using a Lindberg three-zone clamshell furnace. Temperature was measured inside the vessel in the region of the capsules using internal inconel-sheathed, chromel/alumel thermocouples, placed at each end of the gold capsules. Temperature variance along the length of the capsules was <5°C. Samples were reacted for times ranging from 7 to 42 days at 300 ± 5 °C and 100 ± 1 MPa. After specified reaction times vessels were removed from the furnace and quenched in ice-water, cooling from 300°C to 20°C in less than 2 min, at which time pressure was released slowly to avoid bursting capsules. At each time interval, a capsule was removed from the pressure vessel, dried, and weighed to check for leakage.

4.2 Analytical Methods

4.2.1 Organic

After removing the capsules from the vessel and weighing, each capsule was immediately placed in rubber-stopper-sealed 10 ml glass vial that was then evacuated to 0.003 Pa. After evacuation of the vials, the capsules were pierced with a needle inserted through the rubber stopper to release gaseous products. Analysis of the volatiles was performed on a Varian CP-3800 GC equipped with a thermal conductivity detector (TCD) and a flame ionization detector (FID), using a Supelco SPB-1000 capillary column. GC operating parameters were as follows: the initial temperature was held at 100 °C for 3 min, a temperature ramp was applied (10 °C/min up to 180 °C) and then temperature was held at 180°C for 14 min. Helium was used as carrier gas with a total flow of 30 ml/min. The injector temperature was set to 175 °C, the TC detector was set at 230 °C, and the FI detector was set at 275 °C. Peak identifications were confirmed via injections of primary standards.

Immediately following the GC analysis, the reacted clay samples remaining in the capsules were transferred to 1.5 ml silanized glass vials in which non-volatile reaction products were extracted with 1 ml dichloromethane (DCM) for at least 20 hours at room temperature. Analysis of non-volatiles was performed by gas chromatography-mass spectrometry (GC-MS) on a Shimadzu GC-17A GC using a J&W Scientific DB-5MS fused silica capillary column coupled to a Shimadzu GC-MS-QP5000 quadrapole MS. The GC operating parameters in 2X split mode were as follows: the column was held isothermally at 65 °C for 2 min, temperature was increased at 6 °C/min from 65 °C to 190 °C, and then increased further from 190 to 250 °C at 10 °C/min, using a helium carrier gas with a total flow of 8.4 ml/min. Injector temperature and interface temperatures were set to 300 °C. The quadrupole MS method used linear scanning from 45-330 atomic mass units (amu), with a scan rate of 0.1 sec/scan, in electron impact ionization mode.

4.2.2 Mineral

After the minerals were extracted, the cations were exchanged with K^+ and approximately 60 mg of clay was used to make oriented mounts for mineral X-ray diffraction (XRD). Ethylene glycol was used to expand the smectite interlayers for determining the % expandable clay, which is one measure of reaction progress. The characterization of mixed-layered clay minerals is described in Moore and Reynolds (1997). The analyses were performed using a Siemens D5000 spectrophotometer with CuKa radiation.

5 Results

Of the total amount of C supplied, 15% was converted to complex organic compounds by montmorillonite after 5 weeks (Williams et al 2005). Reaction with illite produced the same compounds observed in the experimental blanks (methanol reacted without clay). The most compelling result of this study was the production of small but increasing quantities of aromatic and polycyclic aromatic organic compounds with both of the smectite-type clays, but not with illite (Table 1). This suggests that the most important reactive sites are on the interlayer surfaces of the smectite. All aromatic and polycyclic aromatic hydrocarbons formed were ^{13}C labeled. However, a small amount of alkanes (<10%) contained ^{12}C, indicating a source other than the reactant methanol. Methyl esters extracted from both smectites were predominantly ^{12}C with a ^{13}C functional methyl group. It is likely that the source of ^{12}C for the alkanes was the inorganic carbon in the starting material (Canfield 2006).

The increase in organic product complexity with illitization of montmorillonite corresponds with the increasing silicate layer charge, increasing cation exchange, and with changes in charge density during illitization. The classes of organic compounds formed with saponite are similar to those observed with montmorillonite. However, saponite is non-reactive (metastable) under these experimental conditions (Eberl et al 1976) and the quantities of organics produced are generally smaller. Only the continuous smectite to illite transition resulted in consistently increasing amounts of complex hydrocarbons. Also interesting was the gas phase products in these experiments (H_2, CO_2, and CH_4). The molecular hydrogen recovered in the quenched gas phase for experiments where methanol was reacted with illite, saponite, and montmorillonite minerals, and water alone (blank) is shown in Fig. 5. The reaction with montmorillonite shows an initial increase in hydrogen, but as

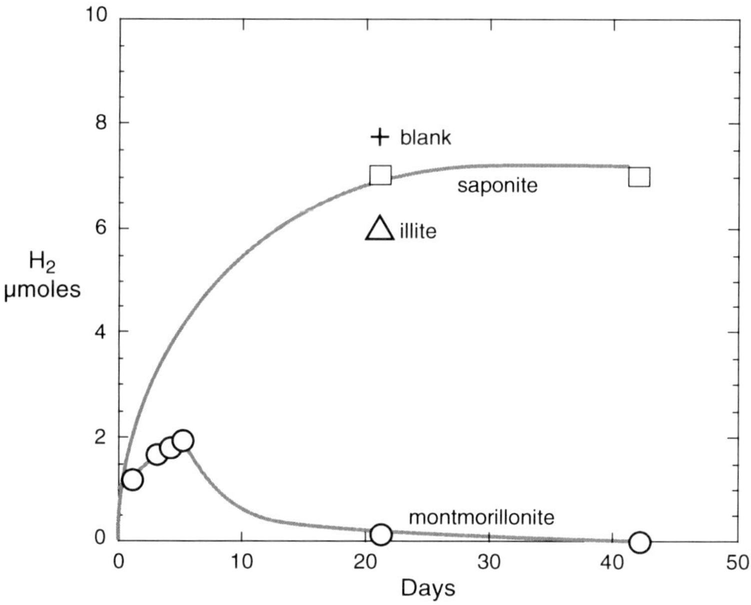

Fig. 5 Hydrogen production as a function of time during reaction of dilute methanol with the three clay minerals and with no minerals (blank)

the mineral begins to change structurally, hydrogen becomes depleted, perhaps owing to its use in organic synthesis. The decrease in hydrogen corresponds with the increase in organic compound concentrations listed in Table 1. This is not reflected in the reactions with illite or saponite, where hydrogen accumulates in the absence of the mineralogical reaction during degradation of methanol. Saponite incorporates Mg rather than Al in the octahedral layers, and since we did not supply Mg, the reaction may have been supply limited. This is an example of a variable that might affect clay surfaces and organic syntheses in nature. Nonetheless, natural saponite will eventually go through recrystallization to chlorite in a mixed-layered sequence analogous to I-S (Chang et al 1986) and we would expect analogous organic syntheses.

5.1 *Summary of Experimental Results*

1. Without clay minerals, methanol decomposes to CH_4, CO_2, H_2, and dimethyl ether.
2. In the presence of montmorillonite, increasingly abundant and complex production of organic molecules over time reflect organic synthesis on the interlayer surfaces in response to mineralogical changes.
3. The smectite interlayer surfaces promote synthesis of organic complexes unlike those promoted by exterior surfaces (demonstrated by lack of similar products with illite).
4. Organic synthesis in hydrothermal systems is significantly influenced by changes in clay mineral structure and must be studied on appropriate temporal and spatial scales.
5. Many of the organic products of reaction with montmorillonite are not stable in the absence of the clay at the pressure and temperature at which they were produced.

6 Discussion

The observed reaction products are numerous and extremely varied in abundance, however by categorizing the products according to the processes by which they formed it may be possible to evaluate some reaction pathways. Early stage reactions involving the formation of light alkanes and alkenes, carbon dioxide, and dimethyl ether were all observed in the control samples incubated with no clay minerals; however, because the abundance of these products was greater in the presence of montmorillonite, it is likely that the clay increased the reaction rates.

McCollom et al (1999) suggested that under hydrothermal conditions montmorillonite may not be a classical catalyst, but rather may act primarily to absorb and thus remove products from the aqueous solution, driving the organic reaction to the product side. While clay absorption may help to explain the effect of the clay minerals on the early stage reactions, late stage reactions involving the formation and modification of aromatic moieties appear to require interaction with the montmorillonite surface as it undergoes an increase in layer charge during illitization.

6.1 Thermodynamic Analysis of the Hydrothermal Organic Synthesis Experiment

Thermodynamic calculations can be used to constrain the range of conditions provided within the clay interlayers (instead of the bulk solvent) that favor the transformation of aqueous methanol to the observed products. This can be done by first defining the basis species of the system as CO_2, H_2O and $H_{2\,(aq)}$. Using aqueous standard-state designations for the volatile compounds permits calculating equivalent concentrations (molalities) of these species in the known amount of solvent, while calculation of fugacities associated with a gaseous standard-state designation would be more problematic because the pressure of gas in the capsule is not known. In the experiments, methanol is consumed and toluene is formed (Table 1). The overall transformation of methanol to toluene can be written as

$$7CH_3OH_{(aq)} \rightarrow C_7H_{8\,(aq)} + 7H_2O + 3H_{2\,(aq)} \qquad (1)$$

which conserves carbon. Reaction 1 and others like it do not represent the mechanisms of the organic transformations but rather provide a foundation for assessing the change in energy in the system as a result of the overall reactions. The change in energy is expressed in terms of the chemical affinity (A), which is given by

$$A = 2.303RT\ log\ (K/Q) \qquad (2)$$

where K stands for the equilibrium constant and Q denotes the activity product of the reaction, $Q = \Pi_{a_i}^{\nu_i}$ (where ai and ni stand for the activity and the reaction coefficient of the ith species). At 300° C and 100MPa, log K for Reaction 1 is equal to 22.4. This value was obtained using the revised HKF equations of state (Tanger and Helgeson 1988; Shock and Helgeson 1990; Shock et al 1992) together with data given by (Amend and Helgeson 1997). In order to calculate the log Q of Reaction 1, it is necessary to turn to the experimental concentrations of methanol and the analytical results for toluene and hydrogen. If we take in a first approximation activity coefficients of unity, the log a of methanol is 1.02 when the starting concentration is 10.4m. Note that the 10.4m methanol solution was prepared using 2098 μmol of methanol; accordingly an upper limit for the molality of toluene at the end of Week 1 in the montmorillonite experiment (Table 1) is 568 * 10.4 / 2098 = 2.8, or log a = 0.45. Likewise, an upper limit for the activity of H_2 (aq) (assuming that the amount present in the quenched gas phase was entirely dissolved at high temperature and pressure) is 0.8 * 10.4 / 2098 = 4.0×10^{-3}, or log $a = -2.4$. Combining these values with log a H_2O = 0 gives log $Q = -13.9$, which is less than the value of log K, so the chemical affinity of Reaction 1 is positive under these conditions. Because the activities of toluene and H_2 (aq) used in these calculations are upper limits, the actual values of log Q are likely to be even lower. Note that decreasing the activity of H_2O would decrease the value of log Q, potentially further driving the reaction forward. Because of the high starting concentration of methanol in the experiments, it can be worthwhile to consider the energetics of Reaction 1 when the activities of both methanol and toluene are equal to low values, which might be more representative of natural conditions. For example, if we take log $a = -6$ for both methanol and toluene, log $a = -2.4$ for hydrogen and log $a = 0$ for H_2O, the value of log Q for Reaction 1 is 28.8, which is greater than the value of log K. Accordingly, under these conditions there is no potential for the conversion of methanol to toluene. Such a potential could develop under dehydrating conditions (log a $H_2O < \approx -0.9$) or more oxidizing conditions (log a $H_2 < \approx -4.5$). Although analytical results for benzene are not listed in Table 1, the results of the analysis presented above are likely to hold as well for the transformation of methanol to benzene. The observation that toluene and other methylated aromatic compounds are dominant reaction products (rather than benzene) can be related to the energetics of the overall reaction of benzene to toluene,

$$7C_6H_{6(aq)} + 3H_{2(aq)} \rightarrow 6C_7H_{8(aq)} \tag{3}$$

which again is written only in order to assess the relative stabilities of the two species and not to describe any possible mechanism for transformations between them. At 300° C and 100 MPa, log K of Reaction 3 is 19.2. Taking logarithms of activities of benzene and toluene both equal to 0.45, and log a = −2.4 for H_2 (*aq*) gives log Q = 1.95; if logarithms of activities of benzene and toluene are both equal to −6 then log Q = 8.4. Hence, conversion of benzene to toluene is favored at the likely oxidation states of the experiments with either high or low concentrations of organic species.

It appears from these calculations that the oxidation states of the experiments are suitable for methylation of aromatic compounds. In addition, the progressive dehydration of the environment within the clay interlayers during the smectite to illite transition may help to drive the initial steps of formation of aromatic compounds, and this effect could be more pronounced when the total concentrations of organic species are significantly lower than those used in the experiments described.

6.2 Reaction Pathways of the Hydrothermal Organic Synthesis Experiment

6.2.1 Early-Stage Reactions: Hydrogen Production

There is as yet no known reaction mechanism that explains the conversion of aqueous methanol to hexmethylbenzene in these experiments. We can however begin to build a plausible reaction pathway by assessing what is presently known about high

T and P reaction mechanisms that involve our starting material, methanol, and what is known about reactions that produce similar products: H_2, CO_2, CH_4, dimethyl ether (C_2H_6O) and the aromatic organic compounds.

In all experiments containing aqueous methanol, both those containing clay minerals and those without, the dominant reaction product was dimethyl ether. Often accounting for up to one quarter of the total carbon, dimethyl ether was formed by the following dehydration reaction:

$$2CH_3OH \rightarrow CH_3OCH_3 + H_2O \tag{4}$$

Although dimethyl ether accounts for a high fraction of the total carbon present at the beginning of the experiment, it is not clear how this species participates in the synthetic pathway to the more complex organic reaction products. The observation that each of the experiments converged towards comparable amounts of dimethyl ether suggests that the abundance may be controlled by metastable equilibrium of Reaction 4, presumably through a Brønsted acid-catalyzed SN2 (bimolecular nucleophilic substitution) mechanism, in which methanol displaces a water molecule from a protonated methanol molecule (Smith and March 2007). This is supported by the comparable amounts of dimethyl ether observed in samples incubated both with and without clay minerals.

Hydrogen was initially produced in all experiments (Fig. 5). In particular, data from the capsules with illite and saponite (the non-expandable clays) are consistent with the following reaction (in the absence of clay), by which methanol is oxidized to carbon dioxide and water reduced to H_2:

$$CH_3OH + H_2O \rightarrow CO_2 + 3H_2 \tag{5}$$

In the blank capsules containing only aqueous methanol and no clay, the reaction products H_2 and CO_2 were also observed (Table 2). One proposed reaction pathway for Reaction (5) includes intermediate products formaldehyde, methylene glycol, and formic acid (Fig. 6); however, none of these intermediate species were detected in our experiment. Formic acid has been shown to quickly decompose to CO_2 and H_2 under similar T and P conditions (McCollom and Seewald 2003a; Seewald et al 2006) and similarly rapid reactions could explain why we did not see the proposed intermediates.

The reduction of methanol to methane:

$$CH_3OH + H_2 \rightarrow CH_4 + H_2O \tag{6}$$

has also been proposed for the speciation of single carbon compounds under hydrothermal conditions (Seewald et al 2006). Seewald et al (2006) were not able to demonstrate a metastable equilibrium for this reaction (9), but our experiments do have CH_4 as a product. One explanation for this might be that the clay mineral surfaces direct different reaction pathways than observed in aqueous solutions (irrespective of organic concentrations). If organic components are attracted to the clay interlayer surfaces, their activity may increase thus promoting reactions that do not occur in the absence of the minerals influence.

Table 2 Volatile reaction products (gas phase at room temperature) from aqueous methanol reacted with and without montmorillonite. Various experiment outcomes are summarized after one week, demonstrating the importance of the recrystallization process on the quantity of organic products (all reported in μmols)

Experiment	Time (d)	H_2	CO_2	CH_4	C_2H_4	C_2H_6	C_3H_6	C_3H_8	n-C_4H_{10}	*iso*-C_4H_{10}	n-C_5H_{12}
K-montmorillonite	1	1.2	5.3	0.7	0.034	0.013	0.009	0.0077	0.0077	0.004	
300 °C, 100 MPa	2	1.7	7.8	1.8	0.044	0.063	0.0048	0.026	0.02	0.012	
	3	1.8	9.1	2.8	0.054	0.15	0.0053	0.069	0.046	0.036	
	4	2	10	3.6	0.059	0.18	0.0095	0.079	0.051	0.042	
	6	2	11	4.3	0.055	0.21	0.0069	0.095	0.061	0.051	
	7	2.7	10.2	6.4	0.051	0.22	0.01	0.12	0.069	0.052	0.0017
	21	0.2	12	25	0.12	1	0.031	0.64	0.29	0.22	0.0077
	42	0.04	21	71	0.16	3	0.053	2	0.94	0.66	0.021
K-montmorillonite 300 °C, 50 MPa	7	3.5	19	7.8	0.026	0.22	0.023	0.068	0.02		
Pre-reacted clay 60%I - 40%S 300 °C, 100 MPa	7	2.1	7.6	5.8	0.011	0.096	0.0037	0.024	0.0065		
Blank 300°C, 100 MPa	7	159	60	12		0.031		0.01	0.043		
Blank 300°C, 50 MPa	7	198	66	5.8		0.01					

Blank = aqueous methanol without clay

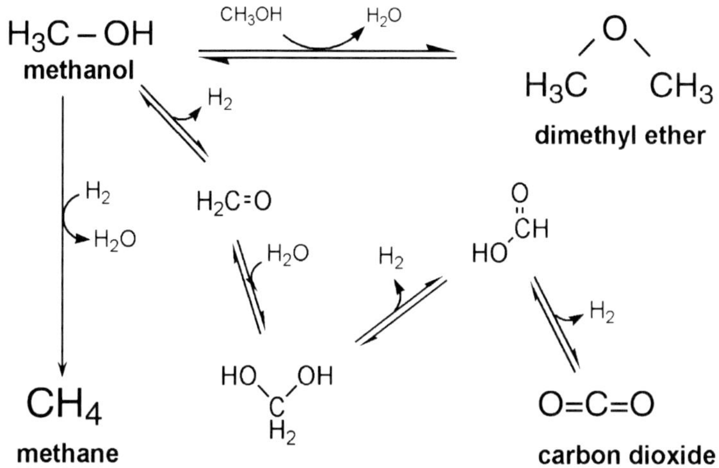

Fig. 6 Schematic representation of the primary reactions of methanol at 300 °C and 100 MPa. These reactions occur in samples containing aqueous methanol, as well as those containing aqueous methanol and clay minerals

In the experiments with the expandable clay montmorillonite, H_2 was also produced in the initial stages of the reaction. However, in contrast to the other experiments, after ~10 days H_2 began to be consumed in the capsules. Under hydrothermal vent conditions, Fisher-Tropsch-type syntheses are believed to occur; these reactions involve the production of organic compounds from reactions of CO_2 and H_2 in the presence of a metal catalyst (Anderson 1984). The first step in such a process is the water-gas shift reaction:

$$CO_2 + H_2 \rightarrow CO + H_2O \tag{7}$$

This reaction does occur at the temperatures and pressures present at hydrothermal vents, and it could account not only for the consumption of H_2 but also provide a reactive C source for the more complex organic products observed. However, there was no *CO* detected in any sample at any point in the reaction progress. It is possible that *CO* is reactive enough that it did not persist under our experimental conditions. If so, Reaction (7) is a possible mechanism for consuming H_2 in the early stages of the montmorillonite experiments.

While our results are generally consistent with a variety of proposed reaction mechanisms that can generate some of our products, the intermediate compounds necessary to confirm a specific reaction mechanism were not identified, either because of low abundance or rapid reaction kinetics. Future experiments could be sampled more often to test for intermediate organic products that may be present for shorter reaction times, and may help to elucidate the reaction pathways.

6.2.2 Late-Stage Reactions: Clay–Surface Interactions

The specific mechanism by which the initial aromatic ring structures are formed remains poorly understood. Small amounts of C_2 through C_5 alkanes and alkenes are formed during the experiments with montmorillonite and it is likely that these molecules are the starting materials in the formation reactions of the aromatic products. A variety of condensation reactions including the elimination of water from alcohols to form ethers and alkenes, the Aldol condensation, the Cannizzaro reaction and the Diels-Alder reaction (as well as others) have been shown to occur in near critical and supercritical water; though, in general, the mechanisms under these conditions are unknown (for a review see Kruse and Dinjus 2007a; Kruse and Dinjus, 2007b). It has been shown that the adsorption affinity for light alkanes on montmorillonite is directly proportional to carbon number, with adsorption affinity increasing with each additional carbon (Cheng and Huang 2004). It is possible that alkanes formed during the early stage of the experiment are attracted to the interlayer surface of the clay mineral where they could then be rapidly dehydrogenated to alkenes. Once a variety of alkenes are formed, it is possible that cyclic compounds could be formed at the clay surface. Two possible mechanisms for the formation of benzene are presented in Fig. 7. In one case, alkanes are progressively built from the addition of methanol to methane, followed by the dehydrogenation to form cyclohexane and subsequently benzene. In the other case, the alkenes formed by the dehydrogenation of ethane participate in condensation reactions to form cyclohexane. If this reaction proceeds by a Diels-Alder-like mechanism, the alkene must exist in the s-cis conformation. Assuming the reaction occurs at the clay interlayer surface, this could be favored because the alkene is constrained to having both carbons added on the same side of the double bond. In addition, Diels-Alder reactions are known to proceed more rapidly at higher pressures.

Notably, there is a size similarity between the alkylated benzene product and the silicate structure that introduces speculation about a surface-mediated reaction path, directed by cation charge distributions associated with the silicate lattice. The smectite siloxane surface exhibits a hexagonal arrangement of silicate tetrahedra. The dimension of the silicate hexagon that forms the basal siloxane sheet of the smectite (Fig. 8) approximates the size of a benzene molecule (~3Å across). A hexamethyl benzene is ~5.8Å with methyl groups attached (Maverick et al, 1978). The similarity in these dimensions draws attention to the possibility that the silicate lattice might act as a template for the synthesis of these ring structures.

Whatever mechanism is acting to produce hexamethyl benzene in our system, it appears to have relatively fast kinetics. This speculation is based on the fact that the concentrations of alkanes and/or alkenes, from which the aromatic rings presumably form, decrease with increasing carbon number, and because no partially aromatized structures were observed.

In zeolite catalyzed methanol reactions a "carbon pool mechanism" has been proposed (Kolboe 1986; Haw and Marcus 2005). In these reactions propene is

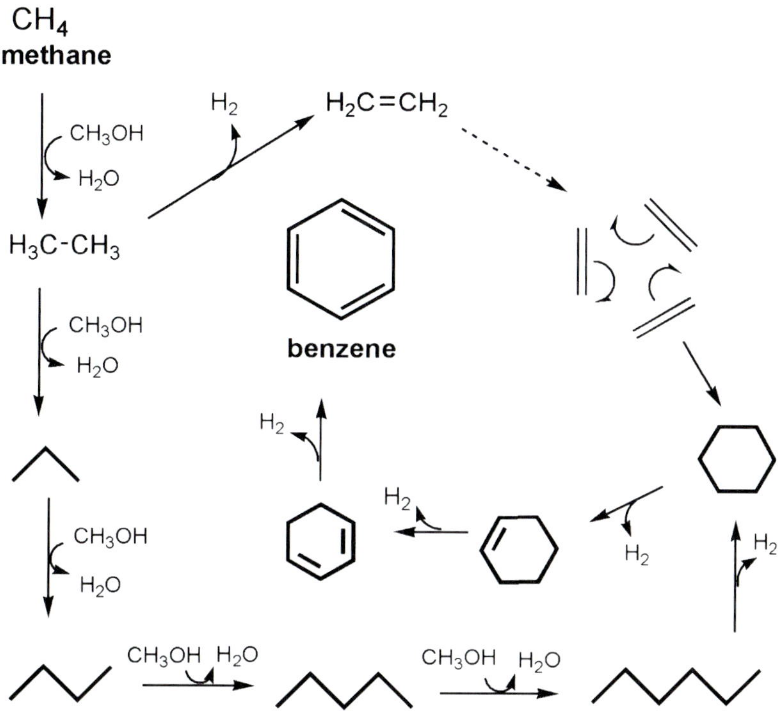

Fig. 7 Schematic representation of two possible pathways from methane to benzene

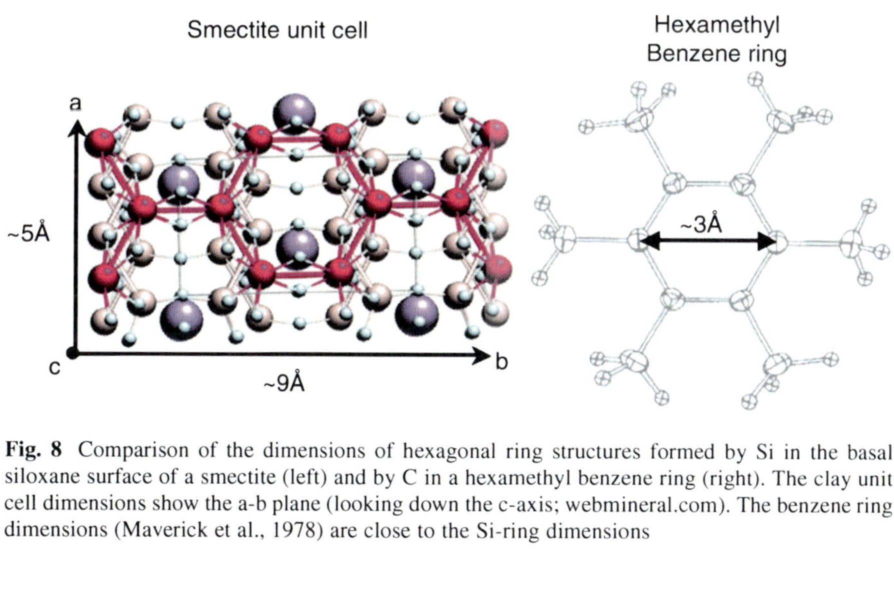

Fig. 8 Comparison of the dimensions of hexagonal ring structures formed by Si in the basal siloxane surface of a smectite (left) and by C in a hexamethyl benzene ring (right). The clay unit cell dimensions show the a-b plane (looking down the c-axis; webmineral.com). The benzene ring dimensions (Maverick et al., 1978) are close to the Si-ring dimensions

the most abundant initial product from methanol; propene trimerizes to form trimethylcyclohexane, which then transfers hydrogen to other propenes, leaving trimethylbenzene. Methyl benzenes are methylated by methanol to hexamethyl benzene, and can be further methylated by side-chain alkylation to an ethyl benzene or isopropyl benzene. The alkylated benzenes then eliminate an ethylene or propene, respectively. The remaining methyl benzene can then be methylated back to hexamethyl benzene by methanol, completing the catalytic cycle. Side-chains may also grow to the point where a new ring can close, adding to the structure.

Given that our system produces propene and hexamethyl benzene, a similar mechanism may occur within the interlayers of the expandable clay minerals. Hexamethyl benzene is the dominant extraction product for all smectite reacted samples, accompanied by progressively decreasing concentrations of methyl benzenes and decreasing numbers of methyl groups. A maximum of 8 carbons have been observed attached to any single benzene molecule, and multiple structural isomers have been observed for benzenes with 6 or fewer attached carbons. This suggests that side-chain alkylation to ethyl and propyl groups is occurring. However, alkenes have been observed only in small amounts, on average one order of magnitude below that of the corresponding alkanes. *n*-Hexane is also observed in trace amounts, suggesting that alkane ring closure could be part of the mechanism for the formation of aromatic ring structures. Once formed, it appears that benzene (and the later formed naphthalene and anthracene) mainly reacts with methyl oxonium ion ($CH_3OH_2^+$) in a series of electrophilic aromatic substitutions (S_EAr mechanism; Smith and March 2007; Fig. 9), progressively methylating the aromatic ring. While much less common, methanol substitutions may also result in attached hydroxyl groups, with the loss of methane; or attached methoxy groups, with the subsequent loss of hydrogen.

6.2.3 Hydrogen Consumption in Montmorillonite-Containing Experiments

Montmorillonite-containing samples showed an initial increase in production of H_2 during the first week of reaction, then a marked decrease in H_2 to nearly undetectable levels over the remaining weeks of reaction (Fig. 5; Table 2). It is notable that the disappearance of H_2 occurs in the same samples that yield the highest quantities of complex carbon products. Proceeding from reduced carbon species, three H_2 molecules should be formed for every one of any of the 6-membered aromatic ring structures produced. If the reaction proceeds along the pathway where methyl groups from methanol are added stepwise to an alkane (CH_4) in a dehydration reaction, then taking Reaction (9) into account leaves the following net reaction that produces H_2:

$$6CH_3OH \rightarrow C_6H_6 + 3H_2 + 6H_2O \quad (8)$$

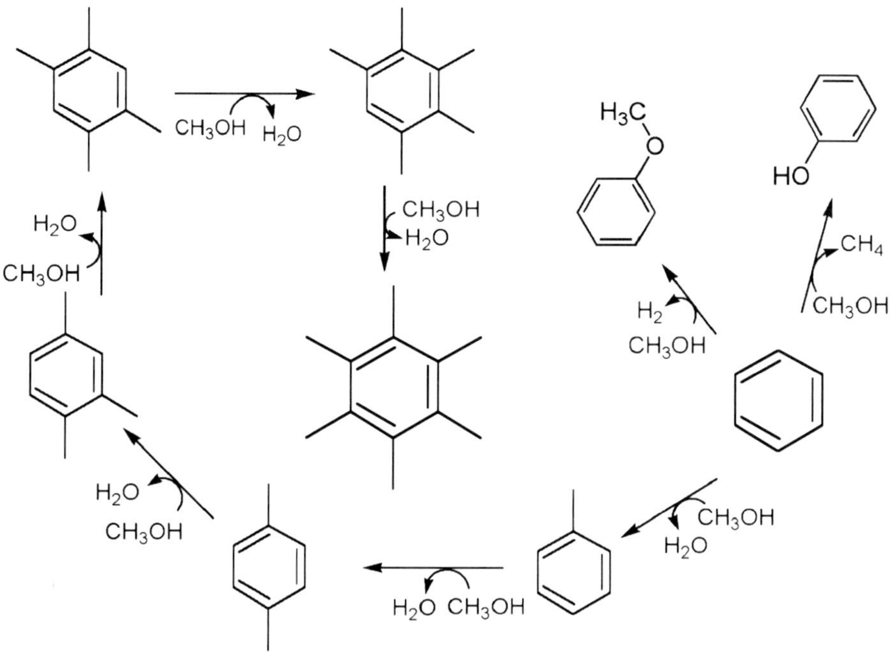

Fig. 9 Schematic representation of major reactions involving benzene. Benzene is progressively methylated through a series of dehydration reactions with methanol to hexamethyl benzene. Less abundant phenols and methoxy phenols are also produced from reactions with methanol, generating H_2 and CH_4. These hydroxyl groups could be similarly methylated through dehydration reactions

If benzene is formed along the alternate pathway (via condensation of alkenes, Fig 6.) then we might then expect to find a surplus of H_2 along such a pathway. However, the opposite has been observed in the samples containing the highest levels of aromatic compounds; H_2 is not found at the end of these experiments (performed in triplicate).

Given the possibility that Fischer-Tropsch-type syntheses are responsible for the observed reaction products, we would anticipate a surplus H_2 due to the oxidation of methanol from Reaction (5). Furthermore, the reduction of methanol to methane via Reaction (6) seems even more likely as the first step in the reaction pathway, because it generates a positive feedback loop. To assess the plausibility of this path we may evaluate the amounts of methane produced relative to the aromatic compounds.

Admittedly, complete mass balance cannot be demonstrated in these experiments because of the large number and range of aromatic reaction products. However, estimates based on the most abundant of these products, hexamethyl benzene, can be used to assess the plausibility of a reaction pathway. In the experiments with the highest yield overall, (montmorillonite reacting for 6 weeks) we observed roughly 5 μmol hexamethyl benzene and roughly 70 μmol methane.

A reasonable estimate based on the semi-quantitative analysis performed would place hexamethyl benzene at ~20% of the overall composition of aromatic products, indicating that the total amount of aromatic compounds produced was near 25 μmol. Assuming that all aromatic originate from single methanol molecules, via stepwise methanol addition (Reaction 8) we could expect 3 moles of H_2 per mole benzene, for a total of 75 μmol CH_4. This is roughly comparable to the measured 70 μmol of CH_4, indicating that this reaction pathway is plausible.

6.2.4 Oxidation State

One variable that we did not attempt to control in these experiments was oxidation state or oxygen fugacity (fO_2). The oxidation state in the experimental capsule is likely to be an important constraint on the reaction pathway because it dictates the activity of *H2* in the experiment. The oxidation state of the fluid and the activity of dissolved hydrogen are related through the dissociation of water:

$$H_2O = H_2 + 0.5O_2 \tag{9}$$

One of the more common methods for controlling fO_2 in hydrothermal experiments is through the use of mineral assemblages that act as fO_2 buffers at equilibrium. An equilibrium oxygen fugacity can be calculated at a given temperature from the equilibrium constant, the activities of the minerals, and the activity of H_2O. Common minerals used as redox buffers in hydrothermal experiments simulating seafloor black smokers include a combination of pyrite (FeS_2), pyrrhotite (FeS), and magnetite (Fe_3O_4) (PPM). But all of these minerals have been shown to catalyze organic syntheses, which would complicate our interpretation of the role of clay minerals.

We anticipate that the organic products from clay mineral reactants would differ under more reduced conditions. In modern seafloor hydrothermal systems, the PPM assemblage has been proposed to control fO_2 (Shock 1990b; Shock et al 1995). There are three other mineral assemblages that buffer fO_2 to conditions common in nature (Fig. 10). In future experiments, the oxygen fugacity may be fixed experimentally at the equilibrium fO_2 values for the hematite–magnetite (HM), quartz–fayalite–magnetite (QFM), and iron–wustite (IW) mineral assemblages; a progression from relatively oxidized to relatively reduced conditions on earth. These conditions will certainly influence the organic products since hydrogen clearly plays an important role in the hydrothermal organic synthesis.

7 Summary

The experimental results showed that organic molecules (up to C_{20}) are formed when aqueous methanol reacts with smectite clays under seafloor hydrothermal conditions. While we cannot as yet define reaction mechanisms for the production

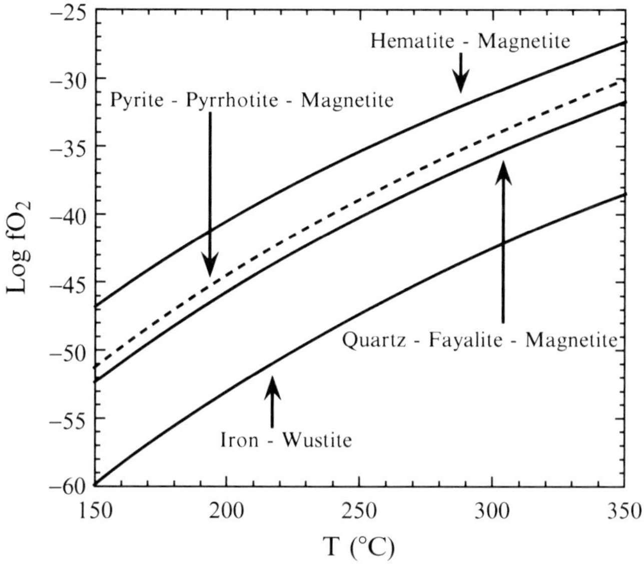

Fig. 10 Calculated log fO_2 values for mineral assemblages commonly used to buffer oxidation state in hydrothermal experiments

of these complex organic compounds, we can begin to assess what is possible in this system from a variety of geochemical perspectives.

Thermodynamics provides a theoretic framework that constrains the conditions present in the experimental system. Calculations using realistic activities for starting materials and for the gas-phase products (CO_2 and CH_4) from the experiments suggest there is a thermodynamic drive to form compounds like benzene and toluene under conditions with a slightly reduced activity of water. Benzene and toluene are only superficially similar to the aromatic products formed in the experiment, and more realistic calculations will require evaluation of the thermodynamic state properties for hexamethyl benzene at elevated temperature and pressure. Nonetheless, the clay interlayer space could very likely exhibit reduced water activity under these conditions, especially as the interlayer collapses during illitization and excludes H_2O.

The products of the reactions with and without clay minerals provide some insight to the reactions taking place in these experiments. In general, the experiments without clay minerals only form the expected thermal decomposition products (CO_2, CH_4, H_2, and dimethyl ether). In the presence of the smectite clays however, more complicated organic molecules form and persist over the course of several weeks. The initial stages of reaction in the smectite experiments produce a variety of reasonable starting materials (dimethyl ether, as well as other small alkanes and alkenes) that could participate in condensation reactions to form aromatic products. These particular experiments were not sampled often enough to detect potential intermediate products that could help to confirm specific reaction pathways. The specific reaction pathways are also very likely dependent on the redox conditions present in the experiment as H_2 is a product for some reactions and a reactant for others. Future experiments to assess reaction mechanisms will need to evaluate intermediate organic products and control the oxygen fugacity (fO_2).

The fact that the production of complex aromatic products requires the presence of an expandable clay mineral and the speculation that reactions which can form aromatic products could be surface-mediated, points to a key role for minerals in the synthesis of organic compounds under hydrothermal conditions. This suggests that the smectites actively incubate the organic compounds and protect them from thermal destruction. It is not clear whether this occurs because the layer-charge in the smectite actually mediates the reactions and stabilizes ionic intermediates, or if conditions within the interlayer space change the aqueous-phase chemistry enough to promote particular reaction pathways. The mineralogical conversion of smectite to illite however corresponds with increasing yields of polycyclic aromatic compounds. It is interesting that only small quantities of alkane products are observed. This suggests either that the alkanes are reactants in the synthesis of aromatic products, or they are released from the protective interlayer of the smectite as illite forms, leaving aromatic complexes behind (Kennedy et al 2002).

If these compounds are released from the smectite incubator, their survival depends on the chemical conditions of the external environment. If the organic products are released into 300°C seawater, the compounds very likely may decompose. However, steep temperature gradients exist in seafloor hydrothermal environments, so the organic compounds might be formed and quickly transported into cooler and more alkaline fluid where their chance of survival is potentially enhanced. These hydrothermal systems are potentially like those on the early Earth where pre-biotic chemistry began. Obviously, we do not yet understand the link between the abiotic synthesis of organic molecules in clays and the production of biomolecules. Nonetheless, the results are compelling and suggest two possibilities: first, complex organic molecules may have been present for early organisms to utilize, and second, early organisms could have followed a "template" derived from the abiotic synthesis of organic molecules in clay minerals. The birth of complex organic molecules from "the primordial womb" could be a precursor for the assembly of primordial biomolecules.

Acknowledgments This material is based upon work supported by the National Science Foundation under grants OCE-0210954 and EAR-0208456 (LBW); OCE0137191 (JRH); EAR-0847616 to (JMD) and OCE- 0826588 (ES).

References

Akiya N, Savage PE (2002) Roles of water for chemical reactions in high temperature water. Chem Rev. 102: 2725–2750

Alargov DK, Deguchi S, Tsujii K, Horikoshi K (2002) Reaction behaviors of glycine under super and subcritical water conditions. Origins of Life and Evolution of the Biosphere 32:1–12

Alt JC, Honnorez J (1984) Alteration of the upper oceanic crust, DSDP Site 417: mineralogy and chemistry. Contrib. Mineral. Petrol. 87:149–169

Alt JC, Honnorez J, Laverne C, Emmermann R (1986) Hydrothermal alteration of a 1 km section through the upper oceanic crust, Deep sea drilling project hole 504B: Mineralogy, Chemistry, and evolution of seawater-basalt interactions. J. Geophys. Res. 91: B10: 10309–10335

Alt JC, Jiang WT (1991) Hydrothermally precipitated mixed-layer illite-smectite in recent massive sulfide deposits from the seafloor. Geology 19: 570–573

Alt, JC (1999) Very low-grade hydrothermal metamorphism of basic igneous rocks. pp 169-201 in: Low-grade Metamorphism (M. Frey and D. Robinson, editors). Blackwell Science Ltd, Oxford, UK

Amend JP, Helgeson HC (1997) Group additivity equations of state for calculating the standard molal thermodynamic properties of aqueous organic species at elevated temperatures and pressures. Geochim. Cosmochim. Acta 61: 11–46

Amend JP, Shock EL (1998) Energetics of amino acid synthesis in hydrothermal ecosystems. Science 281: 1659–1662

An J, Bagnell L, Cablewski T, Strauss CR, Trainor RW (1997) Applications of high-temperature aqueous media for synthetic organic reactions. J. Org. Chem. 62: 2505–2511

Anderson RB (1984) The Fischer-Tropsch Synthesis. Academic Press, New York

Andersson E, Holm NG (2000) The stability of some selected amino acids under attempted redox constrained hydrothermal conditions. Origins of Life and Evolution of the Biosphere 30: 9–23

Ballantine JA, Davies M, Purnell H, Rayanakorn M, Thomas JM, Williams KJ (1981) Chemical conversions using sheet silicates: novel inter-molecular dehydrations of alcohols to polymers. J. Chem. Soc. Chem. Comm. 1981: 427–428

Balogh M, Laszlo P (1993) Organic chemistry using clays. Springer Verlag, New York

Banfield JF, Barker WW (1998) Low-temperature alteration in tuffs from Yucca mountain, Nevada. Clays and Clay Minerals 46: 27–37

Baross JA, Hoffman SE (1985) Submarine hydrothermal vents and associated gradient environments as sites for the origin and evolution of life. Origins of Life and Evolution of the Biosphere 1: 327–345

Benetoli, LOB, de Souza CMD, Da Silva KL, de Souza Jr IG, de Santana H, Paesano Jr A, da Costa ACS, Zaia CTBV, Zaia DAM (2007) Amino Acid Interaction with and adsorption on clays: FT-IR and Mossbauer Spectroscopy and X-ray Diffraction investigations. Orig. Life Evol. Biosph 37: 479–493

Bernal JD (1951) The physical basis of life. Routledge, Kegan Paul, London

Brandes JA, Boctor NZ, Cody GD, Cooper BA, Hazen RM, Yoder HS (1998) Abiotic nitrogen reduction on the early Earth. Nature 395: 365–367

Bujdak J, Rode B (1996) The effect of smectite compositionon the catalysis of peptide bond formation. J. Mol. Evol. 43: 326–333

Bujdak J, Rode B (1999) Silica, alumina and clay catalyzed peptide bond formation: Enhanced efficiency of alumina catalyst. Origins of Life and Evolution of the Biosphere 29: 451–461

Burke K, Kidd WSF (1978) Were Archean continental geotherms much steeper than those of today? Nature 272: 240–241

Cairns-Smith AG (2001) The origin of life: Clays. In: Frontiers of Life (Baltimore D, Dulbecco R, Jacob F, Levi-Montalcini R (eds) Academic Press, London

Cairns-Smith AG (2005) Sketches for a mineral genetic material. Elements 1: 157–161

Canfield BM (2006) Abiotic organic synthesis in simulated subseafloor hydrothermal vent conditions. Ph.D. Dissertation, Dept of Chemistry and Biochemistry, Arizona State University, Tempe, Arizona USA.

Chang HK, Mackenzie FT, Schoonmaker J (1986) Comparisons between the diagenesis of dioctahedral and trioctahedral smectite, Brazilian offshore basins. Clays and Clay Minerals 34: 407–423

Cody GD, Nabil ZB, Filley TR, Hazen RM, Scott JH, Sharma A, Yoder HS Jr. (2000) Primordial carbonylated iron-sulfur compounds and the synthesis of pyruvate. Science 289: 1337–1340

Cody GD, Boctor NZ, Brandes JA, Filley TR, Hazen RM, Yoder HS Jr. (2004) Assaying the catalytic potential of transition metal sulfides for abiotic carbon fixation. Geochim. Cosmochim. Acta, 68: 2185–2196.

Corliss JB, Baross JA, Hoffman SE (1981) An hypothesis concerning the relationship between submarine hot springs and the origin of life on Earth. Oceanol. Acta., No. Sp. 59

Cowan CT, White D (1958) The mechanism of exchange reactions occurring between sodium montmorillonite and various n-primary aliphatic amine salts. Trans. Faraday Soc. 54: 691–697

Cruse AM, Seewald JS (2006) Geochemistry of low-molecular weight hydrocarbons in hydrothermal fluids from Middle Valley, northern Juan de Fuca Ridge. Geochim. Cosmochim. Acta 70: 2073–2092

Decarreau A, Herbillon CF, Manceau A, Naho D, Paquet H, Trauth-Badaud D, Trescases J (1987) Domain segregation in Ni-Fe-Mg-smectites. Clays and Clay Minerals 35: 1–10

Dekov VM, Cuadros J, Shanks WC, Koski RA (2008) Deposition of talk-kerolite-smectite-smectite at seafloor hydrothermal vent fields: Evidence from mineralogical, geochemical and oxygen isotope studies. Chemical Geology 247: 171–194

Delaney J, Kelley D, Lilley M, Butterfield D, Baross J, Wilcock W, Embley R, Summit M (1998) The quantum event of oceanic crustal accretion: Impacts of diking at mid-ocean ridges. Science 281: 222–230

Desprairies A, Tremblay P, Laloy C (1989) Secondary mineral assemblages in a volcanic sequence drilled during ODP Leg 104 in the Norwegian Sea. In: Eldholm O, Thiede J, Taylor F, (eds) *Proceedings of the Ocean Drilling Program, Scientific Results* 1: 397–409

Eberl DD, Srodon J, Northrop HR (1986) Potassium fixation in smectite by wetting and drying. In: Davis JA, Hayes KF (eds) Geochemical Processes at Mineral Surfaces. ACS Symposium Series No. 323, Am. Chem. Soc., Washington, D.C.

Eckert CA, Liotta CL, Bush D, Brown JS, Hallett JP (2004) Sustainable reactions in tunable solvents. Journal of Physical Chemistry B 108:18108–18118.

Epstein S, Krishnamurthy RV, Cronin JR, Pizzarrello S, Yuen GU (1987) Unusual stable isotope ratios in amino acid and carboxylic acid extracts from the Murchison meteorite. Nature 326: 477–479

Ertem G, Ferris JP (1998) Formation of oligomers on montmorillonite: Sites of catalysis. Origins of Life and Evolution of the Biosphere 28: 485–499

Ferris JP, Hill ARJ, Liu R, Orgel LE (1996) Synthesis of long prebiotic oligomers on mineral surfaces. Nature 381: 59–61

Ferris JP (2005) Mineral catalysis and prebiotic synthesis: montmorillonite-catalyzed formation of RNA. Elements 1: 145–149

Fox SW, Windsor CR (1970) Synthesis of amino acids by the heating of formaldehyde and ammonia. Science 170: 984–986

Haw JF, Marcus DM (2005) Well-defined (supra)molecular structures in zeolite methanol-to-olefin catalysis. Topics in Catalysis 34: 41–48

Haymon RM, Kastner M (1986) The formation of high temperature clay minerals from basalt alteration during hydrothermal discharge on the East Pacific Rise axis at 21°N. Geochimica et Cosmochimica Acta 50: 1933–1939

Hennet RJ-C, Holm NG, Engel MH (1992) Abiotic synthesis of amino acids under hydrothermal conditions and the origin of life: a perpetual phenomenon? Naturwissenschaften 79: 361–365

Holloway JR (1997) Laboratory scale hydrothermal vent system experiments: Preliminary results *EOS*, 78: F774

Holloway JR, O'Day, PA (2000) Production of CO_2 and H_2 by diking-eruptive events at mid-ocean ridges: Implications for abiotic organic synthesis and global geochemical cycling. Int. Geol. Rev. 42: 673–683

Holm NG (1992) Why are hydrothermal systems proposed as plausible environments for the origin of life? In: Holm (ed), Marine Hydrothermal Systems and the Origin of Life, Kluwer, Norwell, Massachusetts

Howard KJ, Fisk MR (1988) Hydrothermal alumina-rich clays and boehmite on the Gorda Ridge. Geochim. Cosmochim Acta 52: 2269–2279

Huber C, Wachterhauser G (1997) Activated acetic acid by carbon fixation on (Fe, Ni)S under primordial conditions. Science: 276: 245–247

Hutcheon I, Shevalier M, Abercrombie HJ (1993) pH buffering by metastable mineral-fluid equilibria and evolution of carbon dioxide fugacity during burial diagenesis. Geochim. Cosmochim. Acta 57: 1017–1027

Imai EI, Honda H, Hatori K, Matsuno K (1999) Autocatalytic synthesis of oligoglycine in a simulated submarine hydrothermal system. Origins of Life and Evolution of the Biosphere 29: 249–259

Islam MD, Kaneko T, Kobayashi K (2001) Determination of amino acids formed in suprcritical water flow reactor simulating submarine hydrothermal systems. Anal. Sci. 17: 1631–1634

Jackson ML (1979) Soil Chemical Analysis- Advanced course. 2nd ed., Published by author, Madison, Wisconsin

Johnston CT (1996) Sorption of organic compounds on clay minerals: a surface functional group approach. In: Sawhney BL (ed) Clay Minerals Society Workshop Lectures Vol 8., Organic pollutants in the environment. Clay Minerals Society, Boulder, Colorado

Katritzky AR, Nichols DA, Siskin M, Murugan R, Balasubramanian M (2001) Reactions in high-temperature aqueous media. Chem. Rev. 101: 837–892

Kelley DS, Baross JA, Delaney JR (2002) Volcanoes, fluids and life at mid-ocean ridge spreading centers. Annual Review of Earth and Planetary Sciences 30: 385–491

Kohyama N, Susumu S, Sudo T (1973) Iron-rich saponite (ferrous and ferric forms) Clays and Clay Minerals 21: 229–237

Kolboe S (1986) Methanol reactions on ZSM-5 and other zeolite catalysts-autocatalysis and reaction mechanism. Acta Chem. Scand. 40: 711–713

Kruse A, Dinjus E (2007a) Hot compressed water as reaction medium and reactant: 2. Degradation reactions. The Journal of Supercritical Fluids 41: 361–379

Kruse A, Dinjus E (2007b) Hot compressed water as reaction medium and reactant: Properties and synthesis reactions. The Journal of Supercritical Fluids 39: 362–380

Lewan MD (1997) Experiments on the role of water in petroleum formation. Geochim. Cosmochim. Acta, 61: 3691–3723

MacEwan DMC (1944) Identification of the montmorillonite group of minerals by X-rays. Nature 154: 577–578

Marshall WL (1994) Hydrothermal synthesis of amino acids. Geochim. Cosmochim. Acta. 58: 2099–2106

Maverick E, Trueblood KN, Bekoe DA (1978) Hexamethylbenzene-Tetracyanoethylene (1:1) complex at 113 K: Structure and energy calculations. Acta Cryst. B34: 2777–2781

Meunier A (1994) Mechanism of fluid palaeocirculations in fractured rocks determined by vein alteration studies. In: Hydrocarbon and Petroleum Geology of France, Mascle A, (ed), Springer-Verlag, Berlin, 365–378

Meunier A, Mas A, Beaufort D, Patrier P, Dudoignon P (2008) Clay minerals in basalt-Hawaiite rocks from Mururoa Atoll (French Polynesia). II. Petrography and geochemistry. Clays and Clay Minerals 56: 6: 730–750

McCollom TM (2003) Formation of meteorite hydrocarbons from thermal decomposition of siderite ($FeCO_3$). Geochim. Cosmochim. Acta, 67: 311–317

McCollom TM, Seewald, JS (2001) A reassessment of the potential for reduction of dissolved CO_2 to hydrocarbons during serpentinization of olivine. Geochim. Cosmochim. Acta, 65: 3769–3778

McCollom TM, Seewald JS (2003a) Experimental constraints on the hydrothermal reactivity of organic acids and acid anions: I. Formic acid and formate. Geochim. Cosmochim. Acta 67: 3625–3644

McCollom TM, Seewald JS (2003b) Experimental study of the hydrothermal reactivity of organic acids and acid anions: II. Acetic acid, acetate, and valeric acid. Geochim Cosmochim Acta 67: 3645–3664

McCollom TM, Ritter G, Simoneit RT (1999) Lipid synthesis under hydrothermal conditions by Fischer-Tropsch-Type reactions. Origins of Life and Evolution of the Biosphe 29: 153–166

McCollom TM, Seewald JS Simoneit BRT (2001) Reactivity of monocyclic aromatic compounds under hydrothermal conditions. Geochim. Cosmochim. Acta. 65: 455–468
McMurty GM, Wang C-H, Yeh H-W (1983) Chemical and isotopic investigations into the origin of clay minerals from the Galapagos hydrothermal mounds field. Geochimica et Cosmochimica Acta 47: 475–489
Moore DM, Reynolds RC (1997) X-ray diffraction and the identification and analysis of clay minerals. 2nd ed. Oxford University Press, New York
Mortland MM, Raman KV (1968) Surface acidities of smectites in relation to hydration, exchangeable-cation and structure. Clays and Clay Minerals 16 :393–398
Mortland MM (1970) Clay-organic complexes and interactions. Adv. Agron. 22: 75–117
Ogasawara H, Yoshida A, Imai EI, Honda H, Hatori K, Matsuno K (2000) Synthesizing oligomers from monomeric nucleotides in simulated hydrothermal environments. Origins of Life and Evolution of the Biosphere 30: 519–526
Ogata Y, Imai EI, Honda H, Hatori K, Matsuno K (2000) Hydrothermal circulation of seawater through hot vents and contribution of interface chemistry to prebiotic synthesis. Origins of Life and Evolution of the Biosphere 30: 527–537
Orgel LE (1998) Polymerization on the rocks: theoretical introduction. Origins of Life and Evolution of the Biosphere 28: 227–234
Pichler T, Ridley WI, Nelson EP (1999) Low temperature alteration of dredged volcanics from the Southern Chile ridge: additional information about early stages of seafloor weathering. Marine Geology 159: 272–285
Pinnavaia TJ (1983) Intercalated clay catalysts. Science 220: 365
Pinnavaia, T. J. and Mortland, M.M (1986) Aspects of clay catalysis. In: Cairns-Smith AG, Hartman H (eds) Clay Minerals and the Origin of Life, Cambridge University Press
Pizzarello S, Feng X, Epstein S, Cronin JR (1994) Isotopic analyses of nitrogenous compounds from the Murchison Meteorite - ammonia, amines, amino-acids, and polar hydrocarbons. Geochim. Cosmochim. Acta 58: 5579–5587
Porter TL, Eastman MP, Whitehorse R, Bain E, Manygoats K (2000) The interaction of biological molecules with clay minerals: A scanning force microscopy study. Scanning 22: 1–5
Proskurowski G, Lilley MD, Seewald JS, Früh-Green GL, Olson EJ, Lupton JE, Sylva SP, Kelley DS (2008) Abiogenic Hydrocarbon Production at Lost City Hydrothermal Field. Science 319: 5863: 604 – 607
Rushdi AI, Simoneit BRT (2001) Lipid formation by aqueous Fischer-Tropsch type synthesis over a temperature range of 100-400°C. Origins of Life and Evolution of the Biosphere 31: 103–118
Russell MJ, Hall AJ (1997) The emergence of life from iron monosulphide bubbles at a submarine hydrothermal redox and pH front. J. Geol. Sci. London. 154: 377–402
Savage PE (1999) Organic chemical reactions in supercritical water. Chem. Rev. 99: 603–621
Sherwood-Lollar B, Lacrampe-Couloume G, Voglesonger K, Onstott TC, Pratt LM, Slater GF (2008) Isotopic signatures of CH_4 and higher hydrocarbon gases from Precambrian Shield sites: A model for abiogenic polymerization of hydrocarbons. Geochimica et Cosmochimica Acta 72: 4778–4795
Shock EL (1990a) Do amino acids equilibrate in hydrothermal fluids? Geochim. Cosmochim. Acta 54: 1185–1189
Shock EL (1990b) Geochemical constraints on the origin of organic compounds in hydrothermal systems. Origins of Life and Evolution of the Biosphere 20: 331–367
Shock EL (1992) Chemical environments in submarine hydrothermal systems. In: Holm NG (ed) Marine Hydrothermal Systems and the Origin of Life, a special issue of Origins of Life and Evolution of the Biosphere, 22: 67–107
Shock EL (1996) Hydrothermal systems as environments for the emergence of life. In: Bock GR and Goode JA (eds) Evolution of Hydrothermal Ecosystems on Earth (and Mars?) John Wiley & Sons, New York
Shock EL, Helgeson HC (1990) Calculation of the thermodynamic and transport properties of aqueous species at high pressures and temperatures: Standard partial molal properties of organic species. Geochim. Cosmochim. Acta 54: 915 – 945

Shock EL, Schulte MD (1990) Amino-acid synthesis in carbonaceous meteorites by aqueous alteration of polycyclic aromatic hydrocarbons. Nature 343: 728–731

Shock EL, Schulte MD (1998) Organic synthesis during fluid mixing in hydrothermal systems. J. Geo. Res., 103: 28513–28527

Shock EL, McCollom TM, Schulte MD (1995) Geochemical constraints on chemolithoautotrophic reactions in hydrothermal systems. Origins of Life and Evolution of the Biosphere 25: 141–159

Shock EL, Oelkers EH, Johnson JW, Sverjensky DA, Helgeson HC (1992) Calculation of the thermodynamic properties of aqueous species at high pressures and temperatures: Effective electrostatic radii, dissociation constants, and standard partial molal properties to 1000 °C and 5 kbar. J. Chem. Soc., Faraday Trans. 88: 803 – 826

Schoonen MAA, Xu Y, Bebie J, (1999) Energetics and kinetics of the prebiotic synthesis of simple organic acids and amino acids with the FeS-H_2S/FeS_2 redox couple as a reductant. Origins of Life and Evolution of the Biosphere 29: 5–32

Schulte MD, Shock EL (1995) Thermodynamics of strecker synthesis in hydrothermal systems. Origins of Life and Evolution of the Biosphere 25: 161–173

Seewald JS (1994) Evidence for metastable equilibrium between hydrocarbons under hydrothermal conditions. Nature 370: 285–287

Seewald JS (2001) Aqueous geochemistry of low molecular weight hydrocarbons at elevated temperatures and pressures: Constraints from mineral buffered laboratory experiments. Geochim Cosmochim. Acta 65: 1641–1664

Seewald JS (2003) Organic-inorganic interactions in petroleum producing sedimentary basins. Nature 246: 327–333

Seewald JS, Zolotov MY, McCollom T (2006) Experimental investigation of single carbon compounds under hydrothermal conditions. Geochimica et Cosmochimica Acta 70: 446–460

Simoneit BRT (1993) Aqueous high-temperature and high-pressure organic geochemistry of hydrothermal vent systems. Geochim. Cosmochim. Acta 57: 3231–3243

Simoneit BRT (1995) Evidence for organic synthesis in high temperature aqueous media—facts and prognosis. Origins of Life and Evolution of the Biosphere 25: 119–140

Siskin M, Katritzky AR (2001) Reactivity of organic compounds in superheated water: General background. Chem. Rev. 101: 825–835

Smith MB, March J (2007) March's Advanced Organic Chemistry: Reactions, Mechanisms, and Structure, 6th Ed., Wiley-Interscience, Hoboken, N.J.

Soma Y, Soma M (1989) Chemical reactions of organic compounds on clay surfaces. Environ. Health Persp. 83: 205–214

Srodon J, Eberl DD, Drits VA (2000) Evolution of fundamental-particle size during illitization of smectite and implications for reaction mechanism. Clays and Clay Minerals 48: 446–458

Staudigel H, Hart SR (1983) Alteration of basaltic glass: Mechanism and significance of the oceanic crust-sea water budget. Geochimica et Cosmochimica Acta, 47: 337–350

Staudigel H, Hart SR, Richardson SH (1981) Alteration of the oceanic crust: processes and timing. Earth and Planet. Sci. Lett. 52: 311–327

Stetter KO (1996) Hyperthermophiles in the history of life. In: Bock GR, Goode, JA (eds) Evolution of Hydrothermal Ecosystems on Earth (and Mars?) John Wiley & Sons, New York

Thomas JM (1982) Sheet silicate intercalates: new agents for unusual chemical conversions. In: Whittingham MS, Jacobson AJ (eds) Intercalation Chemistry, Academic Press, New York

Tsukahara H, Imai EI, Honda H, Hatori K, Matsuno K (2002) Prebiotic oligomerization on or inside lipid vesicles in hydrothermal environments. Origins of Life and Evolution of the Biosphere 32:13–21

Varma RS (2002) Clay and clay-supported reagents in organic synthesis. Tetrahedron 58: 1235–1255

Vogelsonger KM, Holloway JR, Dunn EE, Dalla-Betta PJ, O'Day PA (2001) Experimental abiotic synthesis of methanol in seafloor hydrothermal systems during diking events. Chemical Geology 180: 129–139

Wächtershäuser G (1990) Evolution of the first metabolic cycles. Proceedings of the National Academy of Sciences 87: 200–205

Washington J (2000) The possible role of volcanic aquifers in prebiologic genesis of organic compounds and RNA. Origins of Life and Evolution of the Biosphere 30: 53–79

Whitney G, Northrup HR (1988) Experimental investigation of the smectite to illite reaction: Dual reaction mechanisms and oxygen isotope systematics. American Mineralogist, 73: 77–90

Williams LB, Canfield B, Holloway JR, Williams P (2002) Smectite incubation of organic molecules in seafloor hydrothermal systems. The V.M. Goldschmidt Conference, Davos, Switzerland, A837

Williams LB, Canfield B, Voglesonger KM, and Holloway JR (2005) Organic molecules formed in a primordial womb. Geology 33: 913–916

Woese CR (1987) Bacterial Evolution. Microbiology Reviews 51: 221–271

Wolery TJ (1993) EQ3NR, a Computer Program for Geochemical Aqueous Speciation-Solubility Calculations: Theoretical Manual, User's Guide, and Related Documentation. Lawrence Livermore National Laboratory, California

Yanagawa H, Kobayashi K (1992) An experimental approach to chemical evolution in submarine hydothermal systems. Origins of Life and Evolution of the Biosphere 22:147–159

Yokoyama S, Koyama A, Nemoto A, Honda H, Imai EI, Hatori K, Matsuno K (2003) Amplification of diverse catalytic properties of evolving molecules in a simulated hydrothermal environment. Origins of Life and Evolution of the Biosphere 33: 589–595

Zamaraev KI, Romannikov VN, Salganik RI, Wlassoff WA, Khramtsov VV (1997) Modeling of the prebiotic synthesis of oligopeptides: silicate catalysts help to overcome the critical stage. Origins of Life and Evolution of the Biosphere 27: 325–337

Part II
Evidence and Record of Earliest Life on Earth

Towards a Null Hypothesis for Stromatolites

Martin D. Brasier

Abstract "How on Earth did life begin?" is one of the noblest questions we can ask in science, but it took well over a century from Darwin (1859) to gain an understanding of life in the Precambrian. Why did an understanding take so long? Arguably it was because it was, and still remains, a very big and very difficult problem. Its study now involves the whole of the natural sciences, and progress has been a matter of slow attrition. For most of this time, for example, there has been no concept of the vast duration of Precambrian time, nor any evidence for a distinct biota. Since the seminal work of Stanley Tyler and Elso Barghoorn (1954), each generation has come up with its own favourite solution, only to watch it fall as younger scientists have arrived on the scene. As such, this story provides us with a salutary tale of 'paradigm shifts' that have taken place about every 50 years or so. And this process is ongoing and continuous. It is no surprise then, to find that the majority of uniformitarian interpretations for Precambrian fossil assemblages established over the last 50 years now appear highly questionable. That is, of course, exactly how it should be.

Keywords Gunflint chert • Strelley pool formation • Stromatolites • Cyanobacteria • Self-organising structures

1 Peculiar Contortions

Among the first to notice and wonder about the strange forms of stromatolitic growth was Charles Darwin, back in the 1830s, who included a figure (Fig. 1a) in his published 'Journal of Researches' and made the first ever pertinent observations on splash-zone stromatolites during the voyage of HMS Beagle: "*The rocks of St Paul appear from a distance of a brilliantly white colour. This is partly owing to a coating of a hard glossy substance with a pearly lustre, which is intimately united to*

M.D. Brasier (✉)
Department of Earth Sciences, University of Oxford, Parks Road, Oxford OX1 3PR, UK
e-mail: martinb@earth.ox.ac.uk

S.D. Golding and M. Glikson (eds.), *Earliest Life on Earth: Habitats, Environments and Methods of Detection*, DOI 10.1007/978-90-481-8794-2_5,

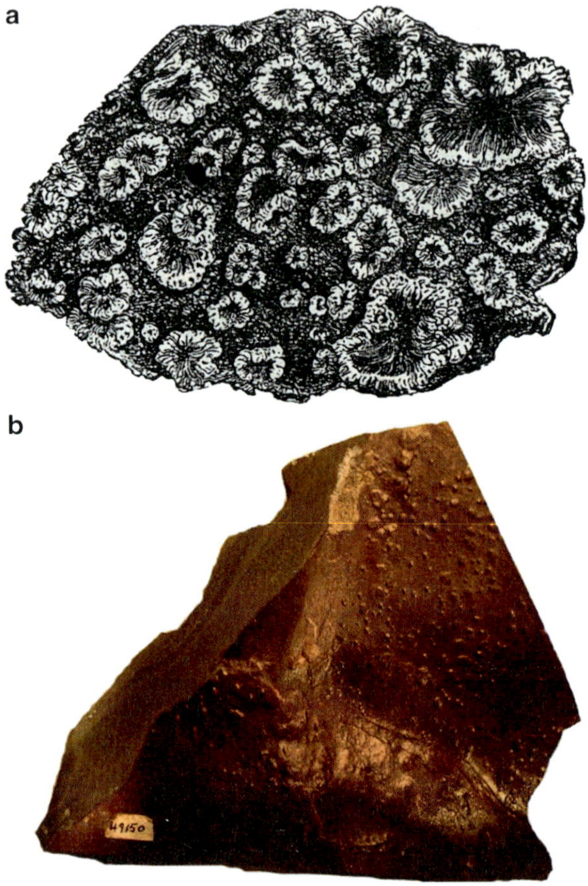

Fig. 1 Peculiar contortions – the discovery of self-organized sedimentary structures. (**a**) A woodcut used by Charles Darwin to illustrate vegetable-like growths of calcium phosphate found growing in the splash zones of oceanic islands; no scale was given. (**b**) One of several specimens of *'Arenicolites sparsus'* collected by J.W. Salter in 1856 from the Ediacaran Longmynd beds of Shropshire and illustrated by him the following year (Salter 1857, pl. 5, Fig. 3). It is arguably among the first Precambrian fossils ever to be figured, named and described. This specimen is now regarded as a microbially-induced sedimentary structure (see McIlroy et al. 2005; Callow and Brasier 2009)

the surface of the rocks. This, when examined with a lens, is found to consist of numerous exceedingly thin layers, its total thickness being about the tenth of an inch. It contains much animal matter, and its origin, no doubt, is due to the action of rain or spray on the birds' dung.… When we remember that lime, either as phosphate or carbonate, enters into the composition of hard parts, such as bones and shells, of all living animals, it is an interesting physiological fact to find substances harder than the enamel of teeth, and coloured surfaces as well polished as those of a fresh shells reformed through inorganic means from dead organic matter – mocking, also, in shape some of the lower vegetable productions." (Darwin 1839). We can only guess that he was thinking about comparisons with fungi and lichens.

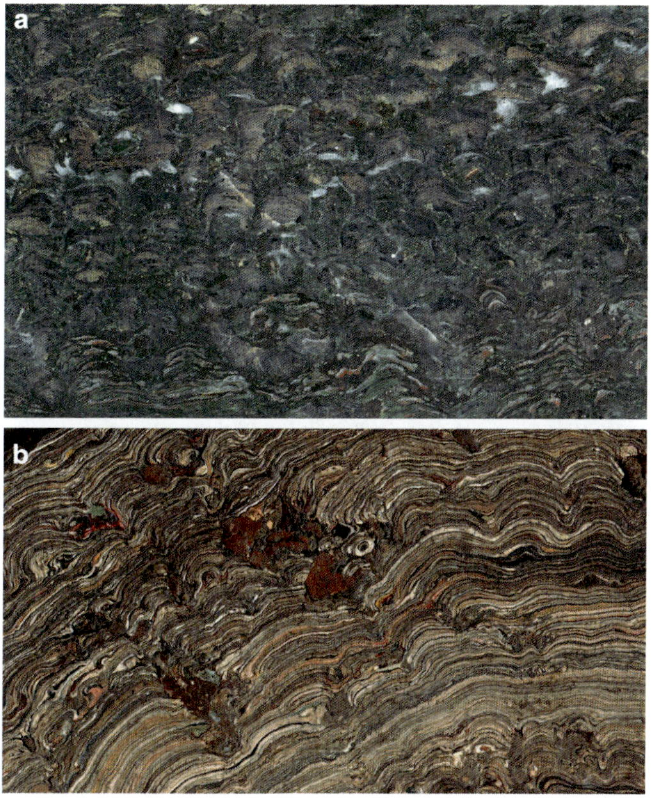

Fig. 2 The discovery and decoding of self-organized sedimentary structures. (**a**) hand specimen of stromatolites from the ~2,000 million year old banded ironstones of the Gunflint chert near Hoyt Lake, Minnesota. Similar structures were interpreted as igneous features by Foster and Whitney (1851) and as paradoxical sedimentary structures by Irving (1883). (**b**) Hand specimen once in the collection of Sir George Taylor, Director of the Royal Botanic Gardens at Kew. The specimen is made from layers of lead-based paint and was likely generated in a spray booth without any participation from biology. Note its similarity to 'a', including the non-isopachous laminae and the inter-columnar spaces filled with matrix (see McLoughlin et al. 2008)

The earliest microbially-mediated sediment to be figured from the Precambrian was discovered as early as 1855, by the great palaeontologist John Salter (see Fig. 1b) in the Ediacaran sediments of the Longmynd in England (Salter 1856, 1857). This material was originally regarded by him as the markings of worm burrows of Cambrian age, but its Precambrian and microbial origin has now become well known (e.g., McIlroy et al. 2005; Callow and Brasier 2009). Interestingly, the specimen in Fig. 1b was indirectly referred to by Darwin (1859) while writing about the Origin of Species, as follows: *"Traces of life have been detected in the Longmynd beds beneath Barrande's so-called primordial [Cambrian] zone"* (Darwin 1859).

From at least 1851, we also find discussions taking place about stromatolitic structures in the ~2,000 million year old Banded Iron Formations from the Great Lakes region of North America (see Fig. 2a). Such ancient rocks were then placed in the Azoic period because of their *"entire absence of organic remains"* (e.g., Foster and

Whitney 1851). At that early time in geological thinking, the Gunflint chert was thought to have been laid down under the great residual heat left over from the fiery origins of the primordial Earth: *"The Azoic period having been one of long continued and violent mechanical action* (ibid., p. 67)." The concern, therefore, was whether the *"beautiful series of intricate convolutions of alternate bands of bright-red and steel-grey"* were really the result of sedimentary processes. *"The flexures are exceedingly intricate and bear no marks of having been the result of original stratification. There is no actual line of separation between the lighter and darker bands"* (ibid., p. 68). Their final suggestion was that they were crumpled by igneous phenomena and had *"risen up, in a plastic state from below"* to pour out onto the deep sea floor.

The geological surveyor Roland Irving was to return to this question in his early monograph on the Archean of the north-western United States. By 1883, it was becoming clearer that the banded iron formations were some kind of paradoxical sedimentary deposit: *"the cherty and jaspery portions, frequently strongly charged with magnetite and other oxides of iron, present often peculiar irregularities and contortions in subordinate bedding, and also often a confused concretionary appearance, and even a brecciated appearance. All of these irregularities are very plainly subordinate to a simple bedding, corresponding entirely to that of the rest of the Animikie series…I anticipate that when we shall have completed our microscopic studies of them we shall get from them some light as to the origin of these confused and much discussed rocks. I may now merely say that… all of these cherts and jasper schists are original, and not the result of a metamorphism upon ordinary sedimentary deposits, though manifestly they are not of eruptive origin, as has been maintained by some"* (Irving 1883).

It was also at about this time that curious case of *Cryptozoon* began to emerge from the fossil record. This cabbage-like structure had been found spread across bedding planes in the upper Cambrian carbonates of New York State. First reported by James Hall (1883), it was originally interpreted as some kind of calcareous algal growth, a view that was developed by Charles Walcott (1883) and then by V.H.E. Kalkowsky (1908), who first introduced the concept of 'stromatolite' as follows: *"organogenic, laminated calcareous rock structures, the origins of which is clearly related to microscopic life"* (translation in Krumbein 1983, but see also Wacey 2010). Stromatolites are so named from the Greek for 'flat stones'. But as can be seen, the definition of Kalkowsky was a genetic one, though direct observation of microbes was seemingly precluded by him. From this strange mismatch there has arisen much misunderstanding, not only about stromatolites, but also about the word 'stromatolite' itself (see Fig. 2b and McLoughlin et al. 2008).

The role of biology in stromatolite growth was widely ignored until Cambridge sedimentologist Maurice Black canoed across the tidal flats of Andros in the Bahamas (Black 1933). He noticed that dense growths of cyanobacteria were forming an 'algal mat' that seemed to survive episodes of sediment deposition by growing upwards through the sediment.

Black's work went largely unnoticed for several decades. By 1954, the search for cyanobacteria-like microfossils in the Gunflint chert stromatolites was starting to bear some fruit in the form of fossilized cells, coccoidal colonies and filaments (Tyler and Barghoorn 1954; Barghoorn and Tyler 1965). There followed an increasing number of descriptions of stromatolites, and of microfossils from stromatolitic cherts, in the succeeding decades (e.g., Walter 1976; Krumbein 1983). During this phase, it was widely assumed that stromatolites were largely formed by the trapping and binding activities of microbes such as living cyanobacteria. Such work was helping to promote and establish the paradigm that cyanobacteria are a search image for the earliest life, and that stromatolites are their constructions (see Schopf and Klein 1992; Schopf 1999).

This cyanobacterial vision of the early biosphere was to reach its zenith during the Viking Missions to Mars in 1976. During those expeditions, the chemical tests for early life were clearly designed to sniff out the evidence for cyanobacteria-like photosynthesis. Those early astrobiologists were looking for kinds of metabolisms that involved the uptake of carbon dioxide and the release of gaseous oxygen by means of photoautotrophic enzymes like chlorophyll. As is now well known, they found nothing to their liking.

The strangeness of distant planets and of early worlds really dates from this moment of realization back in 1976. But it has taken three more decades for the cyanobacterial paradigm – the cyanosphere – to crumble.

The first nail in the coffin came with the discovery of stromatolite-like growth around the chimneys of deep sea black smokers. Clearly, these ecosystems and the stromatolites they contain cannot be photoautrophic, let alone oxygenic. A major step forward was then taken by the MIT group of John Grotzinger and Dan Rothman (1996) who showed by experiment that stromatolitic growth, leading to branching and columnar forms, is the predictable outcome from some kinds of abiogenic crystal growth, much like the growth of calcareous flowstone (cf. Fig. 3a). They suggested that such potentially abiogenic stromatolites had isopachous laminae (Fig. 3b), whereas biologically mediated stromatolites tended to have non-isopachous laminae (Fig. 3c).

These conclusions were to have serious implications for the interpretation of stromatolites from the ca. 3,430 million year old Strelley Pool Formation of Western Australia. First regarded as biogenic (Lowe 1980), they were later rejected by him (Lowe 1994), but further examples and arguments were resurrected by Hoffman et al. (1999) and then more recently by Allwood et al. (2006).

Our own studies (e.g., McLoughlin et al. 2008; Wacey et al. 2008; Wacey 2010), however, reveal that these structures typically show isopachous laminae, and form part of a spectrum that ranges from ripple-like corrugations of linear, through sinuous to linguoid and lunate forms, culminating in asymmetrically conical morphologies (Fig. 4). As such, many may be explained as accretionary flow-stone like bed forms formed under supersaturated conditions on the seafloor. As yet, there is no evidence in these stromatolites for the preservation of microbial fossils in the form of sheaths, filaments or cells.

A further challenge to the paradigm of stromatolites as cyanobacterial markers – and even as biosedimentary structures – has come from experimental work undertaken by Nicola McLoughlin and colleagues (2008). They have managed to grow complex digitate and inclined stromatolites with non-isopachous laminae (cf. Fig. 2b) under conditions that are relevant to the accretion of silcrete and calcrete stromatolites (e.g., on the early land surface), and to the accretion of calcareous tufas and siliceous sinters (e.g., in the early oceans). These laboratory simulated stromatolites show some remarkable similarities with the famous Gunflint stromatolites (Fig. 2a).

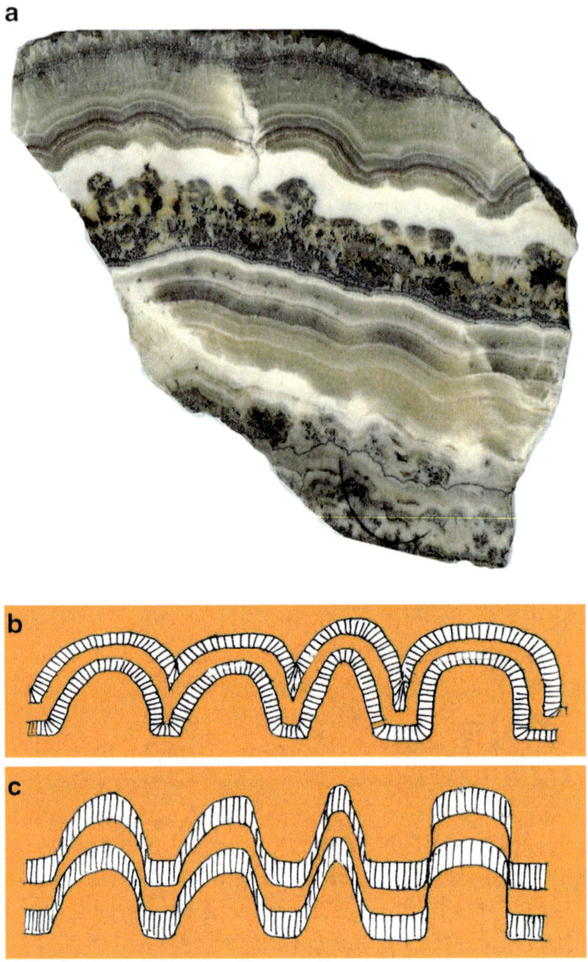

Fig. 3 The decoding of self-organized sedimentary structures. (**a**) Hand specimen of banded agate showing stromatolite-like domes and bush-like dendrites, from an un-named deposit in Wyoming. Note the isopachous nature of the laminae, regarded by Grotzinger and Rothman (1996) as potentially abiogenic. (**b**) Sketch to show the nature of isopachous growth, typically found in agates, malachites and other hydrothermal mineral deposits. (**c**) Sketch to show the nature of non-isopachous laminae, supposedly typical of stromatolites with a biological component

From this and other studies, it emerges that stromatolite morphologies tend to accumulate along the edges of a deposition system where the supply of material is starting to fail – in what can be called the zone of complexity. At best, they may tell us something indirect about viscosity and the presence of extracellular mucilage. But clearly, they can also grow completely without the participation of biology.

A plethora of studies have now shown that stromatolitic morphology should henceforth be regarded as a branch of physical sedimentology. That is to say, whereas ripples are the products of low viscosity accretion, stromatolites may be seen as the products of accretion under more viscous conditions. Clearly, this

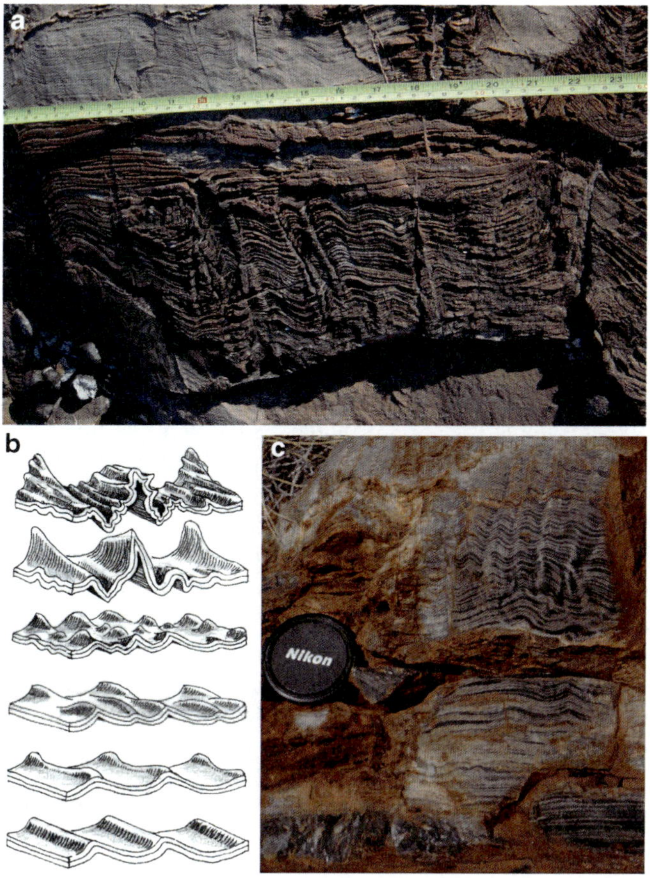

Fig. 4 The decoding of self-organized sedimentary structures from the ~3,430 Ma Strelley Pool Formation. (**a**, **c**) Field photographs showing stacked pseudocolumns of linear, ripple-like features, commonly regarded as 'stromatolites'. (**b**) Reconstructions of the geometry of the so-called stromatolites from the Strelley Pool Formation, showing the continuous spectrum from linear ripples to oversteepened lunate ripples and pseudoconical structures

viscosity may take the form of either abiologically- or biologically-induced crystal precipitates and gels. Both systems can produce self-organized structures of domes and columns under non-equilibrium conditions.

2 Implications

Several major conclusions can be drawn from these case histories. Clearly there is a need for a geological understanding of the context for early life on Earth or Mars, not merely a biological understanding. This means the acquirement of a thorough training in metamorphic, igneous and sedimentary petrology, and the study of context at a range of scales, including the premier disciplines of geological mapping and fabric mapping. Only in that way can scientists yet hope to form an opinion as to whether the context and burial history of the host rock is consistent with claims made for early life.

The second conclusion is equally serious. It concerns our facing up to a non-biological rather than biological origin for candidate morphological (or biogeochemical) signals from the early Earth or Mars. As can now be seen with the sad cases of *Eozoon* (Brasier 2009), *Eoleptonema* (Brasier et al. 2002, 2005) and *Cryptozoon* (cf. McLoughlin et al. 2008), we can no longer afford to ignore the ways in which abiology can simulate earthly biology, let alone unearthly biology. The main problem here is that morphological complexity has for long been taken to be a keystone characteristic for the earliest fossils (e.g., Buick et al. 1981; Schopf 1999). A basic understanding of self-organizing structures (SOS) and complexity is therefore an essential step if the early fossil record is to be correctly decoded. Unfortunately, it is emerging that complex structures do not require complex causes, as shown nearly a century ago by d'Arcy Thompson (1917). As seen with stromatolites, they can arise naturally in physico-chemical systems within the realms of 'chaotic' behaviour (Grotzinger and Rothman 1996).

In Fig. 5, attention is therefore drawn to a range of physico-chemical gradients that can lead to the formation of macroscopic stromatoloids (A) and ripples (B) as well as to microfossil-like structures generated by the growth of dendrites (E), 'coffee-ring' effects (F), polygonal crystal rims (G) and spherulites (H). In each of the systems shown, a move to the right of the diagram results in a loss of symmetry but a gain in morphological or temporal complexity towards the 'chaotic domain' (see Stewart and Golubitsky 1992). This leads to a 'symmetry-breaking cascade', wherein the 'symmetry group' falls and the level of information rises. Symmetry-breaking is a truly conspicuous phenomenon during the growth and recrystallization of spherulites, leading to a natural assemblages of structures that can range from spheroidal (broadly rotational symmetry), to dendritic (reflectional to slide symmetry), to arcuate (no clear symmetry; Fig. 5h). Such symmetry-breaking cascades appear to arise when localised changes in the ionic concentrations of the constituent chemicals (e.g., iron oxide, carbon) fall below a critical threshold, so that the higher levels of symmetry became unstable. In this way, the margins of crystal growth can provide a rich harvest of pseudofossil structures, ranging from polygonal to dendritic to filamentous (e.g., snowflakes, moss agate, pyrolusite 'moss'; Fig. 5e–h) and from spherulitic/botryoidal to dendritic to filamentous (e.g., hydrothermal cherts and jaspers; Fig. 5h). Such complex systems have also been simulated by computational experiments and digital automata (Fig. 5c–d), replicating the self-organization seen within stromatolites and dendrites (Grotzinger and Rothman 1996; Wolfram 2002). Brasier et al. (2006) have reviewed the problems of spheroids, filoids, septate filoids, stromatoloids, wisps and fluffs and the challenges that they present for decoding the earliest fossil record.

Given such a challenge, it therefore seems wise to remain cautious and regard many Archean microfossils, stromatolites and carbon isotopic values (older than, say, ca. 3,000 million years) as open to question until their origin by likely alternative, abiological, processes has been falsified (Brasier et al. 2002, 2006; McLoughlin et al. 2008; Wacey 2010). It is this null hypothesis that makes astrobiology into a science.

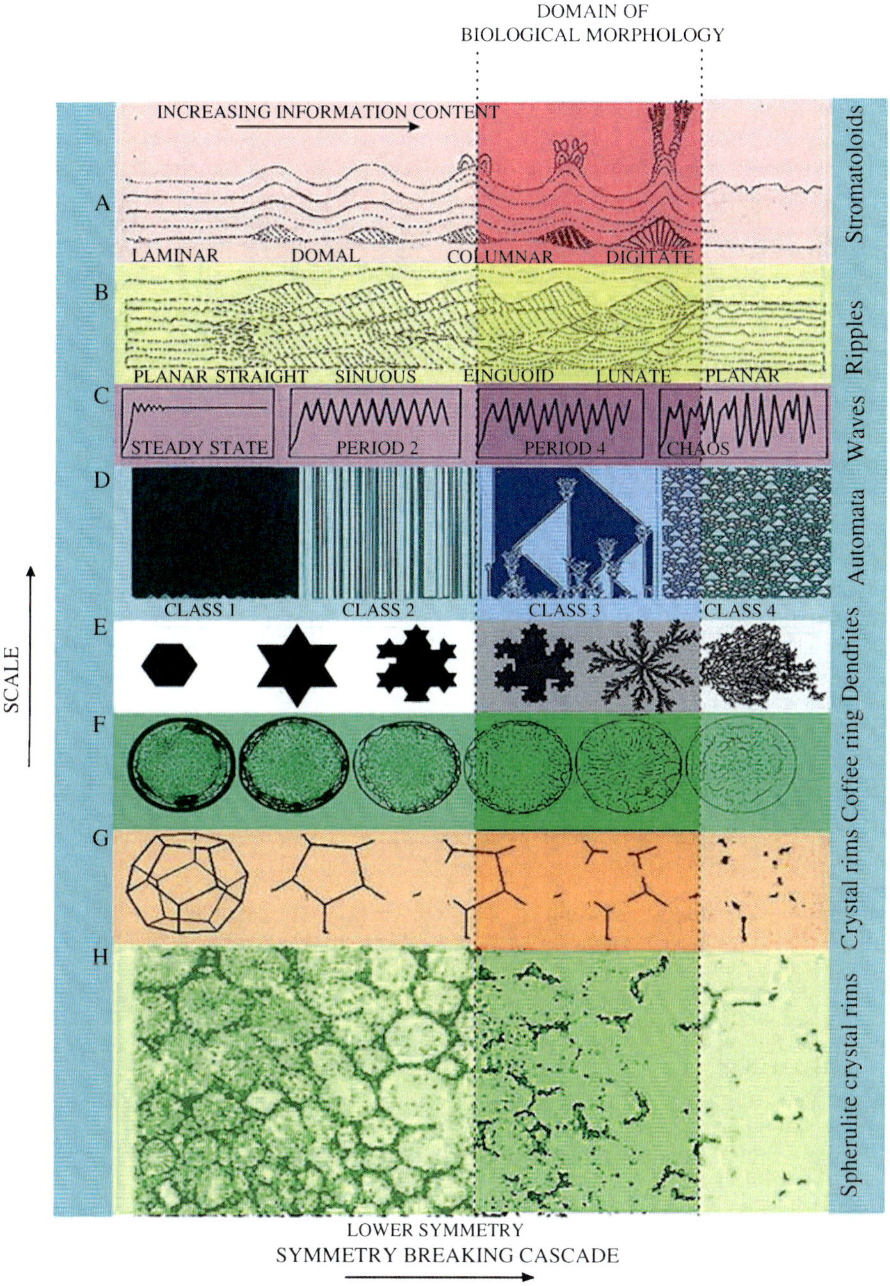

Fig. 5 Self-organising structures (SOS) that can arise naturally in physico-chemical systems within the realms of chaotic behaviour. Symmetry is lost as one moves to the right but morphological complexity increases. In well preserved microfossil assemblages morphological variation of the fossil assemblages is usually less than co-occurring abiogenic structures and so should occupy a more restricted domain ('domain of biological morphology') within the morphospace (Adapted from Brasier et al. 2006)

References

Allwood AC, Walter MR, Kamber BS et al (2006) Stromatolite reef from the Early Archaean era of Australia. Nature 441:714–718

Barghoorn ES, Tyler SA (1965) Microorganisms from the Gunflint Chert. Science 147:563–577

Black M (1933) Algal sediments of Andros Island, Bahamas. Phil Trans R Soc Lond B 222:165–192

Brasier MD (2009) Darwin's lost world. The hidden history of animal life. Oxford University Press, Oxford

Brasier MD, Green OR, Jephcoat AP et al (2002) Questioning the evidence for Earth's oldest fossils. Nature 416:76–81

Brasier MD, Green OR, Lindsay JF et al (2005) Critical testing of Earth's oldest putative fossil assemblage from the 3.5 Ga Apex Chert, Chinaman Creek Western Australia. Precambr Res 140:55–102

Brasier MD, McLoughlin N, Wacey D (2006) A fresh look at the fossil evidence for early Archaean cellular life. Phil Trans R Soc Lond B 361:887–902

Buick R, Dunlop JSR, Groves DI (1981) Stromatolite recognition in ancient rocks: an appraisal of irregularly laminated structures in an early Archaean chert-barite unit from North Pole, Western Australia. Alcheringa 5:161–181

Callow RHT, Brasier MD (2009) A solution to Darwin's dilemma of 1859: exceptional preservation in Salter's material from the Ediacaran Longmyndian Supergroup, England. J Geol Soc 166:1–4

Darwin C (1839) Voyages of the adventure and beagle, volume III – Journal and remarks, 1832–1836. Henry Colburn, London

Darwin C (1859) On the origin of species by means of natural selection, or the preservation of favoured races in the struggle for life (usually shortened to: On the Origin of Species). John Murray, London

Foster JW, Whitney JD (1851) Report on the geology of the Lake Superior land district. Part II, Washington, DC

Grotzinger JP, Rothman DH (1996) An abiotic model for stromatolite morphogenesis. Nature 383:423–425

Hall JD (1883) Cryptozoon (proliferum) ng. (and sp). New York State Museum Annual Report, 36(1883), p 6 and caption

Hoffman HJ, Grey K, Hickman AH et al (1999) Origin of 3.45 Ga coniform stromatolites in Warawoona group, Western Australia. Geol Soc Am Bull 111:1256–1262

Irving R (1883) Copper bearing rocks of Lake Superior. Monograph of the United States Geological Survey

Kalkowsky VHE (1908) Oolith und Stromatolith im Norddeutschen Buntsandstein. Zeitschrift der Deutschen Geologischen Gesellschaft 60:84–125

Krumbein WE (1983) Stromatolites: the challenge of a term in space and time. Precambr Res 20:493–531

Lowe DR (1980) Stromatolites 3,400-Myr old from the Archean of Western Australia. Nature 284:441–443

Lowe DR (1994) Abiological origin of described stromatolites older than 3.2 Ga. Geology 22:387–390

McIlroy D, Crimes TP, Pauley JC (2005) Fossils and matgrounds from the Neoproterozoic Longmynd Supergroup, Shropshire, UK. Geological Magazine 142:441–455

McLoughlin N, Wilson L, Brasier MD (2008) Growth of synthetic stromatolites and wrinkle structures in the absence of microbes: implications for the early fossil record. Geobiology 6:95–105

Salter J (1856) On fossil remains of Cambrian rocks of the Longmynd and North Wales. Q J Geol Soc Lond 12:246–251

Salter J (1857) On annelide-burrows and surface markings from the Cambrian rocks of the Longmynd. Q J Geol Soc Lond 13:199–206

Schopf JW (1999) The cradle of life. Princeton University Press, New York
Schopf JW, Klein C (eds) (1992) The Proterozoic biosphere: a multidisciplinary study. Cambridge University Press, Cambridge
Stewart I, Golubitsky M (1992) Fearful symmetry. Is God a geometer. Penguin Science, London
Thompson D'AW (1917) On Growth and Form. Cambridge University Press, Cambridge
Tyler SA, Barghoorn ES (1954) Occurrence of preserved plants in pre-Cambrian rocks of the Canadian shield. Science 119:606–608
Wacey D (2010) Stromatolites in the ~3400 Strelley Pool Formation, Western Australia: examining biogenicity from the macro- to the nano-scale. Astrobiology 10:381–395
Wacey D, McLoughlin N, Stoakes CA et al (2008) The ~3.4 Ga Strelley Pool Chert in the East Strelley greenstone belt – a field and petrographic guide. Western Australia Geologic Survey Record
Walcott CD (1883) Pre-Carboniferous strata in the Grand Canyon of the Colorado, Arizona. Am J Sci 26:437–442
Walter MR (1976) Stromatolites. Elsevier, Amsterdam
Wolfram S (2002) A new kind of science. Wolfram Media, Champaign

Trace Element Geochemistry as a Tool for Interpreting Microbialites

Gregory E. Webb and Balz S. Kamber

Abstract Microbialites are critical for documenting early life on earth and possibly elsewhere in the solar system. However, criteria for microbialite identification are controversial. Trace element geochemistry provides two types of information that aid interpretation of putative microbialites. Firstly, because most microbialites consist of hydrogenous precipitates, trace elements can be used to investigate the fluids in which the structures formed, thus aiding identification of environments of formation. For example, rare earth elements preserved in microbialites have proven very useful in discriminating depositional environments. Secondly, microbes utilize and concentrate a wide range of elements, including many metals. Preservation of such elemental enrichments may provide a valuable biosignature. Although research in this field is relatively young, high precision, in situ measurement of metals in microbialites using techniques such as laser ablation-inductively coupled plasma-mass spectrometry, now with spatial mapping, have identified consistent enrichments in biologically important metals in microbialites. Hence, trace element studies are finding increasing utility in studying microbialites, and so long as diagenesis and the degree to which specific precipitates represent microenvironments are taken into account, trace element inventories may provide important information about depositional settings and, potentially, metabolic processes within biofilms.

Keywords Microbialite • Stromatolite • Biogeochemistry • Trace element geochemistry • Rare earth elements • Microenvironment

G.E. Webb (✉)
School of Natural Resource Sciences, Queensland University of Technology, GPO Box 2434, Brisbane, Queensland 4001, Australia
e-mail: ge.webb@qut.edu.au

B.S. Kamber
Department of Earth Sciences, Laurentian University, 935 Ramsey Lake Road, Sudbury Ontario P3E 2C6, Canada

S.D. Golding and M. Glikson (eds.), *Earliest Life on Earth: Habitats, Environments and Methods of Detection*, DOI 10.1007/978-90-481-8794-2_6,

1 Introduction

Microbialites are one of the most important records of early life on Earth (Schopf 1994; Walter 1994) and are among the most likely types of potential evidence of extraterrestrial life (Gibson et al. 2001; Cady et al. 2003; Barbieri and Cavalazzi 2004). Microbialites are 'organo-sedimentary' structures formed by the incidental induction of mineral precipitation within or on microbial biofilms with, or without, trapping and binding of ambient sediments (Burne and Moore 1987; Riding 1991b; Dupraz et al. 2008; c.f., the stromatolite definition of Awramik and Margulis in Walter 1976). A 'biofilm' is a consortium of microbes contained within their own extracellular products attached to a substrate. Most authors differentiate biofilms and microbial mats, restricting the former to microbial consortia occurring on solid surfaces, and the latter to commonly thicker communities that occur on, or within, particulate sediment (Stolz 2000). However, the difference is commonly one of scale with thick and complex benthic 'microbial mats' intergrading continuously with 'biofilms' that may be only micrometers in thickness. Functionally, the two end members are similar and the nature of the relationship between substrate and microbial community is far from clear for many ancient microbialites. Hence, for the purposes of this chapter the two terms can be considered more or less interchangeable, but the structural and chemical complexity increase from biofilms to microbial mats. In any case, the study of microbialites forms a nexus between sedimentology, biology, and increasingly now, biogeochemistry and geomicrobiology.

Microbialites include, but are not limited to, the well known laminated forms called stromatolites, which are the most abundant and conspicuous Precambrian fossils. However, the interpretation of microbialites has long been a contentious issue because they include a diverse collection of trace fossils rather than true body fossils (Buick et al. 1981; Walter 1983, 1994; Lowe 1994; Grotzinger and Rothman 1996; Grotzinger and Knoll, 1999; Lindsay et al. 2003; Awramik and Grey 2005). In other words, microbialites are structures that were made by microbial communities rather than the remains of the microbes themselves. As an analogy, buildings are made by people, and thus traces of humanity, but they are not fossils of people themselves and give little indication of the identity or morphology of the people who built them. However, although buildings would be considered irrefutable evidence of life, structures potentially interpreted as microbialites may not be. The difference is that no natural, abiological process can be conceived that could produce a building, whereas it is feasible that structures similar to some microbialites could be made without the intervention of microbial life (Grotzinger and Rothman 1996; Grotzinger and Knoll 1999; Batchelor et al. 2000; Mcloughlin et al. 2008). A variety of natural processes in different environmental settings are capable of creating laminated structures composed of sediment or mineral precipitates without requiring biological activity. These include botryoidal mineral deposits such as malachite or agate, calcrete (Read 1976) and speleothems (Thrailkill 1976), although some degree of microbial influence may affect the morphology of the last two types of structures (e.g., Loisy et al. 1999; Cacchio et al. 2004). Regardless, many authors have chosen to use the term 'stromatolite' in a non-genetic, morphological sense only, meaning an attached, laminated, sedimentary structure that accretes away from the substrate (e.g., Semikhatov et al. 1979; Grotzinger and Knoll 1999; Perry et al. 2007).

In this chapter, we use the term stromatolite exclusively for structures interpreted as having formed owing to the presence of a microbial mat or biofilm, i.e., stromatolites are a subset of microbialites, and abiological laminated structures are not stromatolites. Of necessity, some structures must remain equivocal, and hence, interpretations of their origins will reflect varying degrees of certainty (Awramik and Grey 2005). Regardless, it is acknowledged that the morphologies of stromatolites are heavily affected by ambient environmental conditions, such as sediment supply, water energy and water chemistry (Grotzinger and Knoll 1995, 1999; Knoll and Semikhatov 1998; Sumner 2000; Turner et al. 2000; Dupraz et al. 2006). Additionally, a continuum exists between biologically constructed and completely abiotic structures (Riding 1991b). The rule of thumb being applied here is that we consider a structure to be a stromatolite if it would not have formed with its own distinctive morphology without the presence of a microbial community. We must acknowledge that microbialites formed by the same community might differ in morphology in different environmental settings, as do for example the skeletons of invertebrates such as scleractinian corals (e.g., see Fig. 4 in Veron 1995). We also acknowledge that benthic microbial mats and biofilms are generally formed by a diverse, commonly stratified community of microbes, not a single biological entity (Golubic 1976; Revsbech et al. 1983; Stolz 1990, 2000). Equally microbial consortia and/or metabolic processes may vary spatially and temporally in the same environment or biofilm (Canfield and Des Marais 1993; Wimpenny and Kinniment 1995; Little et al. 1997; Decho 2000; Konhauser 2007; Buckley et al. 2008). Biofilms can be very complex, stratified structures, both in terms of communities and geochemical pathways (e.g., see nice review in Chapter 6 of Konhauser 2007). Hence, microbialites cannot be treated in the same way as traditional body fossils (i.e., preserved parts of the organisms themselves). Irrespective of those factors, we prefer to use the genetic meaning of stromatolite, as that gets at the heart of evidence for early life on Earth, and irrespective of semantics, it is very much our task to make the distinction of whether a structure was biogenic or not. In favor of our approach from a semantic point of view is the fact that other laminated structures that are not thought to be specifically biogenic have names with genetic connotations. Calcrete, speleothems and seafloor crusts are all terms for laminated sedimentary structures that imply a mode of genesis. Why should stromatolites be any different? It must also be pointed out that although stromatolite-like features have been created in computer models without intrinsic biological factors (Grotzinger and Rothman 1996), and limited experimental work has backed them up (Mcloughlin et al. 2008), those models are capable of producing only simple stromatolite-like morphologies (Batchelor et al. 2000; Dupraz et al. 2006). Many stromatolite morphologies, including some simple forms, such as those with side branching and the wide-spread conical *Conophyton*-like forms, cannot be modeled without using inherent biological attributes (Grotzinger and Knoll 1999; Batchelor et al. 2000, 2004). Additionally, demonstrably abiotic structures that could be confused easily with stromatolites have yet to be identified in natural environments (Allwood et al. 2006).

The fact remains that it is very difficult to establish whether a putative stromatolite was biogenic. Hence, a variety of different criteria have been proposed for establishing the origin of such structures (Buick et al. 1981; Walter 1983). These criteria are discussed in more detail below, but in general, have proven to be very difficult to apply to ancient examples. Modern microbialites, on the other hand, can be attributed to biological origin with great certainty, but that certainty decreases with geological age. Hence, there has been a clash between philosophical approaches in dealing with ancient microbialites. One school of thought championed by many long-time stromatolite workers (e.g., Walter 1976; Awramik and Grey 2005; Awramik 2006) suggests that biogenic morphologies can be identified on the basis of setting, morphology and composition by analogy to living or clearly biogenic Phanerozoic examples. The other school (e.g., Grotzinger and Knoll 1999; Brasier et al. 2002; Lindsay et al. 2003) suggests that all such structures should be considered abiotic unless they can meet exacting, and even unlikely, criteria for biogenicity. Hence, the question comes down to a choice of the null hypothesis. Although the 'everything is abiotic unless proven otherwise' philosophical approach may be overkill for the Proterozoic when microbial life was clearly established and stromatolites dominated many shallow benthic settings (Grotzinger and Knoll 1999), it is a necessity for interpreting the earliest structures that might record the presence of life. Once life was established, however, the null hypothesis might be shifted to reflect the fact that life existed and, hence, is the most reasonable explanation for many stromatolite-like structures. The strict adherence to either null hypothesis may be unwarranted, as pointed out by Awramik and Grey (2005), owing to the fine line between biotic and abiotic mechanisms of precipitation in microbialites (Riding 1991b; Webb 2001a). Regardless, in order to search for life's earliest record, we must establish reliable biosignatures for microbialites to allow specific hypotheses of biogenicity to be tested. The purpose of this chapter is to discuss how trace element geochemistry can be used to test the origins of putative microbialites preserved as carbonate rocks. A variety of other minerals may form microbialites (see Konhauser 2007), including various Fe oxyhydroxides (Bazylinski and Moskowitz 1997; Konhauser et al. 2002; Anderson and Pederson 2003, Fortin and Langley 2005), but we will concentrate on carbonate microbialites only in this chapter.

1.1 How Do Microbialites Form?

Several recent reviews have detailed the types and nature of microbialites and processes responsible for their formation (Monty 1976, 1977; Riding 1991b, 2000; Castanier et al. 1999; Dupraz and Visscher 2005; Dupraz et al. 2008). In general, microbialites form in two fundamental ways: (1) trapping and binding of ambient sediment, and (2) inducing or localizing the precipitation of minerals, typically, but not invariably, the carbonate minerals, calcite, aragonite or dolomite. Regardless of the mode of formation, microbialite genesis is attributed to the presence of an active microbial community or associated organic matter.

In the trapping and binding model of stromatolite genesis (Black 1933; Logan et al. 1964), sediment from the local environment is trapped on and between the erect filaments of cyanobacteria that are coated by sticky extracellular polymeric substances (or exopolymers or exopolysaccharides), which generally are abbreviated to EPS (Fig. 1). Structures formed by trapping and binding were termed agglutinated

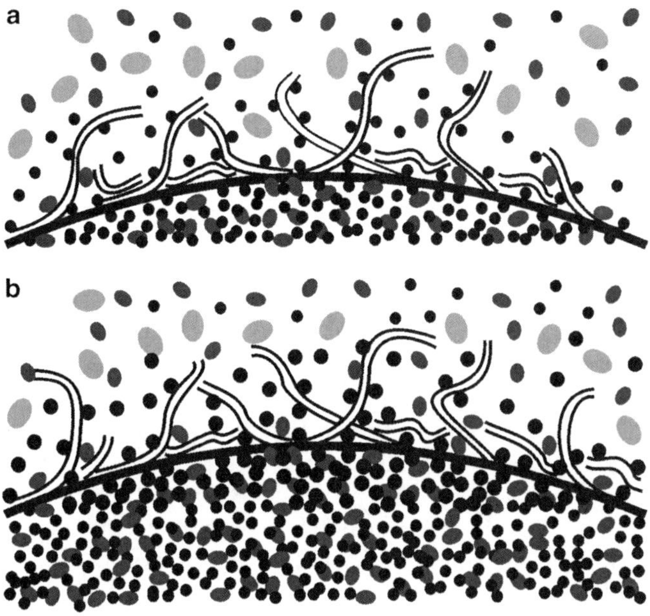

Fig. 1 Schematic diagram of trapping and binding model of stromatolite accretion (after Riding 1991b); (**a**) and (**b**) represent a time series wherein microbial filaments trap sediment and cause its continuing accumulation beneath the growing mat. Note that the coarsest sediment fraction may not be trapped as current or wave energy may be adequate to remove it from the 'sticky' filaments

stromatolites by Riding (1991b) and include the modern stromatolites of Shark Bay, Western Australia, which consist of trapped and bound pelloids, mollusk shells and quartz sand (Logan 1961; Logan et al. 1964), and Bahaman stromatolites, which consist of trapped and bound ooids (Dravis 1983; Dill et al. 1986). Although these structures accreted from ambient particulate sediment, modern examples are lithified relatively quickly on the sea floor (Logan 1961; Reid and Browne 1991; Reid et al. 2000), and rapid lithification is probably a prerequisite for preservation of agglutinated stromatolites in most shallow environments owing to the common reworking of sediment during high energy events (Krumbein et al. 1994). The rapid lithification of Bahaman stromatolites has been attributed to a variety of microbial activities, including even boring by endoliths wherein early (concomitant) cementation within individual microbial borings bridged the gaps between sediment grains, thus cementing them together (Macintyre et al. 2000; Reid et al. 2000). Agglutinated stromatolites may be laminated owing to a variety of biological processes within the microbial mats, including the diurnal behavior of cyanobacteria and ecological succession, and/or episodic sedimentation or cementation events (Monty 1976). Where trapped and bound sediment is coarse, agglutinated stromatolites can be recognized easily, but where trapped and bound sediments are limited to very fine carbonate mud, they may be difficult to distinguish from stromatolites formed primarily by precipitation of microcrystalline carbonate. Even where agglutinated stromatolites occur within relatively coarse ambient sediment, they may be capable of accumulating only the fine sediment that more easily adheres to filaments. Coarse sediment in such environments may be removed from the structures by water agitation (Riding 1991b). Hence, agglutinated micritic sediments could form amidst courser ambient sediment, and thus, petrography alone may not distinguish agglutinated from precipitated microbialites where micritic textures are concerned.

Ever since Black (1933) described intertidal soft microbial mats from Andros Island, Bahamas, as potential stromatolites, the concept of agglutinated stromatolites has been well enshrined in geological literature and thinking (Monty 1977). However, the original stromatolites of Kalkowsky (1908) were clearly rigid, precipitate-dominated structures, and it has become increasingly apparent that mineral precipitates are the more important factor in microbialite genesis (Riding 1991b). Many authors have produced schemes to distinguish the types of precipitates and mineralization processes that form microbial carbonates (Krumbein 1986; Riding 1991b; Webb 1996, 2001a; Dupraz and Visscher 2005; Visscher and Stolz 2005; Dupraz et al. 2008). Riding (1991b) differentiated calcification (i.e., replacement) of EPS, which he termed 'biomineralization,' from mineral precipitation on or exterior to EPS, which he termed 'mineralization,' with the implication that the former was driven by microbial metabolism, whereas the latter was more physico-chemical (i.e., abiotic) in nature, although presumably localized by biofilms. Webb (1996, 2001a), following Lowenstam and Weiner (1989), separated Riding's (1991b) 'biomineralization' into biologically controlled 'enzymatic' secretion of mineral skeletons and the incidental induction of mineral precipitates and suggested that 'enzymatic' versus 'non-enzymatic' processes formed the primary distinction in precipitation mechanisms with microbialites representing non-enzymatic precipitation. Webb (1996, 2001a) noted that non-enzymatic biomineralization represents a continuum from the direct calcification of EPS through to the localization and/or enhancement of precipitation of cements that, once nucleated, are morphologically identical to abiotic precipitates. Hence the degree of biological influence within that continuum correlates to some degree with the microstructure of the resulting microbialite. Greater or more proximal biological influence leads to the formation of identifiable microfossils (i.e., calcimicrobes), whereas less direct influence may enhance cement precipitation rates only or localize precipitation by promoting nucleation, thus forming micritic microbialites and biocementstones that lack identifiable microfossils (Webb 2001a, his Fig. 2). Défarge and Trichet (1995) distinguished 'biominerals' formed by direct association with living bacteria and 'organominerals,' which formed on dead organic matter, and Reitner (1993) pointed out that many precipitates in modern marine 'microbialites' resulted from the degradation of sponges and other multi-cellular eukaryotes (e.g., mucus from bivalves, Braithwait et al. 2000; and corals, Reitner 2005). As a result, Neuweiler and Reitner (1993) and Reitner et al. (1995) suggested that the term 'automicrite' should be used, rather than microbialite or organomineral, where the source of organic matter that played a role in the induction of precipitation was not clear. Perry et al. (2007) re-introduced the term 'organomineral' for basically the same concept as originally suggested by Défarge and Trichet (1995), but with subtle differences (Défarge et al. 2009). In the latest review, Dupraz et al. (2008) restricted biomineralization to

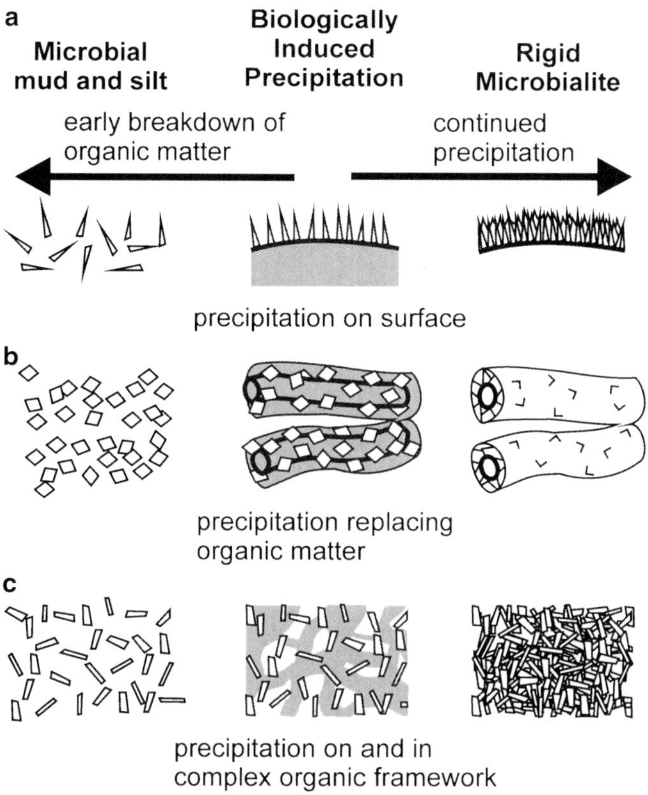

Fig. 2 Schematic diagram showing relative position of mineralization in microbialites. Note that where precipitation does not proceed far enough prior to the degradation of organic matter, precipitates may be distributed as fine sediment (mud and silt). (**a**) Precipitation on organic surfaces may lead to the formation of rigid crusts. (**b**) Precipitation replacing organic matter may produce calcimicrobes or unidentifiable three-dimensional structures. (**c**) Precipitation within complex biofilms or mats may produce felts of more or less randomly oriented crystals that have a poorly organized or massive micritic texture

biologically-controlled mineralization (in the sense of Lowenstam and Weiner 1989; and equivalent to enzymatic precipitation of Webb 1996) and utilized organomineralization to encompass biologically-induced mineralization, which they defined as requiring direct metabolic activity of microbes, and biologically-influenced mineralization, which they considered to be precipitation that is passively driven by ambient water chemistry, but with nucleation on or in living or dead organic matter (see Dupraz et al. 2008, their Fig. 2). Semantics aside, there can be no exact demarcation between precipitates that were formed directly as a consequence of microbial activity and those that precipitate for ambient physicochemical reasons alone; the processes form a continuum. That continuum is affected by the degree to which precipitation occurs within microenvironments (more on this below), and microbes may still be essential for precipitation of some structures even in settings where carbonates would be expected to form for physico-chemical reasons alone (Shiraishi et al. 2008).

Although, the nomenclature associated with mineral precipitates in microbialites is by no means stable, the two fundamental factors associated with mineral precipitation in microbialites are: (1) processes that drive up the local carbonate saturation index, i.e., an 'alkalinity engine,' and/or (2) enhancement of nucleation through provision of organic templates in fresh or degraded EPS or directly on microbial cell surfaces (see the review in Dupraz et al. 2008). A third factor that may also be important is the biological removal of precipitation inhibitors (Webb 2001a), as many organisms produce extracellular products to inhibit calcification (Westbroek et al. 1994) and shallow tropical seawater is already abundantly supersaturated with regards to calcite and aragonite (Lyakhin 1968). This could include biological utilization of P-bearing compounds or biological degradation of Ca^{2+}-binding organic matter.

Cyanobacteria are commonly considered the dominant microbes in stromatolites, but much microbial calcification is driven by heterotrophs (Krumbein 1979; Reitner 1993; Paerl et al. 2001; Dupraz et al. 2008). Important metabolic processes that may increase local carbonate saturation include, but are not limited to, photosynthetic removal of CO_2 (Golubic 1972; Merz 1992; Merz et al. 1995; Arp et al. 2001) and anaerobic sulfate reduction (Krumbein 1979; Lyons et al. 1984; Baumgartner et al. 2006). Many others have been described (e.g., Castanier et al. 1999), and decay processes, such as generation of NH_3 due to ammonification, also drive pH higher (Berner 1968; Krumbein 1979; Reitner 1993). The roles of EPS in regulating and controlling nucleation of minerals in active biofilms and in dead organic matter are very complex, but are becoming better understood (Decho et al. 2005; and see review in Dupraz et al. 2008). Regardless, clear relationships exist between the distribution of organic matter and the nucleation of particular minerals in biofilms (Fig. 2). In some cases, nucleation occurs within organized EPS such that resulting mineral growth is confined to the EPS in the sheaths of cyanobacteria (Merz et al. 1995) or possibly capsular mucilage of coccoid bacteria. This would be equivalent to the 'biomineralization' of Riding (1991b) and 'internal impregnation' of Riding (1991a). Provided mineralization proceeds far enough before the organic matter is degraded and redistributed, the result will be identifiable calcimicrobes, such as *Girvanella* in the case of filamentous microbes or *Renalcis* (Fig. 3a), for example, in the case of colonies of coccoid bacteria. Where mineralization does not proceed very far, crystals may remain isolated, becoming carbonate mud or silt in the local environment (Figs. 2 and 3b). Where mineral precipitation occurs on the surfaces of organic matter, either EPS or cellular surfaces, a variety of more or less organized crusts may form (Figs. 2 and 3c). This would be morphologically equivalent to 'mineralization' of Riding (1991b) and 'external encrustation' of Riding (1991a), although it still may represent microbially-induced precipitation. As the complexity of the microbial biofilm or mat increases, the interplay of mineralization within or on pristine or heavily degraded EPS also becomes increasingly complex making micritic microbialites with complex or poorly organized structures that might be mistaken for the results of trapping and binding of carbonate mud (Figs. 2 and 3d).

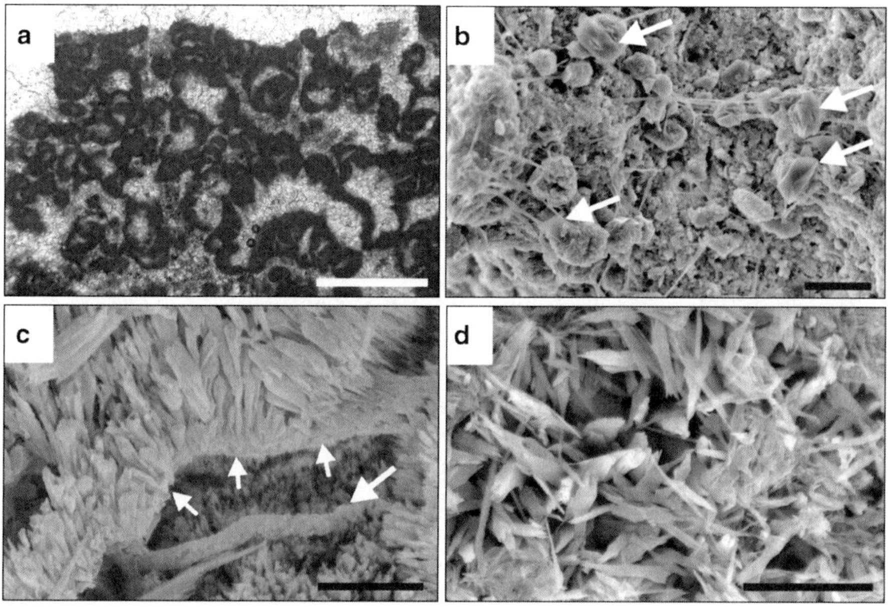

Fig. 3 Examples of microbialite microstructure types shown in Fig. 2. (**a**) Thin section photomicrograph of calcimicrobe *Renalcis* wherein capsular EPS was calcified around presumably coccoid bacteria colonies, from Devonian reef complex, Canning Basin, Western Australia; scale bar = 500 μm. (**b**) Scanning electron microscope photomicrograph of isolated calcite crystals nucleated on microbes or EPS masses (now shriveled) in surface biofilm from modern microstromatolite in beachrock, Heron Island, Australia; scale bar = 20 μm. (**c**) Scanning electron microscope photomicrograph of crust formed by nucleation of aragonite crystals on organic surface (now missing) at *small arrows*. Also note calcified filament (*large arrow*) occupying cavity below aragonite crust within modern microstromatolite in beachrock, Heron Island, Australia; scale bar = 10 μm. (**d**) Scanning electron microscope photomicrograph of randomly oriented aragonite crystals precipitated in situ in complex, irregular biofilm in beachrock, Heron Island, Australia; scale bar = 10 μm (b, c, and d after Webb et al. 1999)

1.2 Biosignatures in Microbialites

The difficulty in selecting and applying criteria for the identification of microbialites has led to a need to identify additional biosignatures in ancient rocks (Awramik and Grey 2005; Ohmoto et al. 2008). Buick et al. (1981) listed eight specific criteria to be applied in identifying stromatolites. These criteria can be reduced to four sets of criteria that apply to other types of microbialites as well. They include: (1) the structures must have formed in an appropriate environment (synsedimentary structures in a sedimentary environment for example); (2) the gross morphology is consistent with biological stromatolites (or other microbialites); (3) the micromorphology is consistent with biogenic structures (in the case of stromatolites, laminae thicken over crests and show wavy to crinkly texture at various wavelengths); and (4) the structures contain fossil microbes with preserved community-related distributions consistent with their having constructed the structure (as opposed to being distributed as incidentally trapped microbes from the ambient environment). Identification of the depositional environment for putative microbialites can be accomplished using basic sedimentology and geochemistry (including trace element geochemistry), and specific examples of those applications are discussed below. Gross morphology is a more complicated issue. As previously noted, computer simulations are capable of reproducing some stromatolite morphologies (both gross structure and lamination) without the need for biological factors (Grotzinger and Rothman 1996). However, abiological benthic marine structures that could be confused easily with biogenic stromatolites using the criteria above have not been demonstrated convincingly. Hence, 'abiotic stromatolites' remain primarily in the realms of computer models. Micromorphology consistent with biogenic structures can be a difficult criterion to apply owing to subsequent diagenetic alteration, especially including recrystallisation. However, even before that, the texture of stromatolite laminae may vary depending on the relative timing of organic decay relative to initial calcification (Knoll and Semikhatov 1998).

The fourth set of criteria is also complicated and, importantly, many unquestionable stromatolites do not preserve life-oriented fossil microbes (Walter 1994; Grotzinger and Knoll 1999). This partly reflects stromatolite construction from metastable carbonate minerals that readily recrystallize during diagenesis (Golubic and Hofmann 1976; Walter et al. 1988; Grotzinger and Knoll 1999; Bartley et al. 2000), but also the rapid recycling of organic matter within microbial biofilms (Bartley 1996). Microfossils can be preserved in such rocks, including the calcimicrobes discussed previously, but even where calcimicrobes occur, they may become unrecognizable and grade into micrite laterally at fine spatial scales due to subsequent diagenetic alteration or inadequate calcification from the outset (Merz 1992; Pratt 1995; Turner et al. 2000; Webb 2001b). Preservation of microfossils in stromatolites generally reflects a fortuitous diagenetic event, such as early silicification (e.g., Lee and Golubic 2000). Hence, in most microbialites it is unrealistic to expect microbe fossils to be present. Some microbialites even grow in such a way as to preclude the formation of microbial fossils in the first place. Agglutinated microbialites may be too course to preserve microfossils and even fine-grained microbialites may not contain them. For example, the microstructure of Holocene thrombolites that formed in cryptic settings in cavities in reef rock in the Great Barrier Reef continually evolved during the growth of the structure (Webb et al. 1998). Initial deposits were thin, smooth sheets that replaced degraded EPS layers at the base of a biofilm. However, through time, these films acted as nucleation sites for larger high-Mg calcite scalenohedra, which increased in size during the life of the film, thus obscuring the original smooth sheets (Fig. 4). The result is an irregular crystalline microstructure that is very similar to what are presumed to be abiotic high-Mg calcite cements (Macintyre 1977). No physical trace of microbes was preserved in the structures, and the growth of scalenohedra halted once the biofilms died, thus preserving the full sequence of microstructures from top to base. The high degree of preservation suggests that the structures were not altered by subsequent diagenetic processes (i.e., early marine cementation) and that ongoing precipitation of the scalenohedra was induced by the overlying biofilm (Webb et al. 1998). Modern micritic microbialites also form from aragonite.

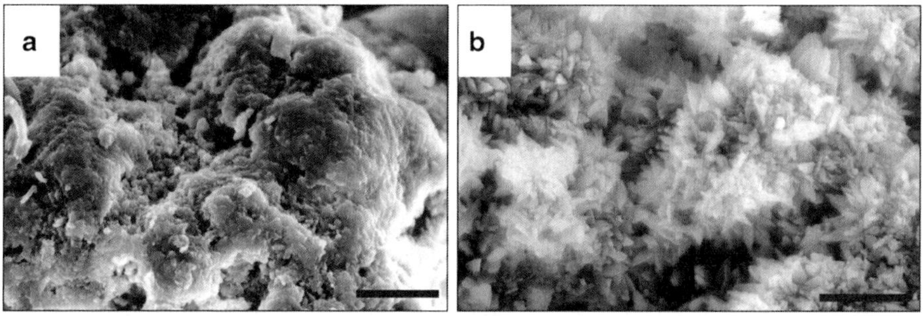

Fig. 4 Scanning electron microscope photomicrographs showing the end-members of microstructure in cryptic microbialites formed in cavities in reef framework, Holocene, Heron Reef, Australia (after Webb et al. 1998). (**a**) Smooth, amorphous to very finely crystalline Mg-calcite replacing the basal EPS layer of the surface biofilm; scale bars = 100 μm. (**b**) Coarser scalenohedral Mg-calcite nucleated from smooth crusts within cavities well below the surface of the microbialite. Consistent preservation and distribution of both calcite morphologies within the microbialites suggest that scalenohedra growth was induced beneath the biofilm by an active microbial alkalinity engine and that precipitation ceased after growth of the microbial community terminated. If the scalenohedral Mg-calcite represented secondary inorganic cement it would be expected to occur on the most recent smooth crusts as well

Aragonite micro-stromatolites in beach rock contain spaces that once hosted microbes, but the lack of well defined edges on these spaces would hinder their recognition in the rock record, especially after recrystallization (Fig. 3d). Hence, adherence to the criterion of preserved microbes in, or out, of growth position automatically excludes many otherwise demonstrably biogenic structures. On the other hand, some abiotic microstructures are capable of mimicking fossil microbes (García-Ruiz 2000), further complicating the criterion. For those reasons, Walter (1994) de-emphasized criteria relating to preservation of microbial fossils and replaced them with a chemical criterion; microbialites should have an appropriate chemical composition (e.g., a calcium carbonate polymorph in a normal marine setting). Regardless, while some morphological features can be used to definitively establish the biogenicity of some microbialites (e.g., calcimicrobes at the microscopic scale up to coniform morphologies at macroscopic scales), additional biosignatures are required.

2 Trace Element Geochemistry and Microbialites

A variety of chemical biosignatures have been investigated in ancient, putative microbialites. Where preservation is adequate, C isotopes have been used to distinguish co-occurring microbial from non-microbial components of limestones (Des Marais et al. 1989; Guo et al. 1996; Schidlowski 2001; Stephens and Sumner 2002; Andres et al. 2005). However, preservation is commonly not adequate to preserve such signatures in the carbonate portion of Precambrian stromatolites owing to diagenetic alteration. Sulfur isotopes are also useful (Cameron 1982; Shen et al.

2001; Shen and Buick 2004; Philippot et al. 2007) and organic biomarkers are valuable potential biosignatures in modern and ancient rocks (Brocks et al. 1999, 2003; Neilan et al. 2002; Derenne et al. 2008). However, in both cases preservation problems can render them useless and this is especially a problem for older Archean sequences that commonly have undergone at least low grade metamorphism.

Trace element geochemistry can be used to help establish biogenicity of putative microbialites in two main ways. Firstly, trace element geochemistry is invaluable for helping establish the type of fluid from which a mineral precipitated, and thus, for establishing the depositional setting of a putative microbialite (i.e., the first set of criteria of Buick et al. 1981 and Walter 1994). Secondly, bioactive trace elements may be sequestered and preserved in microbialites providing direct evidence not only of the presence of biofilms or microbes, but of specific metabolic pathways (Webb and Kamber 2004). Transition metals bind tightly to the organic ligands in EPS (Geesy and Jang 1989), and their ability to bridge between two or three adjacent exopolymer molecules may provide structural strength to biofilms by increasing the cohesiveness of the EPS (Parker et al. 1996, Decho 2000). Additionally, most of the inner transition elements (specifically from V to Zn) and heavier elements such as Mo are cofactors or parts of cofactors for enzymes or structural elements in proteins, thus being essential trace nutrients for organisms (Silver 1983, 1997; Ledin 2000; Rosen 2002; Cavet et al. 2003; Leão et al. 2007). Microbes in particular utilize a wide variety of metals in their diverse metabolic pathways aside from use as redox reagents, including for example, V in nitrogenase (i.e., VNase, an enzyme used for nitrogen fixation), Mn in thylakoidal water-splitting oxygen-evolving complex, Fe in hydrogenase, Co in cobalamin and carbonic anhydrase, Ni in [NiFe]-hydrogenase and as a cofactor in methyl-CoM reductase, Cu in thylakoidal plastocyanin, Zn in α-, β-, and γ-carbonic anhydrase, and Mo in heterocystous nitrogenase (Nealson 1983; Ankel-Fuchs and Thauer 1988; Butler 1998; Fraústo da Silva and Williams 2001; Cavet et al. 2003; Morel and Price 2003; Giordano et al. 2005; Jaun and Thauer 2007). Even Cd can be used in carbonic anhydrase (Lee et al. 1995). Microbes also sequester toxic elements so as to detoxify their local environments (Silver 1997; Tebo and Obraztsova 1998; Ledin 2000; Fraústo da Silva and Williams 2001; Ehrlich 2002; Worms et al. 2006; Konhauser 2007), and Decho (2000) noted that the concentration of transition metals in biofilms generally decreases in the direct vicinity of cells. Importantly, different groups of microbes, both prokaryote and eukaryote, utilize various metals in different ways and in different fundamental ratios (e.g., Quigg et al. 2003). Hence, the accumulation of such elements out of equilibrium with the ambient environment may provide a biosignature, and the ratios of particular metals may provide information about the types of organisms that were involved. Additionally, the abundances of such metals where concentrated in ancient hydrogenous precipitates may tell us much about the evolution of marine biogeochemical cycles owing to the differing availability of some elements under differing redox regimes (Ehrlich 2002; Saito et al. 2003; Anbar et al. 2007; Konhauser et al. 2009).

Preserved trace element inventories in microbialites may reflect four major sources (Fig. 5), which are: (1) biological enrichment in the organic matter of the

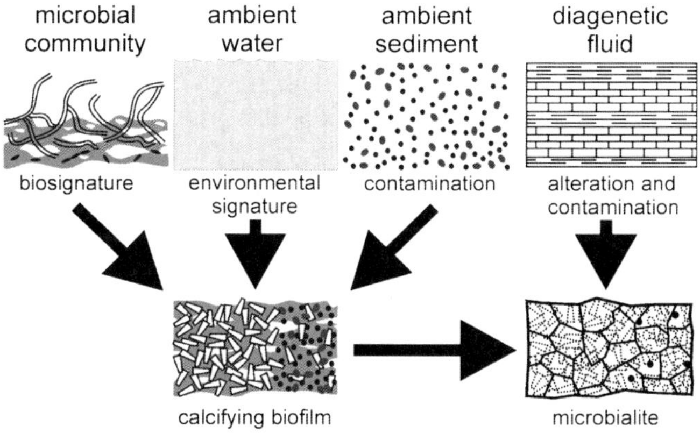

Fig. 5 Schematic diagram showing potential sources of trace elements in microbialites. Trace elements enriched by the microbial community may contain preservable biosignatures, whereas trace elements taken up from the ambient fluids may provide environmental signatures. Trace elements in ambient sediments are generally contaminants. Although such sediment contamination may be identifiable on the basis of correlation with immobile elements such as Th, and Zr, the high metal concentration common in siliciclastic muds may mask any original biosignatures or environmental signatures. Secondary alteration within diagenetic fluids may also contaminate or remove biosignatures and/or environmental signatures

biofilm itself; (2) the ambient water body; (3) trapped and bound ambient sediment, including oxyhydroxides; and (4) younger diagenetic fluids. (1) The first category of elements are those that are concentrated in organic matter (living tissue or EPS) during the life of the microbes, either as nutrients or as toxic elements that are mineralized and sequestered in EPS (e.g., Silver 1997; Ledin 2000; Decho 2000; Ferris 2000; Rosen 2002; Konhauser 2007; Pereira et al. 2009). A variety of microbes show selectivity in sequestration of metals within EPS (e.g., Micheletti et al. 2008), and these elements are of primary interest as potential biosignatures. The degree to which elements that are enriched in a biofilm might be preserved in the resulting microbialite depends on a variety of factors, including their initial concentration, the degree to which organic matter was trapped and preserved during mineral precipitation, the ability of the elements to fit directly into the mineral lattice or within lattice defects, the degree of subsequent diagenesis, and the susceptibility of the element to be remobilized during diagenesis. Trace elements may become incorporated into the preservable carbonate of a microbialite in several ways. Organic matter may be preserved at nanoscales within the carbonate precipitates themselves (Benzerara et al. 2006), just as in eukaryote skeletons (e.g., Johnston 1980; Weiner et al. 1983; Przeniosło et al. 2008). Alternately, elements that are enriched within the biofilm at the time of mineral precipitation may be preserved in the mineral at higher concentrations than would occur from the ambient fluid alone. Finally, elements may form other insoluble complexes or minerals that may become trapped within the growing carbonate (e.g., insoluble sulfides precipitating as a result of sulfate reduction). In general, if the elements are enriched in the local environment of the biofilm, they may be enriched in the resulting microbialite. (2) The second group of elements reflects the chemistry of the ambient water body. Although important bioactive elements may be purposefully enriched within the biofilm, ligands in EPS also passively concentrate metallic cations from the ambient environment (Geesy and Jang 1989; Stone 1997; Webb and Kamber 2000). These elements also may be incorporated directly into the growing mineral precipitates but in ratios that reflect their occurrence in the ambient fluid. They may include metals that are not known to have biological significance or function, but they still may be useful for interpreting the depositional setting of a given microbialite (e.g., the rare earth elements – REE; Webb and Kamber, 2000). (3) The third group of elements includes constituents of minerals in the ambient sediment load or elements adsorbed to particulate matter including clays and organic flocculants in the overlying water column. These elements are contaminants in terms of recognizing and interpreting biosignatures. Unfortunately, the bioactive inner transition elements commonly have relatively high concentrations in fine siliciclastic detritus (Veizer 1983), sulphides and oxihydroxides of Fe and Mn. Hence, contamination of microbialites may mask both biosignatures and signatures of the ambient water body, thus restricting such studies to relatively pure carbonates (Webb and Kamber 2000; Nothdurft et al. 2004). The common presence of immobile elements, such as Zr and Th, in fine siliciclastic detritus typically allows for its recognition in a given trace element analysis (Parekh et al. 1977; Kamber et al. 2004). Oxi-hydroxides of Fe and Mn may be more difficult to recognize but can be monitored via anomalous concentrations of Fe and Mn. Monitoring is especially beneficial where trace element mapping (Hellstrom et al. 2008) can be achieved (more on this below). (4) The final category of elements includes those that may be introduced subsequently with diagenetic fluids. Deep basinal brines may carry transition elements in very high concentrations (e.g., mineralizing fluids) and potentially could completely overprint any original geochemical signatures. At the same time, diagenetic fluids could mobilize and remove elements from the original precipitates, thus altering preserved elemental distributions.

2.1 The Use of Trace Elements to Help Reconstruct Environments of Deposition

Rare earth element (REE) geochemistry has proven very useful in distinguishing the origin of many hydrogenous precipitates, including carbonate, chert, phosphate and banded iron formations (BIF) and some authors have suggested that REEs may provide a bacterial biosignature in BIFs (Takahashi et al. 2005, 2007). The latter possibility is discussed in the section on biosignatures below. Marine REE concentrations are controlled by the balance of different input sources (e.g., terrestrial input from continental weathering, hydrothermal input) and elemental scavenging processes, initially in estuaries and later in the ocean, related to depth, salinity, and oxygen levels (Elderfield 1988; Piepgras and Jacobsen 1992; Bertram and Elderfield 1993; Holser

1997; Kamber and Webb 2001). Combined with Y, the shale-normalized (subscript n) REE have a very distinctive pattern in shallow, oxygenated seawater that is characterized by: (1) light REE (LREE) depletion relative to heavy REE (HREE); (2) superchondritic Y/Ho ratios that range between 40 and 90 (Bau 1996); (3) positive La_n, Gd_n and Lu_n anomalies, of which La_n is the most significant; and (4) a negative Ce_n anomaly in the modern ocean that is related to the oxygenation of Ce and removal on particle surfaces in oxygenated water (Elderfield 1988; Piepgras and Jacobsen 1992; Zhang and Nozaki 1996, 1998). Positive Eu_n anomalies also occur in modern seawater, but they are variable and depend on the shale normalization utilised. They were much larger and more consistent in Archean seawater owing to greater hydrothermal input (Graff 1978; Derry and Jacobsen 1990; Kamber and Webb 2001) and lack of deep ocean ventilation. Archean seawater is also characterized by the lack of the negative Ce_n anomaly, which is consistent with an anoxic ocean in the Archean (Fryer 1977; German and Elderfield 1990; Van Kranendonk et al. 2003).

Fresh water REE_n patterns are more diverse, generally having relatively smooth shale-normalized patterns that: are less LREE depleted than are marine waters – enriched in HREE, middle REE (MREE) or more rarely LREE; and have chondritic Y/Ho ratios (Goldstein and Jacobsen 1988; Elderfield et al. 1990; Sholkovitz 1992, 1995; Lawrence et al. 2006; Lawrence and Kamber 2006). However, the pH and amount and type of organic matter within the water may alter the balance of LREEs to HREEs (Sholkovitz 1992; Tang and Johannesson 2003). Additionally, where fine particulate sediment and colloids occur in fresh water it should impart shale-like patterns on REE distributions where it is sampled. Hydrothermal waters are also diverse and are commonly divided into high- and low-temperature categories. Both types have REE&Y distributions that differ from those of marine and fresh waters, commonly having LREE enrichment, well developed positive Eu_n anomalies, and chondritic Y/Ho ratios (Bau and Dulski 1999; Barrat et al. 2000; Wheat et al. 2002). Hence, REE&Y have been used to investigate the source, depth and oxygenation levels of ancient marine precipitates including BIF (Derry and Jacobsen 1990; Alibert and McCulloch 1993; Bau and Möller 1993; Bau and Dulski 1996; Kato et al. 1998; Bolhar et al. 2004; Alexander et al. 2008; Frei et al. 2008), chert (Kato et al 2002; Sugitani et al. 2002; Hofmann and Bolhar 2007) and carbonates (Kamber and Webb 2001; Yamamoto et al. 2004; Komiya et al. 2008). Microbialites in particular have been shown to serve as robust REE&Y proxies for the water in which they grew from the Holocene to the Archean (Webb and Kamber 2000; Kamber and Webb 2001; Van Kranendonk et al. 2003; Kamber et al. 2004; Nothdurft et al. 2004; Yamamoto et al. 2004; Cabioch et al. 2006; Olivier and Boyet 2006; Bolhar and Van Kranendonk 2007). The ability to reconstruct the source waters of mineral precipitates allows testing of independently derived hypotheses (e.g., on the basis of sedimentological context) regarding the settings of putative microbialites.

The broad utility of trace element studies in early life research can be illustrated using the controversial early Archean rocks of the island of Akilia, southwest Greenland. Mojzsis et al. (1996) used the depletion of ^{13}C in graphite contained within phosphate from metamorphic rocks that they interpreted to be metamorphosed BIFs from Akilia to suggest that the carbon was organically fixed, thus representing the oldest traces of life on Earth (>3.85 Ga). However, the rocks that contain the apatite have been intensely metamorphosed and their protoliths are equivocal. Although Mojzsis et al. (1996) did not suggest that the graphite was initially part of a microbialite, similar constraints exist on the interpretation of carbon isotope ratios as a biosignature for microbialites in general; the ^{13}C depletion can be interpreted as a biosignature only if the rocks that contained the original carbon were initially sediments and the carbon was deposited with them (e.g., the first set of biogenicity criteria mentioned above). Mojzsis and Harrison (2002) used REE&Y geochemistry to test the hypothesis that the graphite-bearing Akilia rocks were originally hydrogenous BIF and concluded that they were. However, Bolhar et al. (2005) used low detection limit solution inductively coupled plasma-mass spectrometry (ICP-MS) to acquire more complete REE data sets so as to specifically compare the Akilia rocks to better preserved metasediments from the nearby early Archean Isua Greenstone Belt (IGB). They concluded on the basis of the $REE\&Y_n$ patterns that whereas the early Archaean IGB BIFs contained unequivocal evidence of their seawater origin, the Isua rocks that contained the depleted ^{13}C signature in graphite lack critical seawater-like characteristics, such as a positive La anomaly. Hence, Bolhar et al. (2005) concurred with Fedo and Whitehouse (2002) that the protolith for the Akilia rocks of interest was an ultramafic igneous rock, not a sedimentary rock. Hence, trace element geochemistry suggests that the depleted ^{13}C in the Akilia rocks cannot be used as a biosignature for those rocks.

REE&Y geochemistry also was used to constrain the depositional environment of some of the oldest known stromatolites from the early Archean of the North Pole Dome area, Pilbara Craton, Western Australia. The oldest known putative stromatolites were first described there by Lowe (1980, 1983) and Walter et al. (1980). However, the biogenicity of those structures quickly became a matter of debate (Buick et al. 1981; Lowe 1994). Subsequently, much better preserved and extensive stromatolites were described from slightly higher in the succession in the 3.43 Ga Strelley Pool Formation (Hofmann et al. 1999; Allwood et al. 2006). Hofmann et al. (1999) provided a convincing argument for biogenicity on the basis of morphology and sedimentary setting, but Lindsay et al. (2003, 2005) disputed their interpretation and suggested that the structures were hydrothermal precipitates that were unlikely to be biological in origin. Van Kranendonk et al. (2003) tested that hypothesis and demonstrated that the structures have seawater-like REE&Y patterns with LREE depletion, a superchondritic Y/Ho ratio and a positive La_n anomaly, as do associated particulate carbonate sediments that show identifiable sedimentary structures. The samples lack a negative Ce_n anomaly, as would be expected in anoxic Archean seawater. Comparison with unequivocal hydrothermal precipitates from the nearby associated Panarama volcano showed very different REE&Y patterns consistent with their hydrothermal origin. Hence, the hypothesis that the structures were precipitated from hydrothermal fluids (Lindsay et al. 2003) was refuted on the basis of REE&Y geochemistry. Seawater-like REE&Y patterns have now been documented in marine carbonates throughout the Precambrian and Phanerozoic (Kamber and Webb 2001; Bolhar et al. 2004; Shields and Webb 2004; Alexander et al. 2008). Subsequently, Allwood et al. (2006, 2007) documented additional stromatolites in the Strelley Pool Formation and showed that their morphology changed in relation to the depositional setting in a way entirely consistent with their having been produced by benthic biofilms.

Several studies have been able to discriminate different water masses and/or differences in local sediment inputs on the basis of REE&Y preserved in microbialites. Kamber and Webb (2001) found that stromatolites from deeper settings over the edge of the late Archean (2.52 Ga) Campbellrand carbonate platform of South Africa had $REE\&Y_n$ patterns that reflected open marine water with greater LREE depletion and better developed Eu_n anomalies relative to stromatolites that formed on the shallow platform top where mixing of lagoonal and continent-derived waters was more prevalent. Nothdurft et al. (2004) were able to clearly discriminate the REE&Y signatures of open marine waters preserved in Devonian calcimicrobes from isolated platform reefs on the Lennard Shelf, Canning Basin, Western Australia from near-shore runoff-rich waters that bathed calcimicrobes in continental fringing reefs. A similar pattern was discerned in Holocene microbialites from reefs in Vanuatu (Cabioch et al. 2006). In that case, microbialites from Urélapa Reef, which is isolated offshore of Espiritu Santo, and microbialites in the Tasmaloum Reef, which fringes the volcanic island itself, show a trend of increasing REE concentration, increased positive Eu_n anomaly and decreased Y/Ho ratio from offshore to onshore. Such a trend is consistent with increasing contamination from runoff from weathering volcanic rocks on Espiritu Santo. In a somewhat special case, Frimmel (2009) showed that the REE&Y geochemistry of Neoproterozoic cap carbonates, including stromatolites, in southwestern and central Africa varied systematically with proximity to the interpreted paleocoastline. Cap carbonates were widely deposited immediately above glacial deposits during the Neoproterozoic, and may represent an initial pulse of high alkalinity resulting from the chemical weathering of finely ground rock in high-CO_2 atmosphere during deglaciation (Hoffman et al. 1998), although alternative interpretations have been advanced (e.g., Kennedy et al. 2001; Young 2002; Le Hir et al. 2009). Relatively flat $REE\&Y_n$ patterns that lack well developed La_n anomalies and superchondritic Y/Ho ratios in the near-shore settings were found not to reflect direct contamination by incorporated siliciclastic detritus, but were interpreted to represent the incorporation of near-shore colloids that had high-concentration shale-like REE&Y distributions. Farther offshore the patterns became more seawater-like with LREE depletion, positive La_n anomalies and superchondritic Y/Ho ratios (e.g., see Frimmel 2009, his Fig. 9). Although an isolated, hydrothermally affected lake could not be excluded for carbonates in the Roch Pina Formation, Frimmel (2009) interpreted most of his analyzed carbonates as near-shore marine fringing stromatolite reefs.

The interpretation of lake settings based on REE&Y geochemistry is complicated. Modern river waters commonly have variable, but relatively smooth, unfractionated $REE\&Y_n$ patterns (Goldstein and Jacobsen 1988; Elderfield et al. 1990; Lawrence et al. 2006). However, as noted by Lawrence et al. (2006), there are very few complete, high quality fluvial REE&Y datasets. Lakes and inland seas have even more variable geochemistry owing to differences in size, input sources, stratification and the degree to which aqueous scavenging processes affect the initial source inventories; they too suffer from a paucity of high quality data sets. For example, Lake Tanganyika has seawater-like LREE depletion whereas Lake Naivasha does not (Bolhar and Van Kranendonk 2007, their Fig. 5; data from Barrat et al. 2000; Ojiambo et al. 2003). A stromatolite from the Eocene Green River Formation of the United States was found to have only slight LREE depletion and lacks both a positive La_n anomaly and a superchrondritic Y/Ho ratio (Bolhar and Van Kranendonk 2007). On the basis of REE&Y patterns, Bolhar and Van Kranendonk (2007) interpreted stromatolites in the Fortescue Group of Western Australia (2.78–2.63 Ga) to have been deposited in an epicontinental lacustrine setting. Samples from the Jeerinah Formation have prominent MREE enrichment and lack well developed La_n anomalies and superchrondritic Y/Ho ratios, thus being very unlike contemporary seawater, but somewhat similar to MREE-enriched rivers reported by Elderfield et al. (1990). However, the average Tumbiana Formation $REE\&Y_n$ pattern (Bolhar and Van Kranendonk 2007, their Fig. 8) does have a positive La_n anomaly and LREE depletion, but still lacks a superchondritic Y/Ho ratio. Hence, the Tumbiana patterns are similar to those interpreted as representing both rivers (Elderfield et al. 1990) and near-shore marine carbonates that were contaminated by terrestrially derived colloids (Frimmel 2009). Hence, additional work may be required to interpret and differentiate some lacustrine and estuarine settings. Regardless, definitive isolated epicontinental water bodies have been recognized on the basis of REE&Y geochemistry in stromatolites. Archean (~2.84 Ga) stromatolites from the Masvingo Greenstone Belt in the Mushandike National Park, Zimbabwe have REE&Y patterns with characteristic marine features such as positive La_n and Gd_n anomalies and superchondritic Y/Ho ratios, but they are enriched in LREE and lack the prominent Eu_n anomaly typical of open Archean seawater (Kamber et al. 2004). However, if the REEs are normalized to the local tonalite gneiss weathering source, a LREE depleted $REE\&Y_n$ pattern more typical of aqueous scavenging is produced. Hence, Kamber et al. (2004) interpreted the stromatolites to have grown in a small basin isolated on older Archaean continental crust with little or no confluence with contemporary open marine waters. That conclusion was supported with Pb-isotope data from the stromatolites, which resembled closely the very distinctive character of the Zimbabwe Craton as a whole.

2.2 *Effect of Microenvironments on Trace Element Inventories*

The use of trace element patterns to interpret depositional environments requires that the preserved elemental inventories are representative of those of the ambient environment. However, microbialites to varying degrees represent microenvironments that may or may not reflect the same chemistry as the ambient water bodies in which they formed. Even the EPS sheaths of individual microbes set up diffusion gradients that isolate the living cells from the ambient chemical environment, and the diffusion pathways become more extended and complicated with depth in more complex biofilms and microbial mats. Decho (2000) suggested that differences in the types and distribution of EPS within biofilms set up 'microdomains' with different physical and chemical attributes. Such microdomains regulate pH as well as the concentration of O_2, compounds of N and S, and metals (nutrients and toxins) in a variety of sedimentary environments (e.g., Revsbech et al. 1983; Wimpenny and Kinniment 1995; Stockdale et al. 2009). In more complex biofilms and microbial mats, depth within the structures isolates a variety of microenvironments, most notably evident from differences in oxygenation levels (see review in Konhauser 2007, Chapter 6). Communities of facultative and obligate anaerobes may occur within millimeters of biofilm surfaces in fully oxygenated, exposed benthic settings (Revsbech 1989; Wimpenny and Kinniment 1995; Baumgartner et al. 2006). As microbial activity is increasingly confined in pore spaces and cavities (crypts) within larger structures, such as large microbialites or reefs, the fluids bathing the confined biofilms may be considerably evolved from ambient waters that reflect the broader depositional environment. Regardless, the growth of a microbialite (biologically induced or influenced precipitation) requires a continual influx of ambient water into the microenvironment to provide adequate ions for mineral growth. However, the degree to which the setting represents a microenvironment with effective diffusion barriers may determine the degree to which the preserved geochemical proxies reflect the chemistry of waters in the 'external' ambient environment.

Microbially mediated microenvironments have been well documented in modern marine settings. For example, Nothdurft et al. (2005) documented the formation of brucite [$Mg(OH)_2$] in microenvironments in normal shallow seawater. Brucite generally forms in solutions with pH above nine (Jordan and Rammensee 1996) and is significantly undersaturated in shallow seawater. However, it precipitates in microenvironments confined within the abandoned spaces of living coral skeletons in modern reefs (Nothdurft et al. 2005; Buster and Holmes 2006) owing to chemical evolution of the seawater within the pore system driven by the activities of microbial endobionts. Although pore waters were exchanged frequently owing to tidal and wave-pumping, phototrophs within the coral skeleton (cyanobacteria and filamentous green algae) removed CO_2, thus increasing the pH of the semi-confined fluid within the cavities and supporting precipitation of aragonite and lowering of the Ca^{2+} activity. Then pH may be driven higher by microbial degradation of organic matter, including ammonification, or by utilization of HCO_3^- for cyanobacterial photosynthesis using carbonic anhydrase (Badger and Price 2003), thus liberating excess OH^- (Nothdurft and Webb 2009). Regardless of the exact mechanism, the brucite could not have grown in ambient seawater without extensive evolution of the local water chemistry by microbes within the semi-confined cavities of the coral skeleton.

Another significant example of microbially-mediated microenvironments leading to precipitation of an unexpected mineral in marine carbonates is the occurrence of low-Mg calcite (LMC) within enclosed microbial borings in living coral skeletons (Nothdurft et al. 2007). LMC is not favoured to precipitate in marine carbonates owing to the high Mg:Ca ratio of seawater (e.g., Bischoff and Fyfe 1968; Folk 1974; Berner 1975). LMC generally precipitates from fresh (meteoric) waters, which mostly have low Mg:Ca ratios (Drever 1997), and thus is generally considered prima facie evidence of meteoric diagenesis. Abiotic marine carbonates are generally either aragonite or high-Mg calcite (HMC). In the case of coral skeletons from the reef flat of Heron Reef, Great Barrier Reef, migrating coccoid cyanobacteria excavated borings within skeletal aragonite at one end, while concomitantly back-filling the borings with LMC. Calcification behind the migrating microbe generated protons that allowed photosynthesis utilizing HCO_3^-, which was generated by dissolution on the other side of the microbe. However, because the Mg:Ca ratios within the microenvironment of the boring reflect those of the dissolving aragonite, rather than ambient seawater, they are very low and calcite is the favoured precipitate within the boring (Nothdurft et al. 2007). A case can be made that the precipitation of calcite in the borings is biologically controlled precipitation as it probably requires direct enzymatic activity by the microbes. If the microbes actively transport Ca^{2+} through the microbe to the site of precipitation utilising Ca-proton exchanging ATPase (e.g., Driessen and Konings 1986) in order to provide protons for converting HCO_3^- to CO_2 and H_2O using carbonic anhydrase (Badger 2003; Badger and Price 2003), the precipitation is being driven by direct enzymatic activity. Regardless, the occurrence of calcite rather than aragonite directly reflects the non-seawater-like Mg:Ca ratio of the fluid within the highly confined microenvironment despite the ambient environment being seawater.

Another complication associated with microenvironments concerns the proximity of precipitation to the 'driver' of that precipitation (i.e., the alkalinity engine). In relatively enclosed environments where the local saturation state for a given mineral has been driven up by microbial activity, abiotic mineral substrates may support mineral growth that is not necessarily proximal to the microbial community that increased the alkalinity. Precipitates produced in this way may be morphologically identical to abiotic cement. Such processes could occur in the interstices of agglutinated stromatolites or fenestrae in micritic stromatolites, but have been demonstrated within abandoned interseptal spaces in modern scleractinian coral skeletons (Nothdurft and Webb 2009). There aragonite cement has a very patchy distribution, commonly nucleating on organic filaments (Fig. 6a), but also growing syntaxially on denuded coral skeleton aragonite (Fig. 6b). The irregular distribution and volume of the cements suggest that they were not the products of passive marine cementation, but that local endolithic-endobiont microbial communities and possibly also boring sponges increased the carbonate saturation index in some interseptal spaces relative to others and the aragonite either nucleated on microbial organic matter or grew syntaxially on existing aragonite surfaces. Without the biological activity (e.g., alkalinity engine), the cement may not have precipitated at all. Hence, in relatively enclosed environments it may be impossible to distinguish between biologically-influenced and inorganic mineralization as defined by Dupraz et al. (2008).

Hence, microbialites represent mineral precipitation in a wide spectrum of microenvironments, encompassing: envelopment by EPS sheaths in open benthic settings; stratified microbial mats; containment within semi-confined cavity systems; and highly confined endolithic settings (Fig. 7). The degree to which the geochemistry of the microbialites reflects the ambient water chemistry may vary accordingly. In the case of REEs, empirical evidence suggests that fluids in the microenvironments in shallow Holocene reef cavities did not evolve away from

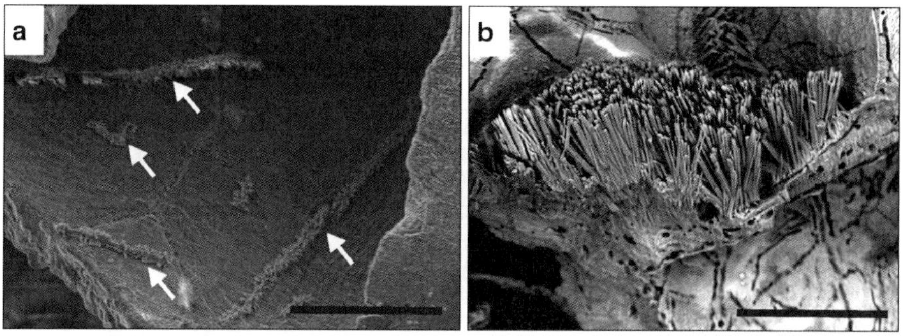

Fig. 6 Scanning electron photomicrographs of aragonite cements within microenvironments contained in porous coral skeletons; scale bars = 100 μm. (**a**) Acicular aragonite cement nucleated on organic filaments (*white arrows*) within abandoned interseptal space. Note that aragonite cements are confined to elongate filaments. (**b**) Abundant aragonite cement nucleated syntaxially on skeletal aragonite of coral within abandoned interseptal space

ambient seawater in regards to elemental ratios (Webb and Kamber 2000). The same appears to be the case for Devonian *Renalcis*, some of which grew in reef cavity systems (Nothdurft et al. 2004). Regardless, compared to eukaryote skeletons (Sholkovitz and Shen 1995; Webb and Kamber 2000; Akagi et al. 2004; Wyndham et al. 2004) and aragonite precipitated directly from water (e.g., Barrat et al. 2000), microbialites are greatly enriched in REEs, thus making them an especially robust seawater proxy. The lack of agreement between theoretical, experimental and empirical REE partition coefficients between fluids and carbonates (Terakado and Masuda 1988; Zhong and Mucci 1995; Barrat et al. 2000; Webb and Kamber 2000; Tanaka and Kawabe 2006) suggests that it currently may not be possible to back-calculate the concentrations of the elements in ancient seawater on the basis of elemental concentrations of REEs in ancient microbialites (e.g., Yamamoto et al. 2004; Komiya et al. 2008), but the elemental ratios are preserved.

2.3 Bioactive Metal Accumulation

Whereas the use of trace element proxies to interpret depositional environments of ancient microbialites requires that elements were incorporated quantitatively from the fluid into the microbialite, the search for biosignatures using bioactive transition elements requires that they be incorporated out of equilibrium with the ambient fluid environment (i.e., enrichment). Despite the importance of trace metals for biological processes, few studies have attempted to analyse potential bioactive trace elements in microbialites, and this partly reflects the analytical difficulty in measuring transition elements at very low abundances in carbonate matrices (Kamber and Webb 2007). Webb and Kamber (2004) used solution ICP-MS to show that Holocene reefal microbialites from the Great Barrier Reef have elevated Cr and V concentrations relative to associated skeletal carbonates, such as scleractinian corals, mollusks and coralline red

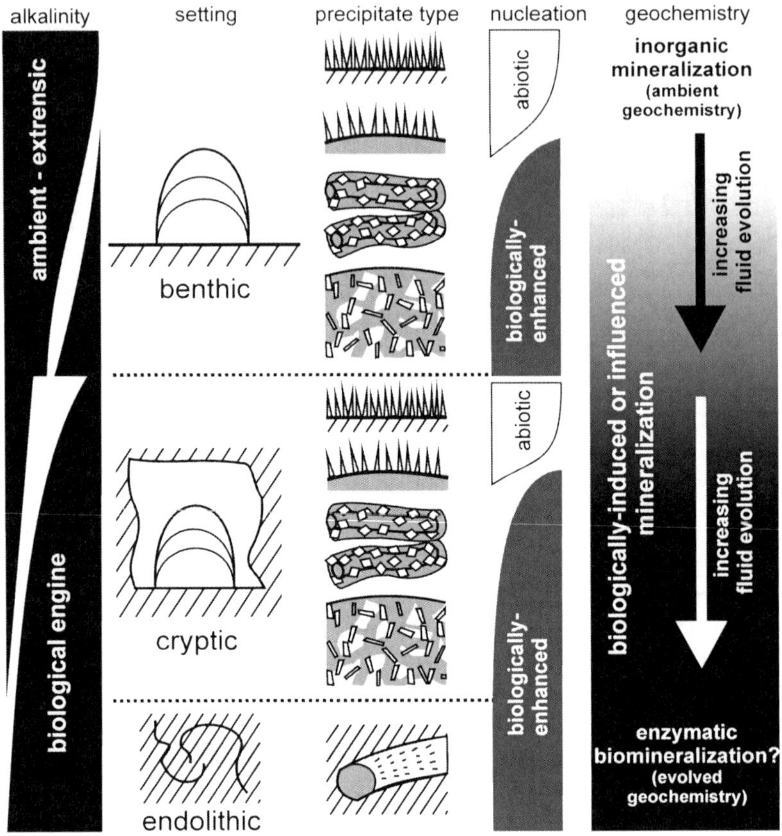

Fig. 7 Diagram showing relationship between depositional settings and microenvironments where microbialites form. The alkalinity source shifts from dominantly extrinsic, ambient fluid sources to more biologically driven sources as the degree of confinement increases down the diagram. Precipitate types are the same as those shown in Fig. 2 with the addition of crusts precipitated on existing mineral substrates (*diagonal lines*). The effectiveness of the biological alkalinity engine increases from mineralization of surfaces to replacement of organic mater and finally precipitation within an enclosed biofilm. Likewise the role of the biological alkalinity engine increases with increased confinement. Nucleation enhancement is dependent primarily on the site of precipitation within a given microbialite. Geochemically, the precipitates range from inorganic minerals to fully enzymatically controlled (biologically controlled) minerals in a continuum that reflects increasing levels of fluid evolution away from ambient fluids with more enclosed settings and more complex biofilms

algae, and in ratios that do not reflect abundances in seawater. Additionally, the high Cr and V concentrations were not sourced from detrital contaminants, and normalization of the elements to nutrient (Ba) and conservative (Sr) elements, suggests that the differences in elemental abundance are robust and that Cr and V concentrations in microbialites vary from those expected in abiotic cements. Although high Cr concentrations may result from microbial mineralization of the element to insoluble forms in order to detoxify the local environment, V enrichment could reflect a variety of microbial activities (Fraústo da Silva and Williams 2001). Elevated V could also reflect incorporation of eukaryote ascidian spicules into the microbialite because ascidians concentrate V in tunichrome blood pigments (Michibata et al. 1998). Ascidians are common inhabitants of reef cavities and their spicules occur in the sampled Heron Reef microbialites. However, Webb and Kamber (2008) subsequently used laser ablation (LA) ICP-MS to analyse the V concentration of the microbialites and, through time series data analysis, found that the V is not concentrated in spicules, but is disseminated throughout the microbial carbonate. Hence, V originally may have been concentrated within the cavities by ascidians and then released during their decay with the increased local V concentration then leading to enhanced concentration in the biofilms and eventual preservation in the microbialites. In that case, the V may serve as a biosignature indicative of ascidians even if spicules were not preserved (Webb and Kamber 2008). However, bacterial processes also may have enhanced the V concentration independently of ascidians. Regardless, enhanced V concentrations could be significant for Precambrian microbialites, which are unlikely to contain traces of ascidians, because VNases may have played an important role in nitrogen fixation prior to the earth attaining an oxidizing atmosphere. In an anoxic ocean V complexes would have been more soluble, and hence bioavailable, than are Mo complexes, which provide Mo for MoNases in the modern oxidized ocean (Fraústo da Silva and Williams 2001). Hence, preservation of elevated V in microbialites may allow recognition of VNase activity in ancient anoxic marine biofilms with implications for interpreting ocean oxygenation and stratification through time (Anbar and Knoll 2002; Huston and Logan 2004).

Although enrichments in transition elements in microbialites from Heron Reef are suggestive of microbial and possibly eukaryote biosignatures, comparison to skeletal carbonates is not optimal because differences in elemental concentrations between microbialites and biologically-controlled skeletons could represent either enhanced concentration in the microbial carbonates or active exclusion of elements from the skeletons. Comparison with coeval early marine cements is a better test of microbial enrichment. Kamber and Webb (2007) applied LA-ICP-MS to investigate transition element concentrations in ancient calcimicrobes and co-occurring cements, along with eukaryote skeletons, from the Late Devonian reef complexes of the Canning Basin, Western Australia. Microbialites from Devonian reefs in the Canning Basin contain typical seawater-like REE&Y patterns (Nothdurft et al. 2004), and calcimicrobes, such as *Renalcis*, retain stable isotope characteristics consistent with their having preserved original isotopic biosignatures (Stephens and Sumner 2002). Hence, the microbialites appear to be very well preserved geochemically. The calcimicrobe *Renalcis* was found to be significantly enriched in V, Sn, Cu and Zn relative to early and late cements, and stromatoporoid sponge skeletons are enriched in V, but not Sn, Cu, or Zn (Kamber and Webb 2007). *Renalcis* also has less significant but recognizable enrichments in Mn, Cd, Co and Cr. No enrichment was found in Mo or Sb. Subsequent trace element mapping (discussed below) showed that elevated Cu and V may be associated with diagenetic Fe-oxyhydroxides, but, regardless, the calcimicrobe *Renalcis* can be differentiated from contemporary carbonates that grew from the same seawater (i.e., cements and stromatoporoid sponge skeletons) and also from later diagenetic cements. Elevated concentrations of vital trace elements in well preserved Devonian *Renalcis* suggest that metabolically significant elements may be preserved in microbial carbonates over geological time frames. However, as noted for the REE, targeted microbialites must be free from significant terrigenous particulates and oxyhydroxides, as they have relatively high concentrations of transition elements (Veizer 1983) and could easily mask any biological signature.

Kamber and Webb (2007) also identified additional sources of contamination relevant to laser ablation techniques. Because LA-ICP-MS commonly utilizes an ablation spot size of less than 200 μm, thin sections are generally necessary in order to target specific microbial carbonates, cements and skeletons. Unfortunately, thin section preparation may introduce contaminants during the grinding and polishing phases, as metal laps, grits and even glasses contain abundant transition elements. Preparation of thin sections on fixed diamond grit surfaces alleviates much of the contamination, but impregnation of samples with organic polymers, such as thin section epoxy resin also helps keep grit from invading pores in the sample. Commercial resins tested so far have not been found to contain abundant metals. However, inks used to mark areas of interest on samples were found to contain abundant metals. Even ink as far away from the ablation target as 1–2 mm could be partially ablated, thus contaminating the helium carrier gas stream with metals. Fortunately, both particulate grit and ink contaminants can generally be identified within an analysis by inspection of the time-series data (Fig. 8), and affected sections of analyses can be discarded. Regardless of contamination issues, LA-ICP-MS techniques are capable of measuring the elements of interest even at very low concentrations (Kamber and Webb 2007; Webb et al. 2009).

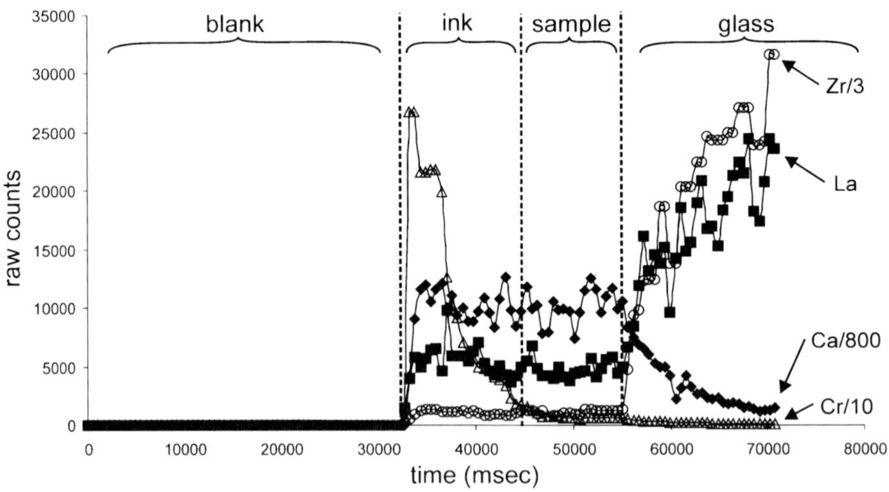

Fig. 8 Time series raw count data for laser ablation of a microbialite sample mounted on glass. Note the low blank for all elements (overlapping). A zone of ink contamination is defined by high Cr counts and ablation into glass is evident from high Zr counts. The zone between the glass and the ink contamination provides a clean sample interval. However, note that the ink does not alter the Ca and La count rates, but that the glass could provide a potential contaminant in regards to La

Takahashi et al. (2005, 2007) suggested that REEs may provide a microbial biosignature in BIFs. They showed experimentally that HREEs preferentially adsorb to carboxylate and phosphate groups on bacterial cell surfaces with milder enrichment in the MREE range in low pH waters. Takahashi et al. (2005) suggested that similar HREE-enriched patterns occurred in biofilms in more alkaline springs but normalized the patterns to account for higher carbonate-complexation in those waters. Ozaki et al. (2006) also found that Eu preferentially adsorbs to bacterial cells, but that competing organic ligands, which are abundant in EPS in natural biofilms (Pereira et al. 2009), significantly affected the degree of the potential enrichment. Hence, although experimental HREE enrichment on bacterial surfaces was demonstrated at low pH, it is unclear whether such a signature could be distinguished in carbonate microbialites that grew in natural seawater. Complicating the issue is the finding by Kamijo et al. (1998) that microbes from soils and river waters preferentially removed the LREE from solution and did not remove HREE. Empirical distribution coefficients between Holocene microbialites and seawater are uniform across the REE mass range (Webb and Kamber 2000) suggesting that those microbialites do not preserve a particular microbial biosignature in the REEs, either HREE-enriched or LREE-enriched. Instead, carbonate microbialites appear to provide robust proxies for seawater REE distributions since the early Archean and in all cases the signatures are HREE enriched. Hence, as both seawater and the potential REE biosignature of Takahashi et al. (2007) are characterized by HREE enrichment, discrimination of the two signatures may require direct comparison of coeval iron and carbonate phases.

3 Diagenetic Disruption of Trace Metal Signatures

Carbonate minerals are metastable at the Earth's surface and many minor and trace elements have been shown to be mobilized and redistributed during carbonate diagenesis (e.g., Kinsman 1969; Brand and Veizer 1980). In some cases changing minor element concentrations reflect differences in lattice spacing in aragonite and calcite (e.g., Mg is more easily accommodated in calcite, whereas Sr is more readily accommodated in aragonite), and thus minor element concentrations can be used to model diagenetic processes in carbonate transformations (Martin et al. 1986; Maliva 1998). Such processes are clearly important in the neomorphism of aragonite to calcite in meteoric environments, and dolomitization is also likely to affect trace element distributions. Veizer (1983) noted that most transition elements have different theoretical 'equilibrium' concentrations in aragonite, calcite and dolomite that are precipitated from seawater, and thus, diagenetic changes from one mineral phase to another may cause changes in the concentrations and ratios of the elements of interest as biosignatures. However, the degree of the shift in concentration during diagenetic transformation depends on the initial concentrations in the mineral phase, elemental concentrations in the diagenetic fluid and the degree of water/rock interaction (i.e., how open versus closed the system is; e.g., Martin et al. 1986). Hence, although the direction of the shift in concentration for a given element in a given diagenetic transformation is likely to be the same, the magnitude of the shift will depend on more than one factor. Regardless, vital transition elements such as Fe, Cu and Zn would be expected to increase in concentration during transformation of both marine HMC and aragonite to meteoric LMC (Veizer 1983), but very little work has been done to document the effects of carbonate diagenesis on the concentrations of specific trace metals. Despite that, the comparison of coeval microbialites, cements and ambient particulate sediments may allow biosignatures to be identified despite some degree of subsequent diagenetic alteration.

The effects of carbonate diagenesis on REE concentrations in limestones and dolomites have been studied to varying degrees and several studies have evaluated the relative partitioning of REEs between fluids and carbonate minerals. However, there has been poor agreement between the results of different workers (Terakado and Masuda 1988; Zhong and Mucci 1995; Lakshtanov and Stipp 2004; Tanaka and Kawabe 2006) and empirical studies are generally in conflict with experimental results (e.g., Scherer and Seitz 1980; Webb and Kamber 2000). Observations from natural diagenetic series of limestones and dolomites have also presented somewhat variable results. Schieber (1988) documented the remobilization of REEs from Proterozoic shale and emplacement in proximal interbedded limestones, but Cretaceous limestones that are rhythmically inter-bedded with marl retained their seawater-like REE&Y signatures despite their very close proximity to the marl (Bellanca et al. 1997). The different behaviors probably reflect differences in the nature of the diagenetic fluids and degree of water-rock interaction. Although dolomite may form as a primary carbonate mineral and some dolomites are formed by microbes (Vasconcelos et al. 1995; Wright 1999; Warthmann et al. 2000; Van Lith et al. 2003; Wacey et al. 2007; McKenzie and Vasconcelos 2009), dolomitization is a common diagenetic process that has resulted in a varying abundance of dolomite through the Phanerozoic with increasing dolomite through the Precambrian (e.g., Given and Wilkinson 1987; Zenger 1989; Burns et al. 2000). Dolomitization has been shown not to have altered the original marine REE&Y patterns of limestones of Proterozoic and Carboniferous age (Banner et al. 1988; Frimmel 2009) although dolomitization in basinal brines associated with Mississippi Valley Type mineralization in some cases altered the REE&Y patterns of Devonian microbial carbonates (Nothdurft et al. 2004). Hence, REE distributions may remain intact through dolomitization, but it depends on elemental concentrations in the diagenetic fluid. Regardless, many Precambrian microbialites may have been dolomite originally and therefore less prone to subsequent diagenetic alteration. In any case, once REEs are set in a carbonate lattice they tend to be very stable even into metamorphic temperatures (Cherniak 1998).

The meteoric transformation of marine aragonite and HMC to LMC represents another very common diagenetic process that potentially could alter REE concentrations. Komiya et al. (2008) used LA-ICP-MS to show that early diagenetic rims on Early Cambrian ooids retained the same REE&Y patterns as the ooids, but blocky meteoric spar that replaced dissolved ooids had somewhat different patterns. Webb et al. (2009) used LA-ICP-MS to demonstrate that REEs were largely retained during the meteoric transition of coral aragonite to LMC in the Pleistocene

Key Largo Limestone despite the diagenetic environment representing an open system with a high degree of water/rock interaction (Maliva 1998). The diagenetic fluids represented clean meteoric waters that percolated down through overlying marine carbonates and never came into contact with continental siliciclastic sediments. Hence, the diagenetic fluids carried REEs in ratios similar to seawater. However, the recrystallizing carbonates acted as sinks for REEs in the diagenetic fluid. Hence, diagenesis in fluids with other types of REE distributions could alter the REE patterns of limestones during their initial stabilization phase. The nature of the diagenetic fluids and degree of water-rock interaction is critical.

It is likely that REE signatures can survive many types of diagenetic alteration, including instances of both meteoric stabilization and dolomitization, especially in extensive or isolated carbonate platforms where diagenetic fluids represent primarily meteoric water or seawater, but contamination could potentially occur. Hence, diagenetic vetting is required for any studies of potential trace element biosignatures in microbialites. However, it remains to be demonstrated how stable many of the other elements are during diagenesis. Empirical studies can document the trace element relationships between differing contemporary carbonates where they occur in very close proximity, and thus provide a means of interpreting potential biosignatures, but coordination of sedimentology, petrography and geochemistry is critical.

4 Outlook

There are at least three areas of research where progress is required to take the emerging field of trace element geochemistry in microbial carbonates to a higher level of interpretation.

The first area concerns the integration of sedimentological information, particularly basin reconstruction, with planning of geochemical analysis and interpretation of geochemical data. Many Archean carbonate occurrences are small and isolated and there is limited scope to put the carbonate rocks into a depositional context. Nonetheless, an important question of greenstone belt-hosted carbonate is the level of connectivity of the greenstone basin with the open ocean. For the purpose of addressing that question it is necessary to combine information from different supracrustal lithologies. A case in point is the late Archean southern Zimbabwe craton. Several of its greenstone belts contain stromatolites (e.g. Abell et al. 1985a, b) but opinions vary regarding the origin (and by implication the setting) of the greenstones themselves. Some structural geologists (e.g., Kusky and Kidd 1992) argued that the greenstone belts at least in southern Zimbabwe represent open marine oceanic plateaus, while others (e.g. Blenkinsop et al. 1993) made a case for deposition on pre-existing continental crust. When considering all evidence, including the fact that many mafic metavolcanic rocks contain inherited zircon (Wilson et al. 1995) and a high level of contamination with radiogenic Pb (Chauvel et al. 1993), it is unavoidable to conclude that the greenstone belts formed as epicontinental basins. Any geochemical information extracted from their hydrogenous sediments must take this depositional setting into account before extrapolating findings to the global ocean. Kamber et al. (2004) showed how Pb- and Sr-isotopes of stromatolitic limestone can be combined with REE data to assess the level of connectivity of greenstone basins with the open ocean. Using the previously published data of Bolhar et al. (2002) it was possible to show that throughout the lifetime of the Belingwe greenstone belt, the degree of restriction of the basin varied greatly. Other isolated Archean stromatolite occurrences, such as the Steep Rock Group (W Ontario, Canada), must be put into a similar context before their chemical signatures can be interpreted.

Many Proterozoic basins contain much thicker sedimentary successions than Archean greenstone belts. These lend themselves to basin reconstruction into which the occurrences of microbial carbonates can be tied. This wealth of sedimentological information awaits full geochemical exploitation. For example, the early Neoproterozoic Little Dal group of the Mackenzie Mountains, Northwestern Canada, contains calcimicrobial reefs, whose development responded to basin-wide environmental change (e.g. Turner et al. 1997). Such very well-preserved carbonate build-ups will yield much information regarding the evolution of the Earth surface environment prior to the late Neoproterozoic appearance of the Ediacaran fauna. Provided such successions can be dated with sufficient accuracy, the geochemical approach that has been demonstrated for Phanerozoic and Archean carbonates will lead to new insight regarding the time immediately prior to the appearance of abundant multicellular life.

The second area for which progress is required is the careful empirical documentation of element associations in stromatolites of undeniable microbial origin. This will require a series of tests on microbial carbonates of various ages. In the best-preserved Phanerozoic microbial carbonates it should be possible to test whether microtextural distinctions described earlier in this chapter manifest as recognizable elemental distributions. In other examples of ancient stromatolites, it will be necessary to test the effect of taphonomic control on microstructure. Yet another avenue of research is the comparison of macroscopically similar stromatolites that have formed at different times in different geographic locations. This will give an indication of which elemental abundances are controlled by microbial activity and can therefore potentially serve as biosignatures in stromatolites of debated origin, and which elemental signatures may be imparted by geographic settings.

The study of element distribution, particularly transition metals in microbial carbonate is an analytical challenge. Conventionally, a small amount (a few mg) of sample would be obtained either by scratching or drilling into suitable material in a rock slab. The digested carbonate would then be analyzed by solution ICP-MS (e.g., Webb and Kamber 2004). When using low resolution quadrupole mass spectrometers, care must be taken to avoid spurious artifacts of molecular interferences caused by the high concentrations of Ca and C in the matrix. More significantly, however, many studied microbial carbonates are so finely textured and integrated with cements that it is physically very difficult to obtain sufficient material for a meaningful analysis.

Much progress has been made in determining trace element concentrations directly in petrographic thin sections using LA-ICP-MS. This technique has inferior detection limits to solution ICP-MS but it is much less prone to isobaric molecular interference and it also permits very specific analysis of areas as small as ca. 50 × 50 × 25 μm. Kamber and Webb (2007) presented the first study in which this technique was trialed at distinguishing between microbial carbonate, stromatoporoid sponge skeletons and a variety of cements. They used repeat spot analyses to test whether one type of carbonate was systematically enriched or depleted in the elements of interest. Although these tests generally showed coherent behavior, the concentration variability within a particular type of carbonate was found to be large for the transition metals. More recently, it has become computationally feasible to produce two-dimensional trace element maps from a series of parallel adjoining LA-ICP-MS traverses (Hellstrom et al. 2008). This approach allows much more rigorous assessment of trace element distribution in relation to sedimentological texture. Two examples of such maps are discussed below, using the instrumental setup described in Ulrich et al. (2009) who showed an additional example of trace element maps for a late Archaean oolitic grainstone.

The first map (Fig. 9) shows an abiogenic laminated carbonate (a pseudostromatolite) from the early Neoproterozoic Little Dal group. It is strongly laminated and interpreted to have formed as a seafloor precipitate interrupted by episodic deposition of fine siliciclastic mud. Some of the mud-rich bands acted as dissolution surfaces along which stylolites formed (white stippled area in Fig. 9). The strong banding is clearly evident in the elemental maps in which warm colors represent high concentrations and blue colors low concentrations. The suspected mud-rich bands stand out on account of higher Mg, Sc and Th concentrations, which are all enriched in fine siliciclastic detritus. By contrast, Ca, Mn and Y are more concentrated in the seafloor carbonate precipitate. The stylolite is also clearly visible on account of much elevated Sc, which suggests that it may represent the trace element inventory of several bands now condensed into one horizon. This type of map could be used as a guide to spot analysis sites for uncontaminated carbonate (i.e. the areas of deep blue in the lower left of the Sc and Mg maps). In this case, the spatial chemical information is in complete agreement with geometric/sedimentologic arguments that strongly suggest an abiogenic origin.

The second map (Fig. 10) is from one of the same thick sections on which Kamber and Webb (2007) analyzed Devonian *Renalcis* and related carbonates. While the concentration contrast between late cements, stromatoporoid and *Renalcis* was beyond doubt for many elements, the differentiation between *Renalcis* and early interstitial cement (still largely microbial micrite) was less clear. A 2.1 mm by 2.2 mm trace element map (Fig. 10) shows obvious contrast for the LREE (shown are La and Ce), Mn and Zn. In all cases, the petrographically darker clots of *Renalcis* are easily identified on the trace element maps as warm colors of relatively high concentrations of those elements. The empirical observation of high REE concentration in microbial carbonate is supported as is the suggestion that Mn and Zn concentrations may be used to identify the outline of calcified microbial communities. By contrast, the distribution of Cu and V does not follow the LREE. Some of the areas of the highest Cu and V concentration are in the interstitial cement, suggesting that they may have been associated with Fe-oxihydroxides (note that V shows a very

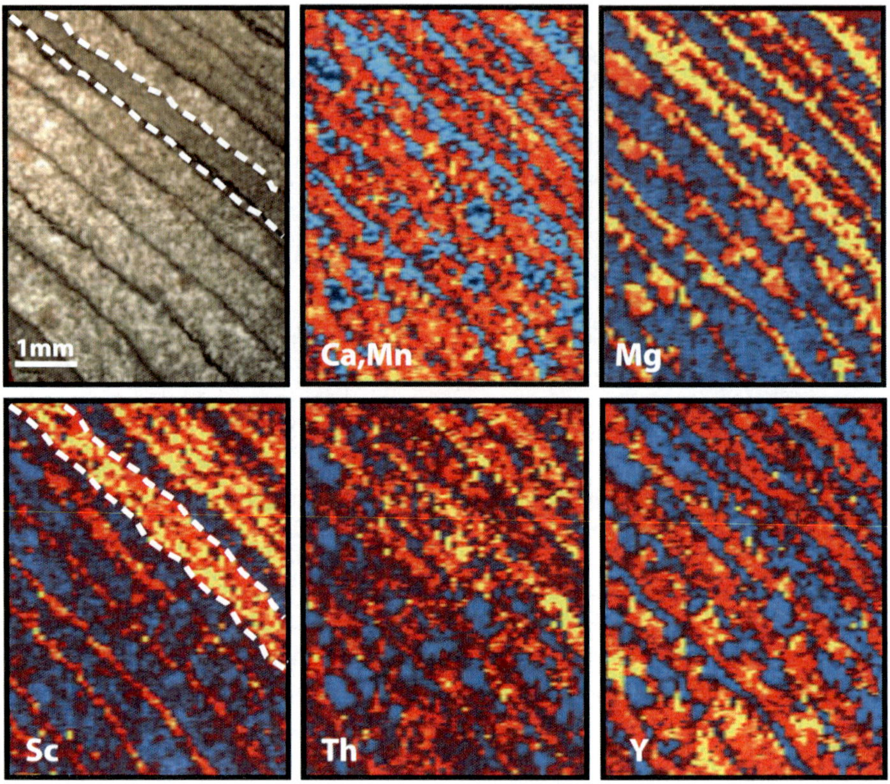

Fig. 9 Semi-quantitative major and trace element maps of an abiogenic stromatolite-like carbonate structure from the Neoproterozoic Little Dal group, Canada (sample RT-A19-501, courtesy of E. C. Turner). The major elements Ca and Mn and the trace metal Y show highest concentrations (orange and yellow colors) in spar-rich petrographically light colored calcite. By contrast, petrographically dark, mud-rich sediment layers show up clearly as high concentration areas of Mg, Sc and Th. Note outline of petrographically dark grey band containing stylolites that show very elevated Sc concentration. Two more such bands occur towards the *top right* corner of the slide. The map was produced from a series of 52 adjacent 80 μm wide LA-ICP-MS traverses (shot at 40 μm s^{-1}) using Iolite (Hellstrom et al. 2008)

strong chemical affinity with Fe). In this example, the decoupling of the LREE, Mn and Zn from Cu and V was not evident from the spot analyses but is easily revealed by trace metal maps. Systematic study of well-preserved microbial carbonates with this new technique will allow close comparison of petrographic information and metal distribution and is expected to yield better trace element criteria for biogenicity. Ulrich et al. (2009) has already used the technique to discuss the effects of dolomitization on trace metal content in oolitic rocks.

Provided that the empirical documentation of trace element distribution yields coherent results, a need will arise for a third area of research, namely, to relate particular sets of trace metal associations, such as Mn and Zn, with a biochemical pathway. The example of Mo and V concentrations in microbial carbonate and their

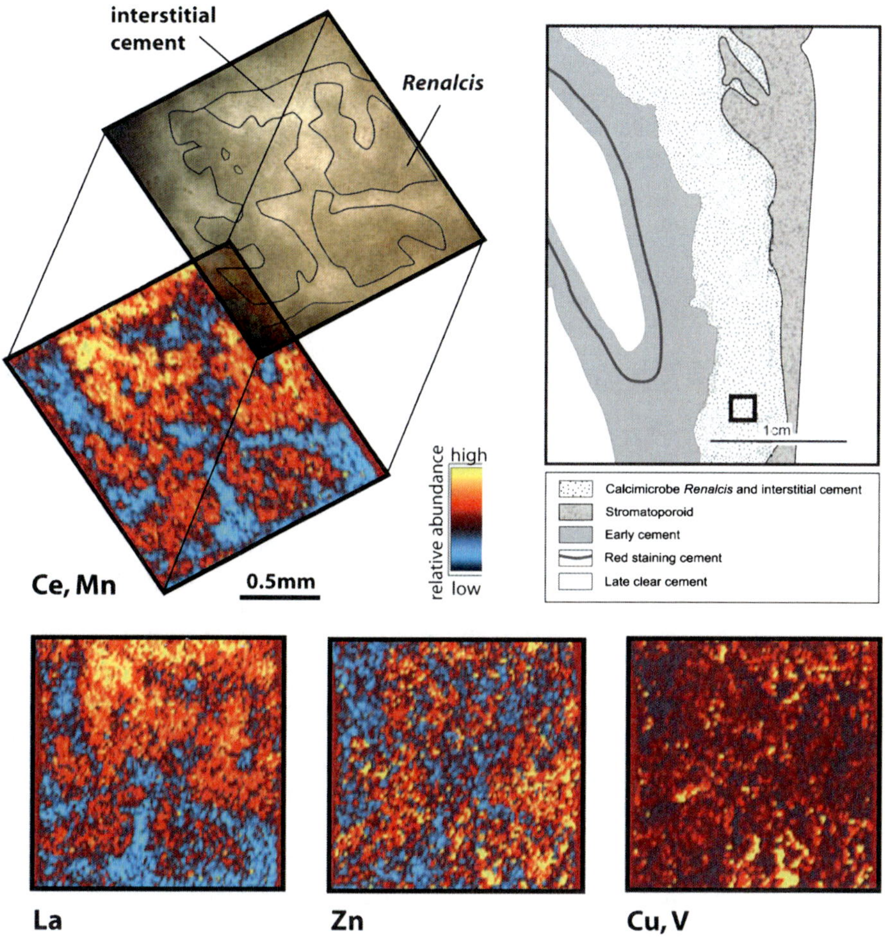

Fig. 10 Semi-quantitative major and trace element maps of the calcimicrobe *Renalcis* and interstitial cement (location of mapped area indicated by *black box* on interpretative sketch of polished petrographic section. Sample is 99-7-254a, which was documented in detail by Kamber and Webb 2007). The composite Ce and Mn concentration highs show very clear correspondence with dark areas of *Renalcis* in the transmitted light photomicrograph. These areas also stand out clearly on La and Zn maps. By contrast, Cu and V distribution highs are not clearly linked to *Renalcis* but could be associated with Fe-oxihydroxides. The map was produced from a series of 70 adjacent 30 μm wide LA-ICP-MS traverses (shot at 15 μm s^{-1})

possible link to nitrogen fixation was discussed earlier as was the need to put modern microbial biochemistry into the context of the evolving Earth surface environment. Progress has already been made at linking certain metal enrichments in hydrothermal minerals with extremophiles (e.g. Holden and Adams 2003) but the systematics arising from study of microbial carbonate are expected to be complex. At this stage, the critical step to be taken is the search for coherent associations of elevated metal contents with particular types of microbial carbonates.

5 Summary

The proper interpretation of microbialites has important implications for understanding early life and the evolution of earth systems through time. However, as incidental biosedimentary structures, microbialites are formed by complex systems that represent a continuum of processes with varying contributions from biological and physical-chemical components. Hence, their identification and interpretation can be challenging. Trace element geochemistry is proving to be a valuable tool for assessing and evaluating the nature and origin of putative microbialites. Trace elements inventories within microbialites represent a variety of sources including: (1) elements enriched either purposefully or incidentally by biological processes within biofilms or microbial mats; (2) elements passively accumulated from the ambient fluid environment; (3) elements associated with trapped and bound particulate matter or sediment; and (4) elements emplaced from diagenetic fluids. The first source of elements, those that are enriched through biological processes, may serve as biosignatures. Recognition of disequilibrium behaviour in such elements could potentially even serve as a biosignature in putative microbialites from an extraterrestrial source. Where the nature of microbialites is better constrained and the activity of specific metals in microbial metabolism is known, some elemental enrichments may provide direct evidence of the types of biological communities, organisms or metabolic pathways that occurred in a given mat or biofilm. Where elemental inventories reflect more the ambient waters of the depositional environment, they may aid interpretation of those environments, including the recognition of different water bodies, and may record changing marine chemistry through time. Rare earth element inventories in ancient microbialites, for example, have already been used to document marine chemistry, oxygenation levels and depositional settings in ancient basins. However, when interpreting ambient physical-chemical conditions from elemental patterns in ancient microbialites it is necessary to acknowledge the degree to which a given microbialite may have been isolated within a microenvironment. The chemistry of fluids within microenvironments may be significantly evolved relative to ambient fluids and precipitation of minerals in most if not all microbialites reflects some degree of microenvironmental forcing (e.g., elevated alkalinity). Elemental inventories introduced from trapped and bound sediment and from subsequent diagenesis, the last two categories, represent significant contaminants in such studies, thus necessitating effective sample vetting. However, where the initial systems are relatively well understood, elements introduced during diagenesis may allow better understanding of the diagenetic processes and fluid sources in a given stratigraphic unit or basin.

In conclusion, the outlook for trace element geochemical analysis of microbialites is good. New analytical techniques such as LA-ICP-MS with spatial mapping are breaking new ground in establishing elemental abundances at very low concentrations, in challenging materials and at the very small spatial scales that are particularly relevant to geomicrobiological studies. Integration of such studies with other geochemical techniques (e.g., stable isotope studies) and basic sedimentology and stratigraphy may allow ancient depositional environments and palaeogeographic settings to be reconstructed with increasing proficiency. However, there is much room for documentation of biosignatures and environmental signatures in modern and unequivocal ancient microbialites to refine interpretations, vetting procedures and analytical approaches. As our baseline understanding of microbialites as recorders of biological communities and geochemical environments increases their analysis will provide critical new data to inform our understanding of the Earth's biogeochemical evolution.

Acknowledgments Thomas Ulrich is thanked for putting together the trace element maps and Elizabeth Turner kindly provided advice regarding Little Dal sedimentology. L.D. Nothdurft provided the SEM photomicrographs in Fig. 6. We wish to thank an anonymous reviewer for many beneficial comments on the manuscript. The analytical work at Laurentian University is supported by the Canada Research Chairs program. GEW is supported by Australian Research Council Discovery Grant (DP0663312).

References

Abell PI, McClory J, Martin A, Nisbet EG (1985a) Archaean stromatolites from the Ngesi Group, Belingwe Greenstone Belt, Zimbabwe: preservation and stable isotopes – preliminary results. Precambr Res 27:357–383

Abell PI, McClory J, Martin A, Nisbet EG, Kyser TK (1985b) Petrography and stable isotope ratios from Archaean stromatolites, Mushandike Formation, Zimbabwe. Precambr Res 27:385–398

Akagi T, Hashimoto Y, Fu F-F, Tsuno H, Tao H, Nakano Y (2004) Variation of the distribution coefficients of area earth elements in modern coral-lattices: species and site dependencies. Geochim Cosmochim Acta 68:2265–2273

Alexander BW, Bau M, Andersson P, Dulski P (2008) Continentally-derived solutes in shallow Archean seawater: rare earth element and Nd isotope evidence in iron formation from 2.9 Ga Pongola Supergroup, South Africa. Geochim Cosmochim Acta 72:378–394

Alibert C, McCulloch MT (1993) Rare earth element and neodymium isotopic compositions of the banded iron-formations and associated shales from Hamersley, Western Australia. Geochim Cosmochim Acta 57:187–204

Allwood AC, Walter MR, Kamber BS, Marshall CP, Burch IW (2006) Stromatolite reef from the Early Archaean era of Australia. Nature 441:714–718

Allwood AC, Walter MR, Burch IW, Kamber BS (2007) 3.43 billion-year-old stromatolite reef from the Pilbara Craton of Western Australia: ecosystem-scale insights to early life on Earth. Precambr Res 158:198–227

Anbar AD, Knoll AH (2002) Proterozoic ocean chemistry and evolution: a bioinorganic bridge? Science 297:1137–1142

Anbar AD, Duan Y, Lyons TW, Arnold GL, Kendall B, Creaser RA, Kaufman AJ, Gordon GW, Scott C, Garvin J, Buick R (2007) A whiff of oxygen before the great oxidation event? Science 317:1903–1906

Anderson CR, Pederson K (2003) *In situ* growth of *Gallionella* biofilms and partitioning of lanthanides and actinides between biological material and ferric oxyhydroxides. Geobiology 1:169–178

Andres MS, Sumner DY, Reid RP, Swart PK (2005) Isotopic fingerprints of microbial respiration in aragonite from Bahamas stromatolites. Geology 34:973–976

Ankel-Fuchs D, Thauer RK (1988) Nickel in biology: nickel as an essential trace element. In: Lancaster JR Jr (ed) The bioinorganic chemistry of nickel. VCH Publishers, New York, pp 93–110

Arp G, Reimer A, Reitner J (2001) Photosynthesis-induced biofilm calcification and calcium concentrations in Phanerozoic oceans. Science 292:1701–1704

Awramik SM (2006) Respect for stromatolites. Nature 441:700–701

Awramik SM, Grey K (2005) Stromatolites: Biogenicity, biosignatures, and bioconfusion. Proc SPIE 5906, 59060P-1. doi:10.1117/12.625556

Badger M (2003) The roles of carbonic anhydrases in photosynthetic CO2 concentrating mechanisms. Photosynth Res 77:83–94
Badger M, Price GD (2003) CO2 concentrating mechanisms in cyanobacteria: molecular components, their diversity and evolution. J Exp Bot 383:609–622
Banner JL, Hanson GN, Meyers WJ (1988) Rare earth element and Nd isotopic variations in regionally extensive dolomites from the Burlington-Keokuk Formation (Mississippian): implications for REE mobility during carbonate diagenesis. J Sediment Petrol 58:415–432
Barbieri R, Cavalazzi B (2004) Astrobiological implications of microbial morphologies in cold fluid-generated carbonates. Adv Space Res 33:1262–1267
Barrat JA, Boulègue J, Tiercelin JJ, Lesourd M (2000) Strontium isotopes and rare-earth element geochemistry of hydrothermal carbonate deposits from Lake Tanganyika, East Africa. Geochim Cosmochim Acta 64:287–298
Bartley JK (1996) Actualistic taphonomy of cyanobacteria: implications for the Precambrian fossil record. Palaios 11:571–586
Bartley JK, Knoll AH, Grotzinger JP, Sergeev VN (2000) Lithification and fabric genesis in precipitated stromatolites and associated peritidal carbonates, Mesoproterozoic Billyahk Group, Siberia. In: Grotzinger JP, James NP (eds) Carbonate sedimentation and diagenesis in the evolving Precambrian world. SEPM Spec Publ 67:60–73
Batchelor MT, Burne RV, Henry BI, Watt SD (2000) Deterministic KPZ model for stromatolite laminae. Physica A 282:123–136
Batchelor MT, Burne RV, Henry BI, Jackson MJ (2004) A case for biotic morphogenesis of coniform stromatolites. Physica A 337:319–326
Bau M (1996) Controls on fractionation of isovalent trace elements in magmatic and aqueous systems: evidence from Y/Ho, Zr/Hf, and lanthanide tetrad effect. Contrib Mineral Petrol 123:323–333
Bau M, Dulski P (1996) Distribution of yttrium and rare-earth elements in the Penge and Kuruman iron-formations, Transvaal Supergroup, South Africa. Precambr Res 79:37–55
Bau M, Dulski P (1999) Comparing yttrium and rare earths in hydrothermal fluids from the Mid-Atlantic Ridge: implications for Y and REE behaviour during near-vent mixing and for the Y/Ho ratio of Proterozoic seawater. Chem Geol 155:77–90
Bau M, Möller P (1993) Rare earth element systematics of the chemically precipitated component in Early Precambrian iron formations and the evolution of the terrestrial atmosphere-hydrosphere-lithosphere system. Geochim Cosmochim Acta 57:2239–2249
Baumgartner LK, Reid RP, Dupraz C, Decho AW, Buckley DH, Spear JR, Przekop KM, Visscher PT (2006) Sulfate reducing bacteria in microbial mats: changing paradigms, new discoveries. Sediment Geol 185:131–145
Bazylinski DA, Moskowitz BM (1997) Microbial biomineralization of magnetic iron minerals: microbiology, magnetism and environmental significance. In: Banfield JF, Nealson KH (eds) Geomicrobiology: interactions between microbes and minerals. Reviews in Mineralogy, vol 35. Mineralogical Society of America, Washington, DC, pp 181–223
Bellanca A, Masetti D, Neri R (1997) Rare earth elements in limestone/marlstone couplets from the Albian-Cenomanian Cismon section (Venetian region, northern Italy): assessing REE sensitivity to environmental changes. Chem Geol 141:141–152
Benzerara K, Menguy N, López-García P, Yoon T-H, Kazmierczack J, Tyliszczak T, Guyot F, Brown GE Jr (2006) Nanoscale detection of organic signatures in carbonate microbialites. PNAS 103:9440–9445
Berner RA (1968) Calcium carbonate concretions formed by the decomposition of organic matter. Science 159:195–197
Berner RA (1975) The role of magnesium in the crystal growth of calcite and aragonite in seawater. Geochim Cosmochim Acta 39:489–504
Bertram CJ, Elderfield H (1993) The geochemical balance of the rare earth elements and neodymium isotopes in the oceans. Geochim Cosmochim Acta 57:1957–1986
Bischoff JL, Fyfe WS (1968) Catalysis, inhibition, and the calcite-aragonite problem. I. the aragonite-calcite transformation. Am J Sci 266:65–79
Black M (1933) The algal sediments of Andros Islands, Bahamas. Phil Trans R Soc Lond B 222:165–192

Blenkinsop TG, Fedo CM, Eriksson KA, Martin A, Nisbet GE, Wilson JF (1993) Ensialic origin for the Ngezi Group, Belingwe greenstone belt, Zimbabwe. Geology 21:1135–1138

Bolhar R, Van Kranendonk MJ (2007) A non-marine depositional setting for the northern Fortescue Group, Pilbara Craton, inferred from trace element geochemistry of stromatolitic carbonates. Precambr Res 155:229–250

Bolhar R, Hofmann A, Woodhead JD, Hergt JM, Dirks P (2002) Pb- and Nd-isotope systematics of stromatolitic limestones from the 27 Ga Ngezi Group of the Belingwe Greenstone Belt: constraints on timing of deposition and provenance. Precambr Res 114:277–294

Bolhar R, Kamber BS, Moorbath S, Fedo CM, Whitehouse MJ (2004) Characterisation of early Archaean chemical sediments by trace element signatures. Earth Planet Sci Lett 222:43–60

Bolhar R, Kamber BS, Moorbath S, Whitehouse MJ, Collerson KD (2005) Chemical characterization of earth's most ancient clastic metasediments from the Isua Greenstone Belt, southern West Greenland. Geochim Cosmochim Acta 69:1555–1573

Braithwait CJR, Taylor JD, Glover EA (2000) Marine carbonate cements, biofilms, biomineralization, and skeletogenesis: some bivalves do it all. J Sediment Res 70:1129–1138

Brand U, Veizer J (1980) Chemical diagenesis of a multicomponent carbonate system. 1. Trace elements. J Sediment Petrol 50:1219–1236

Brasier MD, Green OR, Jephcoat AP, Kleppe AK, Van Kranendonk MJ, Lindsay JF, Steele A, Grassineau N (2002) Questioning the evidence for Earth's oldest fossils. Nature 416:76–81

Brocks JJ, Logan GA, Buick R, Summons RE (1999) Archean molecular fossils and the rise of eukaryotes. Science 285:1033–1036

Brocks JJ, Buick R, Logan GA, Summons RE (2003) Composition and syngeneity of molecular fossils from the 2.78 to 2.45 billion-year-old Mount Bruce Supergroup, Pilbara Craton, Western Australia. Geochim Cosmochim Acta 67:4289–4319

Buckley DH, Baumgartner LK, Visscher PT (2008) Vertical distribution of methane metabolism in microbial mats of the Great Sippewissett Salt Marsh. Environ Microbiol 10:967–977

Buick R, Dunlop JSR, Groves DI (1981) Stromatolite recognition in ancient rocks: an appraisal of irregularly laminated structures in an Early Archaean chert-barite unit from North Pole, Western Australia. Alcheringa 5:161–181

Burne RV, Moore LS (1987) Microbialites: organosedimentary deposits of benthic microbial communities. Palaios 2:241–254

Burns SJ, McKenzie JA, Vasconcelos C (2000) Dolomite formation and biogeochemical cycles in the Phanerozoic. Sedimentology 47(Suppl 1):49–61

Buster NA, Holmes CW (2006) Magnesium content within the skeletal architecture of the coral *Montastraea faveolata*: locations of brucite precipitation and implications to fine scale data fluctuations. Coral Reefs 25:243–253

Butler A (1998) Acquisition and utilization of transition metal ions by marine organisms. Science 281:207–210

Cabioch G, Camoin G, Webb GE, Le Cornec F, Molina MG, Pierre C, Joachimski MM (2006) Contribution of microbialites to the development of coral reefs during the last deglacial period: Case study from Vanuatu (South-West Pacific). Sediment Geol 185:297–318

Cacchio P, Contento R, Ercole C, Cappuccio G, Martinez MP, Lepidi A (2004) Involvement of microorganisms in the formation of carbonate speleothems in the Cervo Cave (L'Aquila-Italy). Geomicrobiol J 21:497–509

Cady SL, Farmer JD, Grotzinger JP, Schopf JW, Steele A (2003) Morphological biosignatures and the search for life on Mars. Astrobiology 3:351–368

Cameron EM (1982) Sulphate and sulphate reduction in early Precambrian oceans. Nature 296:145–148

Canfield DE, Des Marais DJ (1993) Biogeochemical cycles of carbon, sulfur and free oxygen in a microbial mat. Geochim Cosmochim Acta 57:3971–3984

Castanier S, Le Métayer-Levrel G, Perthuisot J-P (1999) Ca-carbonates precipitation and limestone genesis – the microbiogeologist point of view. Sediment Geol 126:9–23

Cavet JS, Borrelly GPM, Robinson NJ (2003) Zn, Cu and Co in cyanobacteria: selective control of metal availability. FEMS Microbiol Rev 27:165–181

Chauvel C, Dupré B, Arndt NT (1993) Pb and Nd isotopic correlation in Belingwe komatiites and basalts. In: Bickle MJ, Nisbet EG (eds) The geology of the Belingwe Greenstone Belt, Zimbabwe. Geological Society of Zimbabwe Special Publication, Rotterdam, pp 167–174

Cherniak DJ (1998) REE diffusion in calcite. Earth Planet Sci Lett 160:273–287

Decho AW (2000) Exopolymer microdomains as a structuring agent for heterogeneity within microbial biofilms. In: Riding RE, Awramik SM (eds) Microbial sediments. Springer, Berlin, pp 9–15

Decho AW, Visscher PT, Reid RP (2005) Production and cycling of natural microbial exopolymers (EPS) within a marine stromatolite. Palaeogeogr Palaeoclimatol Palaeoecol 219:71–86

Défarge C, Trichet J (1995) From biominerals to "organominerals": the example of the modern lacustrine calcareous stromatolites from Polynesian atolls. In: Allemand D, Cuif J-P (eds) Biomineralization 93: Proceedings of the 7th international symposium on biomineralization. Bull l'Inst océanogr, Monaco, Num spéc 14(2):265–271

Défarge C, Gautret P, Reitner J, Trichet J (2009) Defining biominerals: comment on 'defining biominerals and organominerals: direct and indirect indicators of life' by Perry et al. (2007, Sedimentary Geology, 201, 157–179). Sediment Geol 213:152–155

Derenne S, Robert F, Skrzypczak-Bonduelle A, Gourier D, Binet L, Rouzaud J-N (2008) Molecular evidence for life in the 3.5 billion year old Warrawoona chert. Earth Plat Sci Lett 272:476–480

Derry LA, Jacobsen SB (1990) The chemical evolution of Precambrian seawater: evidence from REEs in banded iron formations. Geochim Cosmochim Acta 54:2965–2977

Des Marais DJ, Cohen Y, Nguyen H, Cheatham M, Cheatham T, Munoz E (1989) Carbon isotopic trends in the hypersaline ponds and microbial mats at Guero Negro, Baja California Sur, Mexico: implications for Precambrian stromatolites. In: Cohen Y, Rosenberg E (eds) Microbial mats: physiological ecology of benthic microbial communities. American Society of Microbiology, Washington, DC, pp 191–205

Dill RF, Shinn EA, Jones AT, Kelly K, Steinen RP (1986) Giant subtidal stromatolites forming in normal salinity water. Nature 324:55–58

Dravis JJ (1983) Hardened subtidal stromatolites, Bahamas. Science 219:385–386

Drever JI (1997) The geochemistry of natural waters, 3rd edn. Prentice Hall, Upper Saddle River, NJ

Driessen AJM, Konings WN (1986) Calcium transport in membrane vesicles of *Streptococcus cremoris*. Eur J Biochem 159:149–155

Dupraz C, Visscher PT (2005) Microbial lithification in marine stromatolites and hypersaline mats. Trends Microbiol 13:429–438

Dupraz C, Pattisina R, Verrecchia EP (2006) Translation of energy into morphology: simulation of stromatolite morphospace using a stochastic model. Sediment Geol 185:185–203

Dupraz C, Reid RP, Braissant O, Decho AW, Norman RS, Visscher PT (2008) Processes of carbonate precipitation in modern microbial mats. Earth Sci Rev. doi:10.1016/j.earscirev.2008.10.005

Ehrlich HL (2002) Geomicrobiology, 4th edn. Marcel Dekker, Inc, New York

Elderfield H (1988) The oceanic chemistry of the rare-earth elements. Phil Trans R Soc Lond A 325:105–106

Elderfield H, Upstill-Goddard R, Sholkovitz ER (1990) The rare earth elements in rivers, estuaries, and coastal seas and their significance to the composition of ocean waters. Geochim Cosmochim Acta 54:971–991

Fedo CM, Whitehouse MJ (2002) Metasomatic origin of quartz-pyroxene rock, Akilia, Greenland, and implications for Earth's earliest life. Science 296:1448–1452

Ferris FG (2000) Microbe-metal interactions in sediments. In: Riding RE, Awramik SM (eds) Microbial sediments. Springer, Berlin, pp 121–126

Folk R (1974) The natural history of crystalline calcium carbonate; effect of magnesium content and salinity. J Sediment Petrol 44:40–53

Fortin D, Langley S (2005) Formation and occurrence of biogenic iron-rich minerals. Earth Sci Rev 75:1–19

Fraústo da Silva JJR, Williams RJP (2001) The biological chemistry of the elements: the inorganic chemistry of life, 2nd edn. Oxford University Press, Oxford

Frei R, Dahl PS, Duke EF, Frei KM, Hansen TR, Frandsson MM, Jensen LA (2008) Trace element and isotopic characterization of Neoarchean and Paleoproterozoic iron formations in the Black

Hills (South Dakota, USA): assessment of chemical change during 2.9–1.9 Ga deposition bracketing the 2.4–2.2 Ga first rise of atmospheric oxygen. Precambr Res 162:441–474

Frimmel HE (2009) Trace element distribution in Neoproterozoic carbonates as palaeoenvironmental indicator. Chem Geol 258:338–353

Fryer BJ (1977) Rare earth evidence in iron formations for changing Precambrian oxidation states. Geochim Cosmochim Acta 41:361–367

García-Ruiz JM (2000) Geochemical scenarios for the precipitation of biomimetic inorganic carbonates. In: Grotzinger JP, James NP (eds) Carbonate sedimentation and diagenesis in the evolving Precambrian world. SEPM Spec Publ 67:75–89

Geesy GG, Jang L (1989) Interactions between metal ions and capsular polymers. In: Beveridge TJ, Doyle RJ (eds) Metal ions and bacteria. Wiley, New York, pp 325–358

German CR, Elderfield H (1990) Application of the Ce anomaly as a paleoredox indicator: the ground rules. Paleoceanography 5:823–833

Gibson EK Jr, McKay DS, Thomas-Keprta KL, Wentworth SJ, Westall F, Steele A, Romanek CS, Blee MS, Toporski J (2001) Life on Mars: evaluation of the evidence within Martian meteorites ALH84001, Nakhla, and Shergotty. Precambr Res 106:15–34

Giordano M, Beardall J, Raven JA (2005) CO2 concentrating mechanisms in algae: mechanisms, environmental modulation, and evolution. Annu Rev Plant Biol 56:99–131

Given RK, Wilkinson BH (1987) Dolomite abundance and stratigraphic age: constraints on rates and mechanisms of Phanerozoic dolostone formation. J Sediment Petrol 57:1068–1078

Goldstein SJ, Jacobsen SB (1988) Rare earth elements in river waters. Earth Planet Sci Lett 89:35–47

Golubic S (1972) The relationship between blue-green algae and carbonate deposits. In: Carr NG, Whitton BA (eds) The biology of blue-green algae. Blackwell, Oxford, pp 434–472

Golubic S (1976) Organisms that build stromatolites. In: Walter MR (ed) Stromatolites, developments in sedimentology, vol 20. Elsevier, Amsterdam, pp 113–126

Golubic S, Hofmann HJ (1976) Comparison of modern and mid-Precambrian Entophysalidaceae (Cyanophyta) in stromatolitic algal mats: cell division and degradation. J Paleontol 50:1074–1082

Graff JL (1978) Rare earth elements, iron formations and seawater. Geochim Cosmochim Acta 42:1845–1850

Grotzinger JP, Knoll AH (1995) Anomalous carbonate precipitates: is the Precambrian the key to the Permian? Palaios 10:578–596

Grotzinger JP, Knoll AH (1999) Stromatolites in Precambrian carbonates: evolutionary mileposts or environmental dipsticks? Annu Rev Earth Planet Sci 27:313–358

Grotzinger JP, Rothman DH (1996) An abiotic model for stromatolite morphogenesis. Nature 383:423–425

Guo L, Andrews J, Riding R, Dennis P, Dresser Q (1996) Possible microbial effects on stable carbon isotopes in hot spring travertines. J Sediment Res 66:468–473

Hellstrom J, Paton C, Woodhead JD, Hergt JM (2008) Iolite: software for spatially resolved LA-(quad and MC) ICPMS analysis. Mineral Assoc Can Short Course Ser 40:343–348

Hoffman PF, Kaufman AJ, Halverson GP, Schrag DP (1998) A Neoproterozoic snowball Earth. Science 281:1342–1346

Hofmann A, Bolhar R (2007) Carbonaceous cherts in the Barberton Greenstone Belt and their significance for the study of early life in the Archean record. Astrobiology 7:355–388

Hofmann HJ, Grey K, Hickman AH, Thorpe R (1999) Origin of 3.45 Ga coniform stromatolites in Warrawoona Group, Western Australia. Geol Soc Am Bull 111:1256–1262

Holden JF, Adams MWW (2003) Microbe-metal interactions in marine hydrothermal environments. Curr Opin Chem Biol 7:160–165

Holser WT (1997) Evaluation of the application of rare-earth elements to paleoceanography. Palaeogeogr Palaeoclimatol Palaeoecol 132:309–323

Huston DL, Logan GA (2004) Barite, BIFs and bugs: evidence for the evolution of the Earth's early hydrosphere. Earth Planet Sci Lett 220:41–55

Jaun B, Thauer RK (2007) Nickel and its surprising impact in nature. In: Sigel A, Sigel H, Sigel RKO (eds) Metal ions in life sciences, vol 2. Wiley, Chichester, IL, pp 323–356

Johnston I (1980) The ultrastructure of skeletogenesis in hermatypic corals. Int Rev Cytol 67:171–214

Jordan G, Rammensee W (1996) Dissolution rates and activation of brucite (001): a new method based on the microtopography of crystal surfaces. Geochim Cosmochim Acta 60:5055–5062

Kalkowsky E (1908) Oolith und Stromatolith im norddeutschen Buntsandstein. Z Dtsch Geol Ges 60:68–125

Kamber BS, Webb GE (2001) The geochemistry of late Archaean microbial carbonate: implications for ocean chemistry and continental erosion history. Geochim Cosmochim Acta 65:2509–2525

Kamber BS, Webb GE (2007) Transition metal abundances in microbial carbonate: a pilot study based on *in situ* LA-ICP-MS analysis. Geobiology 5:357–389

Kamber BS, Bolhar R, Webb GE (2004) Geochemistry of late Archaean stromatolites from Zimbabwe: evidence for microbial life in restricted epicontinental seas. Precambr Res 132:379–399

Kamijo M, Suzuki T, Kawai K, Murase H (1998) Accumulation of Yttrium by *Variovorax paradoxus*. J Ferment Bioeng 86:564–568

Kato Y, Ohta I, Tsunematsu T, Watanabe Y, Isozaki Y, Maruyama S, Imai N (1998) Rare earth element variations in mid-Archean banded iron formations: implications for the chemistry of ocean and continent and plate tectonics. Geochim Cosmochim Acta 62:3475–3497

Kato Y, Nakao K, Isozaki Y (2002) Geochemistry of Late Permian to Early Triassic pelagic cherts from southwest Japan: implications for an oceanic redox change. Chem Geol 182:15–34

Kennedy MJ, Christie-Blick N, Sohl LE (2001) Are Proterozoic cap carbonates and isotopic excursions a record of gas hydrate destabilization following Earth's coolest intervals? Geology 29:443–446

Kinsman DJJ (1969) Interpretation of the strontium concentration in carbonate minerals and rocks. J Sediment Petrol 39:486–508

Knoll AH, Semikhatov MA (1998) The genesis and time distribution of two distinctive Proterozoic stromatolite microstructures. Palaios 13:407–421

Komiya T, Hirata T, Kitajima K, Yamamoto S, Shibuya T, Sawaki Y, Ishikawa T, Shu D, Li Y, Han J (2008) Evolution of the composition of seawater through geologic time, and its influence on the evolution of life. Gondwana Res 14:159–174

Konhauser KO (2007) Introduction to geomicrobiology. Blackwell, Malden Mass

Konhauser KO, Hamade T, Raiswell R, Morris RC, Ferris FG, Southam G, Canfield DE (2002) Could bacteria have formed the Precambrian banded iron formations? Geology 30:1079–1082

Konhauser KO, Pecoits E, Lalonde SV, Papineau D, Nisbet EG, Barley ME, Arndt NT, Zahnle K, Kamber BS (2009) Oceanic nickel depletion and the methanogen famine before the Great Oxidation Event. Nature 458:750–754

Krumbein WE (1979) Photolithotropic and chemoorganotrophic activity of bacteria and algae as related to beachrock formation and degradation (Gulf of Aqaba, Sinai). Geomicrobiol J 1:139–203

Krumbein WE (1986) Biotransfer of minerals by microbes and microbial mats. In: Leadbeater BSC, Riding R (eds) Biomineralization in lower plants and animals. Syst Assoc Spec 30:55–72

Krumbein WE, Paterson DM, Stal LJ (1994) Biostabilization of sediments. Bibl Inf Syst Universität Oldenburg, Oldenburg, Germany

Kusky TM, Kidd WSF (1992) Remnants of an Archean Oceanic Plateau, Belingwe Greenstone-Belt, Zimbabwe. Geology 20:43–46

Lakshtanov LZ, Stipp SLS (2004) Experimental study of europium (III) coprecipitation with calcite. Geochim Cosmochim Acta 68:819–827

Lawrence MG, Kamber BS (2006) The behaviour of the rare earth elements during estuarine mixing – revisited. Mar Chem 100:147–161

Lawrence MG, Greig A, Collerson KD, Kamber BS (2006) Rare earth element and yttrium variability in southeast Queensland waterways. Aquat Geochem 12:39–72

Le Hir G, Donnadieu Y, Goddéris Y, Pierrehumbert RT, Halverson GP, Macouin M, Nédélec A, Ramstein G (2009) The snowball Earth aftermath: exploring the limits of continental weathering processes. Earth Planet Sci Lett 277:453–463

Leão PN, Vasconcelos MTSD, Vasconcelos VM (2007) Role of marine cyanobacteria in trace metal bioavailability in seawater. Microb Ecol 53:104–109

Ledin M (2000) Accumulation of metals by microorganisms- processes and importance for soil systems. Earth Sci Rev 51:1–31

Lee S-J, Golubic S (2000) Biological and mineral components of an ancient stromatolite: Gaoyuzhuang Formation, Mesoproterozoic of China. In: Grotzinger JP, James NP (eds) Carbonate sedimentation and diagenesis in the evolving Precambrian world. SEPM Spec Publ 67:91–102

Lee JG, Roberts SB, Morel FMM (1995) Cadmium – a nutrient for the marine diatom *Thalassiosira weissflogii*. Limnol Oceanogr 40:1056–1963

Lindsay JF, Brasier MD, McLoughlin N, Green OR, Fogel M, McNamara KM, Steele A, Mertzman SA (2003) Abiotic Earth – establishing a baseline for earliest life, data from the Archean of Western Australia. In: Lunar and Planetary Institute, Annual Meeting, Lunar Planet Inst Contrib 1156, 1137.pdf

Lindsay JF, Brasier MD, McLoughlin N, Green OR, Fogel M, Steele A, Mertzman SA (2005) The problem of deep carbon – an Archean paradox. Precambr Res 143:1–22

Little BJ, Wagner PA, Lewandowski Z (1997) Spatial relationship s between bacteria and mineral surfaces. In: Banfield JF, Nealson KH (eds) Geomicrobiology: interactions between microbes and minerals. Reviews in Mineralogy, vol 35. Mineralogical Society of America, Washington, DC, pp 123–159

Logan BW (1961) *Cryptozoan* and associated stromatolites from the recent, Shark Bay, Western Australia. J Geol 69:517–533

Logan BW, Rezak R, Ginsburg RN (1964) Classification and environmental significance of algal stromatolites. J Geol 72:68–83

Loisy C, Verrecchia EP, Dufour P (1999) Microbial origin for pedogenic micrite associated with a carbonate palaeosol (Champagne, France). Sediment Geol 126:193–204

Lowe DR (1980) Stromatolites 3,400-Myr old from the Archean of Western Australia. Nature 284:441–443

Lowe DR (1983) Restricted shallow water sedimentation of Early Archean stromatolitic and evaporitic strata of the Strelley Pool Chert, Pilbara Block, Western Australia. Precambr Res 19:239–283

Lowe DR (1994) Abiological origin of described stromatolites older than 3.2 Ga. Geology 22:387–390

Lowenstam HA, Weiner S (1989) On biomineralization. Oxford University Press, Oxford

Lyakhin YuI (1968) Calcium carbonate saturation of Pacific water. Oceanology 8:45–53

Lyons WB, Long DT, Hines ME, Gaudete HE, Armstrong PB (1984) Calcification of cyanobacterial mats in Solar Lake, Sinai. Geology 12:623–626

Macintyre IG (1977) Distribution of submarine cements in a modern Caribbean fringing reef, Galeta Point, Panama. J Sediment Petrol 47:503–516

Macintyre IG, Prufert-Bebout L, Reid RP (2000) The role of endolithic cyanobacteria in the formation of lithified laminae in Bahamian stromatolites. Sedimentology 47:915–921

Maliva RG (1998) Skeletal aragonite neomorphism – quantitative modelling of a two-water diagenetic system. Sediment Geol 121:179–190

Martin GD, Wilkinson BH, Lohmann KC (1986) The role of skeletal porosity in aragonite neomorphism – *Strombus* and *Montastrea* from the Pleistocene Key Largo Limestone, Florida. J Sediment Petrol 56:194–203

McKenzie JA, Vasconcelos C (2009) Dolomite Mountains and the origin of the dolomite rock of which they mainly consist: historical developments and new perspectives. Sedimentology 56:205–219

Mcloughlin N, Wilson LA, Brasier MD (2008) Growth of synthetic stromatolites and wrinkle structures in the absence of microbes – implications for the early fossil record. Geobiology 6:95–105

Merz MUE (1992) The biology of carbonate precipitation by cyanobacteria. Facies 26:81–102

Merz M, Schlue WR and Zankl H (1995) pH measurements in the sheath of calcifying cyanobacteria. In: Allemand D, Cuif J-P (eds) Biomineralization 93: Proceedings of the 7th international symposium on biomineralization. Bull l'Inst océanogr, Monaco, Num. spéc. 14(2):281–289

Micheletti E, Colica G, Viti C, Tamagnini P, De Philippis R (2008) Selectivity in the heavey metal removal by exopolysaccharide-producing cyanobacteria. J Appl Microbiol 105:88–94

Michibata H, Uyama T, Kanamori K (1998) The accumulation mechanism of vanadium by ascidians. Am Chem Soc Symp Ser 711:248–258

Mojzsis SJ, Harrison TM (2002) Origin and significance of Archaean quartzose rocks at Akilia, Greenland. Science 298:917a

Mojzsis SJ, Arrhenius G, McKeegan KD, Harrison TM, Nutman AP, Friend CRL (1996) Evidence for life on Earth before 3,800 million years ago. Nature 384:55–59

Monty CLV (1976) The origin and development of cryptalgal fabrics. In: Walter MR (ed) Stromatolites, developments in sedimentology, vol 20. Elsevier, Amsterdam, pp 193–249

Monty CLV (1977) Evolving concepts on the nature and the ecological significance of stromatolites. In: Flügel E (ed) Fossil algae, recent results and developments. Springer, Berlin, pp 15–35

Morel FMM, Price NM (2003) The biogeochemical cycles of trace metals in the oceans. Science 300:944–947

Nealson KH (1983) Microbial oxidation and reduction of manganese and iron. In: Westbroek P, de Jong EW (eds) Biomineralization and biological metal accumulation. D Reidel, Dordrecht, Holland, pp 459–479

Neilan BA, Burns BP, Relman DA, Lowe DR (2002) Molecular identification of cyanobacteria associated with stromatolites from distinct geographical locations. Astrobiology 2:271–280

Neuweiler F, Reitner J (1993) Initially indurated structures if fine-grained calcium carbonate firmed in place (automicrite). 7th Symposium on Biominerals (abstract vol) Monaco

Nothdurft LD, Webb GE (2009) Earliest diagenesis in scleractinian coral skeletons: implications for palaeoclimate-sensitive geochemical archives. Facies 55:161–201

Nothdurft LD, Webb GE, Kamber BS (2004) Rare earth element geochemistry of Late Devonian reefal carbonates, Canning Basin, Western Australia: confirmation of a seawater REE proxy in ancient limestones. Geochim Cosmochim Acta 68:263–283

Nothdurft LD, Webb GE, Buster NA, Holmes CW, Sorauf JE, Kloprogge JT (2005) Brucite microbialites in living coral skeletons: indicators of extreme microenvironments in shallow-marine settings. Geology 33:169–172

Nothdurft LD, Webb GE, Bostrom T, Rintoul L (2007) Calcite-filled borings in the most recently deposited skeleton in live-collected *Porites* (Scleractinia): implications for trace element archives. Geochim Cosmochim Acta 71:5423–5438

Ohmoto H, Runnegar B, Kump LR, Fogel ML, Kamber BS, Anbar AD, Knauth PL, Lowe DR, Sumner DY, Watanabe Y (2008) Biosignatures in ancient rocks: a summary of discussions at a field workshop on biosignatures in ancient rocks. Astrobiology 8:883–907

Ojiambo SB, Lyons WB, Welch KA, Poreda RJ, Johannesson KH (2003) Strontium isotopes and rare earth elements as tracers of groundwater-lake water interactions, Lake Naivasha, Kenya. Appl Geochem 18:1789–1805

Olivier N, Boyet M (2006) Rare earth and trace elements of microbialites in Upper Jurassic coral- and sponge-microbialite reefs. Chem Geol 230:105–123

Ozaki T, Suzuki Y, Nankawa T, Yoshida T, Ohnuki T, Kimura T, Francis AJ (2006) Interactions of rare earth elements with bacteria and organic ligands. J Alloy Comp 408:1334–1338

Paerl HW, Steppe TF, Reid RP (2001) Bacterially mediated precipitation in marine stromatolites. Environ Microbiol 3:123–130

Parekh PP, Möller P, Dulski P, Bausch WM (1977) Distribution of trace elements between carbonate and non-carbonate phases of limestone. Earth Planet Sci Lett 34:39–50

Parker DL, Schram BR, Plude JL, Moore RE (1996) Effect of metal cations on the viscosity of a pectin-like capsular polysaccharide from the cyanobacterium *Microcystis flos-aquae* $C_{3\text{-}40}$. Appl Environ Microbiol 62:1208–1213

Pereira S, Zille A, Micheletti E, Moradas-Ferreira P, De Philippis R, Tamagnini P (2009) Complexity of cyanobacterial exopolysaccharides: composition, structures, inducing factors and putative genes involved in their biosynthesis and assembly. FEMS Microbiol Rev 33:917–941

Perry RS, Mcloughlin N, Lynne BY, Sephton MA, Oliver JD, Perry CC, Campbell K, Engel MH, Farmer JD, Brasier MD, Staley JT (2007) Defining biominerals and organominerals: direct and indirect indicators of life. Sediment Geol 201:157–179

Philippot P, Van Zuilen M, Lepot K, Thomazo C, Farquhar J, Van Kranendonk MJ (2007) Early Archaean microorganisms preferred elemental sulfur, not sulfate. Science 317:1534–1537

Piepgras DJ, Jacobsen SB (1992) The behavior of rare earth elements in seawater: precise determination of variations in the North Pacific water column. Geochim Cosmochim Acta 56:1851–1862

Pratt BR (1995) The origin, biota and evolution of deep-water mud-mounds. In: Monty CLV, Bosence DWJ, Bridges PH, Pratt BR (eds) Carbonate mud-mounds: their origin and evolution. Int Assoc Sediment Spec Publ 23:49–123

Przeniosło R, Stolarski J, Mazur M, Brunelli M (2008) Hierarchically structured scleractinian coral biocrystals. J Struct Biol 161:74–82

Quigg A, Finkel ZV, Irwin AJ, Rosenthal Y, Ho T-Y, Reinfelder JR, Schofield O, Morel FMM, Falkowski PG (2003) The evolutionary inheritance of elemental stoichiometry in marine phytoplankton. Nature 425:291–294

Read JF (1976) Calcretes and their distinction form stromatolites. In: Walter MR (ed) Stromatolites, developments in sedimentology, vol 20. Elsevier, Amsterdam, pp 55–71

Reid RP, Browne KM (1991) Intertidal stromatolites in a fringing Holocene reef complex, Bahamas. Geology 19:15–18

Reid RP, Visscher PT, Decho AW, Stolz JF, Bebout BM, Dupraz C, Macintyre IG, Paerl HW, Pinckney JL, Prufert-Bebout L, Steppe TF, Desmarais DJ (2000) The role of microbes in accretion, lamination, and early lithification of modern marine stromatolites. Nature 406:989–992

Reitner J (1993) Modern cryptic microbialite/metazoan facies from Lizard Island (Great Barrier Reef, Australia). Facies 29:3–39

Reitner J (2005) Calcifying extracellular mucus substances (EMS) of *Madrepora oculata* – a first geobiological approach. In: Friewald A, Roberts JM (eds) Cold-water corals and ecosystems. Springer, Berlin, pp 731–744

Reitner J, Gautret P, Maurin F, Neuweiler F (1995) Automicrites in a modern marine microbialite. Formation model via organic matrices (Lizard Island, Great Barrier Reef, Australia). In: Allemand D, Cuif J-P (eds) Biomineralization 93: Proceedings of the 7th international symposium on biomineralization. Bull l'Inst océanogr, Monaco, Num. spéc. 14(2):237–263

Revsbech NP (1989) Diffusion characteristics of microbial communities determined by use of oxygen microsensors. J Microbiol Methods 9:111–122

Revsbech NP, Jørgensen BB, Blackburn TH (1983) Microelectrode studies of the photosynthesis and O_2, H_2S, and pH profiles of a microbial mat. Limnol Oceanogr 28:1062–1074

Riding R (1991a) Calcified cyanobacteria. In: Riding R (ed) Calcareous algae and stromatolites. Springer, Berlin, pp 55–87

Riding R (1991b) Classification of microbial carbonates. In: Riding R (ed) Calcareous algae and stromatolites. Springer, Berlin, pp 21–51

Riding R (2000) Microbial carbonates: the geological record of calcified bacterial-algal mats and biofilms. Sedimentology 47(Suppl 1):179–214

Rosen BP (2002) Transport and detoxification systems for transition metals, heavy metals and metalloids in eukaryotic and prokaryotic microbes. Comp Biochem Physiol A 133:689–693

Saito MA, Sigman DM, Morel FMM (2003) The bioinorganic chemistry of the ancient ocean: the co-evolution of cyanobacterial metal requirements and biogeochemical cycles at the Archean-Proterozoic boundary? Inorg Chim Acta 356:308–318

Scherer M, Seitz H (1980) Rare-earth element distribution in Holocene and Pleistocene corals and their distribution during diagenesis. Chem Geol 28:279–289

Schidlowski M (2001) Carbon isotopes as biogeochemical recorders of life over 3.8 Ga of Earth history: evolution of a concept. Precambr Res 106:117–134

Schieber J (1988) Redistribution of rare-earth elements during diagenesis of carbonate rocks from the mid-Proterozoic Newland Formation, Montana, U.S.A. Chem Geol 69:111–126

Schopf JW (1994) The oldest known records of life: early Achean stromatolites, microfossils, and organic matter. In: Bengtson S (ed) Early life on Earth. Nobel Symposium No. 84. Columbia University Press, New York, pp 193–206

Semikhatov MA, Gebelein CD, Cloud P, Awramik SM, Benmore WC (1979) Stromatolite morphogenesis – progress and problems. Can J Earth Sci 19:922–1015
Shen Y, Buick R (2004) The antiquity of microbial sulfate reduction. Earth Sci Rev 64: 243–272
Shen Y, Buick R, Canfield DE (2001) Isotopic evidence for microbial sulphate reduction in the early Archaean era. Nature 410:77–81
Shields GA, Webb GE (2004) Has the REE composition of seawater changed over geological time? Chem Geol 204:103–107
Shiraishi F, Reimer A, Bissett A, de Beer D, Arp G (2008) Microbial effects on biofilm calcification, ambient water chemistry and stable isotope records in a highly supersaturated setting (Westerhöfer Bach, Germany). Palaeogeogr Palaeoclimatol Palaeoecol 262:91–106
Sholkovitz ER (1992) Chemical evolution of rare earth elements: fractionation between colloidal and solution phases of filtered river water. Earth Planet Sci Lett 114:77–84
Sholkovitz ER (1995) The aquatic chemistry of rare earth elements in rivers and estuaries. Aquat Geochem 1:1–34
Sholkovitz E, Shen GT (1995) The incorporation of rare earth elements in modern coral. Geochim Cosmochim Acta 59:2749–2756
Silver S (1983) Bacterial interactions with mineral cations and anions: good ions and bad. In: Westbroek P, DeJong EW (eds) Biomineralization and biological metal accumulation, biological and geological perspectives. D. Reidel, Dordrecht, Holland, pp 439–457
Silver S (1997) The bacterial view of the periodic table: specific functions for all elements. In: Banfield JF, Nealson KH (eds) Geomicrobiology: interactions between microbes and metals. Reviews in mineralogy, vol 35. Mineralogical Soceity of America, Washington, DC, pp 345–360
Stephens NP, Sumner DY (2002) Renalcids as fossilized biofilm clusters. Palaios 17:225–236
Stockdale A, Davison W, Zhang H (2009) Micro-scale biogeochemical heterogeneity in sediments: a review of available technology and observed evidence. Earth Sci Rev 92:81–97
Stolz JF (1990) Distribution of phototrophic microbes in the flat laminated microbial mat at Laguna Figueroa, Baja California, Mexico. BioSystems 23:345–357
Stolz JF (2000) Structure of microbial mats and biofilms. In: Riding RE, Awramik SM (eds) Microbial sediments. Springer, Berlin, pp 1–8
Stone AT (1997) Reactions of extracellular organicligands with dissolved metal ions and mineral surfaces. In: Banfield JF, Nealson KH (eds) Geomicrobiology: interactions between microbes and metals. Reviews in mineralogy, vol 35. Mineralogical Society of America, Washington, DC, pp 309–344
Sugitani K, Yamamoto K, Wada H, Binu-Lal SS, Yoneshige M (2002) Geochemistry of Archean carbonaceous cherts deposited at immature island-arc setting in the Pilbara Block, Western Australia. Sediment Geol 151:45–66
Sumner DY (2000) Microbial versus environmental influences on the morphology of Late Archean fenestrate microbialites. In: Riding R, Awramik S (eds) Microbial sediments. Springer, Berlin, pp 307–314
Takahashi Y, Châtellier X, Hattori KH, Kato K, Fortin D (2005) Adsorption of rare earth elements onto bacterial cell walls and its implication for REE sorption onto natural microbial mats. Chem Geol 219:53–67
Takahashi Y, Hirata T, Shimuzu H, Ozaki T, Fortin D (2007) A rare earth element signature of bacteria in natural waters? Chem Geol 244:569–583
Tanaka K, Kawabe I (2006) REE abundances in ancient seawater inferred from marine limestone and experimental REE partition coefficients between calcite and aqueous solution. Geochem J 40:425–435
Tang J, Johannesson KH (2003) Speciation of rare earth elements in natural terrestrial waters: assessing the role of dissolved organic matter from the modeling approach. Geochim Cosmochim Acta 67:2321–2339
Tebo BM, Obraztsova AY (1998) Sulfate-reducing bacterium grows with Cr(VI), U(VI), Mn(IV), and FE(III) as electron receptors. FEMS Microbiol Lett 162:193–198
Terakado Y, Masuda A (1988) The coprecipitation of rare-earth elements with calcite and aragonite. Chem Geol 69:103–110

Thrailkill J (1976) Speleothems. In: Walter MR (ed) Stromatolites, developments in sedimentology, vol 20. Elsevier, Amsterdam, pp 73–86
Turner EC, James NP, Narbonne GM (1997) Growth dynamics of Neoproterozoic calcimicrobial reefs, Mackenzie Mountains, northwest Canada. J Sediment Res 67:437–450
Turner EC, James NP, Narbonne GM (2000) Taphonomic control on microstructure in early Proterozoiic reefal stromatolites and thrombolites. Palaios 15:87–111
Ulrich T, Kamber BS, Jugo PJ, Tinkham DK (2009) Imaging element-distribution patterns in minerals by laser ablation – inductively coupled plasma – mass spectrometry (LA-ICP-MS). Can Mineral 47:1001–1012
Van Kranendonk MJ, Webb GE, Kamber BS (2003) Geological and trace element evidence for a marine sedimentary environment of deposition and biogenicity of 3.45 Ga stromatolitic carbonates in the Pilbara Craton, and support for a reducing Archean ocean. Geobiology 1:91–108
Van Lith Y, Warthmann R, Vasconcelos C, McKenzie JA (2003) Sulphate-reducing bacteria induce low temperature Ca-dolomite and high Mg-calcite formation. Geobiology 1:71–79
Vasconcelos C, McKenzie JA, Bernasconi S, Grujic D, Tien AJ (1995) Microbial mediation as a possible mechanism for natural dolomite formation at low temperatures. Nature 377:220–222
Veizer J (1983) Trace elements and isotopes in sedimentary carbonates. In: Reeder RJ (ed) Carbonates: mineralogy and chemistry. Reviews in mineralogy, vol 11. Mineralogical Society of America, Washington, DC, pp 265–299
Veron JEN (1995) Corals in space & time. UNSW Press, Sydney, p 321
Visscher PT, Stolz JF (2005) Microbial mats as bioreactors: populations, processes, and products. Palaeogeogr Palaeoclimatol Palaeoecol 219:87–100
Wacey D, Wright DT, Boyce AJ (2007) A stable isotope study of microbial dolomite formation in the Coorong Region, South Australia. Chem Geol 244:155–174
Walter MR (1976) Introduction. In: Walter MR (ed) Stromatolites, developments in sedimentology, vol 20. Elsevier, Amsterdam, pp 1–3
Walter MR (1983) Archean stromatolites: evidence of Earth's earliest benthos. In: Schopf JW (ed) Earth's earliest bioshphere, its origin and evolution. Princeton University Press, Princeton, NJ, pp 187–213
Walter MR (1994) Stromatolites: the main source of information on the evolution of the early benthos. In: Bengtson S (ed) Early life on Earth. Nobel Symposium No. 84. Columbia University Press, New York, pp 270–286
Walter MR, Buick R, Dunlop JSR (1980) Stromatolites 3,400–3,500 Myr old from the North Pole area, Western Australia. Nature 284:443–445
Walter MR, Krylov IN, Muir MD (1988) Stromatolites from Middle and Late Proterozoic sequences in McArthur and Georgina Basins and the Mount Isa Province, Australia. Alcheringa 12:79–106
Warthmann R, van Lith Y, Vasconcelos C, McKenzie JA, Karpoff AM (2000) Bacterially induced dolomite precipitation in anoxic culture experiments. Geology 28:1091–1094
Webb GE (1996) Was Phanerozoic reef history controlled by the distribution of non-enzymatically secreted reef carbonates (microbial carbonate and biologically induced cement)? Sedimentology 43:947–971
Webb GE (2001a) Biologically induced carbonate precipitation in reefs through time, Chapter 5. In: Stanley G (ed) The history and sedimentology of ancient reef systems: topics in geobiology, vol 17. Kluwer/Plenum, New York, pp 159–203
Webb GE (2001b) Famennian mud-mounds in the proximal fore-reef slope, Canning basin, Western Australia. Sediment Geol 145:295–315
Webb GE, Kamber BS (2000) Rare earth elements in Holocene reefal microbialites: a new shallow seawater proxy. Geochim Cosmochim Acta 64:1557–1565
Webb GE, Kamber BS (2004) Biogenicity inferred from microbialite geochemistry. Microbiol Aust 25:34–35
Webb GE, Kamber BS (2008) BAR-25. Biosignatures in Holocene reefal microbialites, Great Barrier Reef, Australia: old microstructural approaches versus new geochemical techniques. Astrobiology 8:906
Webb GE, Baker JC, Jell JS (1998) Inferred syngenetic textural evolution in Holocene cryptic reefal microbialites, Heron Reef, Great Barrier Reef, Australia. Geology 26:355–358

Webb GE, Jell JS, Baker JC (1999) Cryptic intertidal microbialites in beachrock, Heron Island, Great Barrier Reef: implications for the origin of microcrystalline beachrock cement. Sediment Geol 126:317–334

Webb GE, Nothdurft LD, Kamber BS, Kloprogge JT, Zhao J-X (2009) Rare earth element geochemistry of scleractinian coral skeleton during meteoric diagenesis: a sequence through neomorphism of aragonite to calcite. Sedimentology 56:1433–1463

Weiner S, Traub W, Lowenstam HA (1983) Organic matrix in calcified exoskeletons. In: Westbroek P, DeJong EW (eds) Biomineralization and biological metal accumulation, biological and geological perspectives. D. Reidel, Dordrecht, Holland, pp 205–224

Westbroek P, Buddemeier B, Coleman M, Kok DJ, Fautin D, Stal X (1994) Strategies for the study of climate forcing by calcification. In: Doumenge F (ed) Past and present biomineralization processes: considerations of the carbonate cycle. Bull l'Inst océanogr, Monaco, Num spéc 13:37–60

Wheat CG, Mottl MJ, Rudnicki M (2002) Trace element and REE composition of a low-temperature ridge-flank hydrothermal spring. Geochim Cosmochim Acta 66:3693–3705

Wilson JF, Nesbitt R, Fanning CM (1995) Zircon geochronology of Archaean felsic sequences in the Zimbabwe craton: a revision of greenstone stratigraphy and models for crustal growth. In: Ries AC, Coward MP (eds) Early Precambrian processes. Geological Society Special Publication, London, pp 109–126

Wimpenny JWT, Kinniment SL (1995) Biochemical reactions and establishment of gradients within biofilms. In: Lappin-Scott HM, Costerton JW (eds) Microbial biofilms. Cambridge University Press, Cambridge, pp 99–117

Worms I, Simon DF, Hassler CS, Wilkinson KJ (2006) Bioavailability of trace metals to aquatic microorganisms: importance of chemical, biological and physical processes on biouptake. Biochimie 88:1721–1731

Wright DT (1999) The role of sulphate-reducing bacteria and cyanobacteria in dolomite formation in distal ephemeral lakes of the Coorong region, South Australia. Sediment Geol 126:147–157

Wyndham T, McCulloch M, Fallon S, Alibert C (2004) High-resolution coral records of rare earth elements in coastal seawater: biogeochemical cycling and a new environmental proxy. Geochim Cosmochim Acta 68:2067–2080

Yamamoto K, Itoh N, Matsumoto T, Tanaka T, Adachi M (2004) Geochemistry of Precambrian carbonate intercalated in pillows and its host basalt: implications for the REE composition of circa 3.4 Ga seawater. Precambr Res 135:331–344

Young GM (2002) Stratigraphic and tectonic settings of Proterozoic glaciogenic rocks and banded iron formations: relevance to the snowball Earth debate. J Afr Earth Sci 35:451–466

Zenger DH (1989) Dolomite abundance and stratigraphic age: constraints on rates and mechanisms of Phanerozoic dolostone formation, Discussion. J Sediment Petrol 59:162–164

Zhang J, Nozaki Y (1996) Rare earth elements and yttrium in seawater: ICP-MS determinations in the East Caroline, Coral Sea, and South Fiji basins of the western South Pacific Ocean. Geochim Cosmochim Acta 60:4631–4644

Zhang J, Nozaki Y (1998) Behaviour of rare earth elements in seawater at the ocean margin: a study along the slopes of the Sagami and Nankai troughs near Japan. Geochim Cosmochim Acta 62:1307–1317

Zhong S, Mucci A (1995) Partitioning of rare earth elements (REEs) between calcite and seawater solutions at 25°C and 1 atm, and high dissolved REE concentrations. Geochim Cosmochim Acta 59:443–453

A Modern Perspective on Ancient Life: Microbial Mats in Sandy Marine Settings from the Archean Era to Today

Nora Noffke

Abstract Most valuable paleontological studies on the ancient life in the Archean era are on fossils of bacterial cells preserved in chert, or on the famous stromatolites we find in the oldest carbonate or silica rocks. The past decade, a new window into Earth's most antique worlds opened: Modern sandy coastal areas and the biofilms and microbial mats therein! The interaction of the modern microbial mats with the physical sediment dynamics creates sedimentary structures of mm to m sizes. Those 'microbially induced sedimentary structures – MISS' are found in the fossil record as well. Both modern and fossil examples look alike, but much different to stromatolites. We distinguish 17 main types of those MISS, related to their modes of formations. Our studies showed that ancient microbial mats developed in tidal, lagoonal, and shelf sandstones during the Phanerozoic, and Proterozoic ages, and even colonized already sandy marine deposits billions of years ago – during the Archean time. The ancient MISS suggest the existence of cyanobacteria potentially before 3 billion years ago.

Keywords Microbial mat • Archean • Tidal flats • MISS • Early life • Siliciclastic

1 Introduction

It is an intriguing task to reconstruct life in the Archean era, the oldest time period of Earth's rock record, because the sediments that time have been formed in an environment so much different to our own. After their deposition 3.8–2.5 billion years ago, the sediments lithified, and then experienced a multitude of diagenetic and tectonic alterations. Those processes erased a lot of the information those rocks had once contained. However, some information is still there and therefore windows

N. Noffke (✉)
Department of Ocean, Earth & Atmospheric Sciences, Old Dominion University, Norfolk, VA 23529, USA
e-mail: nnoffke@odu.edu

S.D. Golding and M. Glikson (eds.), *Earliest Life on Earth: Habitats, Environments and Methods of Detection*, DOI 10.1007/978-90-481-8794-2_7,

into Earth's most antique past exist. The 2.9 Ga Pongola and Witwatersrand Supergroups, and the 3.2 Ga Barberton Greenstone Belt in South Africa constitute one of the world's most famous paleontological sites for the exploration of oldest life (Lowe and Tice 2007). Precious fossils of tiny bacterial cells preserved in glass-like chert, or beautiful stromatolites give significant insight into the evolution of Earth's prokaryotes (e.g., Schopf et al. 2007; Walter et al. 2007).

In contrast, the siliciclastic rock successions of same age have been less studied. The reason is that sandstones rarely are regarded as sites of good fossil preservation – even sandstones of much younger Earth age. The siliciclastic material is porous and air and water is circulating through the deposits altering any organic matter therein quickly and thoroughly. But this is chemistry. We have to keep in mind that chemical processes are dependent on the chemical milieu they take place. Without doubt, the chemical composition of the Archean oceans and the atmosphere was different to today. Therefore, we must use caution in interpreting structures, texture or fossils that rose from chemical processes. In contrast, the interaction of life with physical processes should give rise to more reliable insight (contributions in Noffke and Paterson 2008). The reason is that the physical laws on Earth have never changed: Water has moved sand grains the same way 3 billion years ago as water in the present does. In consequence, the sedimentary structures that rise from this mechanical interaction of organisms with moving water always must look the same – be it today or billions of years ago. The past decade, the studies on sedimentary structures preserved in sandstones have opened a new path to investigate life's early steps in the Archean era (Noffke 2000, 2009; Noffke et al. 2002, 2003b, 2006a, b, 2008) (Fig. 1).

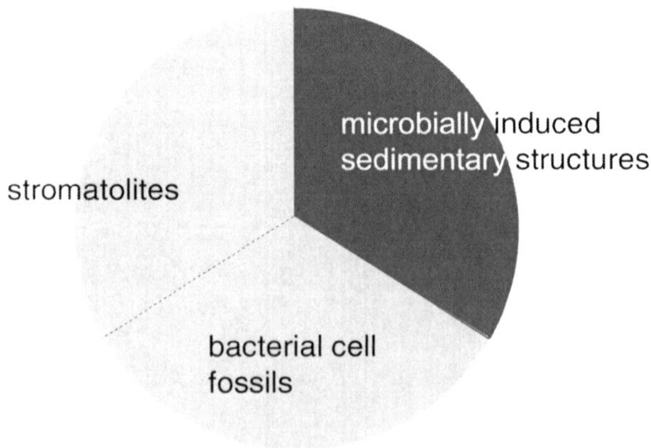

Fig. 1 Microbially induced sedimentary structures (MISS) constitute a new group of biogenic sedimentary structures that allow conclusions on early life in the Archean era. In contrast to fossils of bacterial cells, or to stromatolites, the MISS occur in siliciclastic sedimentary deposits. The structures are distributed throughout Earth history, from the early Archean to the present

Aim of this review is to give a brief summary about those biogenic sedimentary structures in siliciclastic deposits. This contribution likes to demonstrate, how they rise and what they tell us about Archean Earth.

2 Microbially Induced Sedimentary Structures – MISS

In contrast to chert and carbonate rocks, sandstones display sedimentary structures that resemble neither body fossils, nor stromatolites (Fig. 2). Those sedimentary structures come in multitude geometries and morphologies, and in sizes of meter to millimeter scales. Those 'microbially induced sedimentary structures – MISS' are not the product of chemical processes such as in situ carbonate mineral precipitation, but of the biotic-physical interaction of microbial mats with the sedimentary dynamics in marine environments. Due to their so specific genesis, the structures have been placed as own category into the classification of primary sedimentary structures (Noffke et al. 2001b; Schieber et al. 2007). The MISS have been described especially well from terminal Proterozoic sandstone successions, where the role of microbial mats in the preservation of Ediacara fossils has long been suggested. In his pioneering work, Schieber (1989) detected ancient microbial mats in Mesoproterozoic shales of Montana. 'Elephant skin textures', crinkled upper bedding planes, record ancient microbial mats (e.g., Bland 1984; Runnegar and Fedonkin 1992; Gehling 1999, 2000; Hagadorn and Bottjer 1999; Eriksson et al. 2000; Prave 2002; Sarkar et al. 2005). An overview on those early studies on microbial mat-related sedimentary structures was published for the first time in the volume by Hagadorn et al. (1999), followed later by a beautiful photo atlas (Schieber et al. 2007). Simultaneously to the time, when the first elephant skin textures were described, work on modern microbial mats and the sedimentary structures they form was conducted (e.g., Cameron et al. 1985; Gerdes and Krumbein 1987; Gerdes et al. 1991, 1993). In 1996, we decided to introduce the term 'microbially induced sedimentary structures (MISS)' for microbial mat-related sedimentary structures, based on our research on quantitative analyses on those structures in modern sandy tidal flats (Noffke et al. 1996).

The scholarly dry definition of MISS is: "MISS are primary sedimentary structures that rise syndepositionally from the physical interaction of biofilms and microbial mats with the sediment dynamic caused by hydraulic parameters in siliciclastic aquatic environments. Biostabilization counteracts erosion, baffling and trapping responds to deposition of sediment, and binding and growth take place during latencies (the time periods of no or low sediment reworking). Mostly, all biotic-physical interactions overlap in the formation of MISS, which is quantified with modern examples. Whereas primary mineral precipitation does not play any role in the formation of the MISS, secondary mineral accretion induced by the decay of the biofilm or mat constructing microorganisms and their EPS assists in the preservation of those structures. In thin-section, the MISS must include microtextures that are related to, have been caused by, or represent ancient biofilms or microbial mats. MISS occur from the early Archean to the present (cit. from Noffke 2009)."

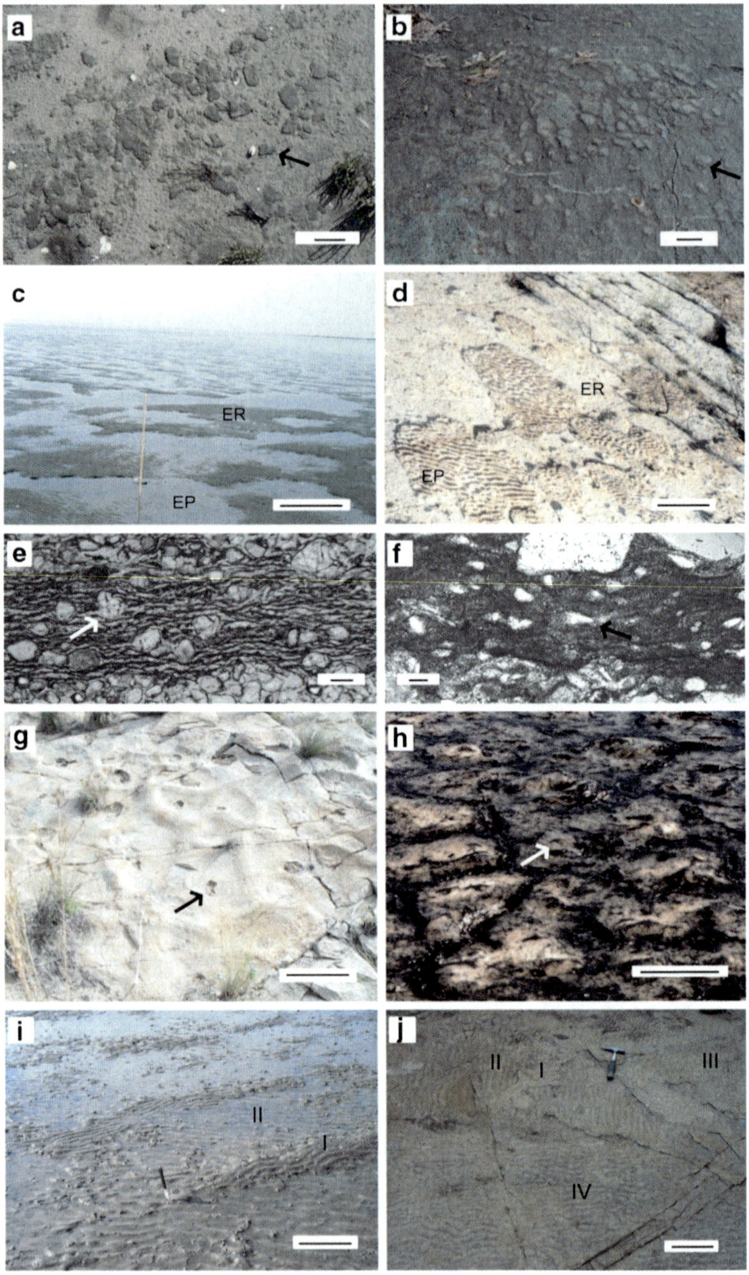

Fig. 2 Microbially induced sedimentary structures in sandy tidal flats – modern examples on the left and fossil analogues on the right. (**a**) Microbial Mat Chips (Fishermans Island, USA); scale: 5 cm. (**b**) fossil mat fragments (2.9 Ga Pongola Supergroup, South Africa); scale: 5 cm. (**c**) Erosional remnants (ER) and pockets (EP) (Mellum Island, North Sea); scale 1 m. (**d**) fossil erosional pockets (EP) situated in mat-stabilized sedimentary surface (ER) (Cretaceous Dakota Sandstone, Colorado, USA); scale 1 m. (**e**) Thin-section through modern microbial mat displaying oriented grains (microbial mat from Mellum Island, North Sea); scale 0.2 mm. (**f**) Fossil microtexture (3.2 Ga Moodies

3 Modern MISS and Their Formation

Classical study site Mellum Island, North Sea. The first examples of MISS have been described in the modern tidal flats of Mellum, an island situated at the North Sea coast of Germany (Gerdes and Krumbein 1987; Gerdes et al. 1991, 1993, 2000; Noffke et al. 1996). Here, microbial mats overgrow quartz-rich sands of fine sand sizes. Microbial mats are dense and coherent organic layers that from the distance look like bluegreen carpets (introduction e.g., Golubic and Knoll 1999; Stolz 2000; Paerl et al. 2000; Stal 2000). In close-up, this carpet is composed of cyanobacteria and their mucilages they secrete (extracellular polymeric substances – EPS), Decho 1990. But microbial mats are not everywhere of same type. Along a transect crossing the tidal flat from the low to the high water line, a lateral succession of different types of microbial mats establishes – each mat type best adapted to its specific site of colonization. Each colonization site is defined by a characteristic pattern of tidal currents that inundates or exposes the tidal flat surface for different time periods (Noffke and Krumbein 1999).

Close to the low water line, the constant reworking of the substrate by hydraulic movement prohibits the formation of 'real' microbial mats. Here, the coccoid cyanobacterium *Merismopedia punctata* forms sotermed 'biofilms' thin organic coatings around the single grains of the sandy surface. Those biofilm-coated grains are swirled around by the agitated water and rarely come to rest on the seafloor. Whenever the waves and currents settle for a moment, all particles suspended in the water sink down – however the biofilm-coated ones at last. The reason for this slowest downward movement is that the ratio between diameter and specific weight of a biofilm-coated grain favours long residence in the water column compared to that of the relatively heavier, sterile grains.

If we ascend the tidal flat along our transect, we reach the intertidal zone, which is affected by the daily tidal currents. Here another cyanobacterial species becomes abundant: *Oscilatoria limosa.* The trichomes (single filaments) of *Oscilatoria limosa* are highly mobile that is they are able to move quickly through the sediments. Because the intertidal zone is reworked with every flood current, this high motility is of great advantage for this cyanobacterium. Indeed, this cyanobacterium can form a dense microbial mat within the few hours an ebb tide lasts (Villbrandt 1992; Noffke 1998)!

In the morphologically highest portions of the tidal flat, the lower supratidal zone, the dominant microbial mat builder is *Microcoleus chthonoplastes*. This tidal zone is typically inundated by sea water only during the spring high tides or during

---

Fig. 2 (continued) Group, South Africa); scale 0.2 mm. (**g**) Oscillation cracks define a polygonal pattern in the surface of a microbial mat. The center of each polygon shows an erupted gas dome, one example marked with an arrow (southern Tunisia); scale about 30 cm; (**h**) Same polygonal structure. *Arrow* indicates a hole, where gas dome erupted. (2.9 Ga Pongola Supergroup, South Africa); scale 25 cm. (**i**) Multidirected ripple marks. Two generations of ripples (I and II) are displayed on this tidal surface (Mellum Island); scale 50 cm. (**j**) Multidirected ripple marks with four generations of ripple marks (I–IV) (2.9 Ga Pongola Supergroup, South Africa); scale 30 cm strong landward winds. *Microcoleus chthonoplastes* is well adapted to the long lasting periods of subaerial exposure of the tidal surface. Its ubiquitous EPS protect the filaments of this cyanobacterium against desiccation or osmotic pressure by quickly changing salinities (e.g., Decho 1990).

On Mellum Island, the classical MISS include 'erosional remnants and pockets', 'multidirected ripple marks', 'gas domes', or 'mat chips' atop the sediments. In vertical sections we see 'sponge pore fabrics', 'microsequences', 'roll-ups', 'oriented grains', and other structures and textures (see examples for MISS in Figure 2).

Classical study site sabkha Bahouane, southern Tunisia. The second classical study site for the MISS are the sabkha-like tidal flats along the coast of southern Tunisia (Noffke et al. 2001a). A sabkha-like tidal flat is inundated only every 2 weeks up to every 3 months – depending on the seasonal precipitation in the hinterland and on the rise and fall of the intrasedimentary groundwater. It is the groundwater that nourishes the predominantly evaporitic mineral formation at the depositional surface of the salt-encrusted sabkha. In this environment, the deepest portions of the tidal flats, sometimes large depressions of several hundreds of meters in diameter, are covered for long weeks by a few cm deep water. At those locations, thick and bluegreen microbial mats can grow, mostly constructed by *Microcoleus chthonoplastes* or *Lyngbya aestuarii* (Gerdes et al. 2000). Thick mats of up to 4 cm thicknesses lead a prospering life in warm water not disturbed even by slightest winds. However, if we leave this mat paradise and walk towards higher locations, we quickly learn that this mat paradise is locally restricted. The higher the morphology, even if the height differences is just centimeters, the longer the microbial mats are exposed to harsh sun light and winds that dry out the deposits so quickly. Here, other cyanobacterial species establish. For example, *Synechnococcus* builds leathery microbial mats and its orange-red pigments that serve so well against the sun light give rise to a rose-coloured, seemingly extraterrestrial monotonous landscape (Gerdes et al. 2000). The classical MISS at this study location include 'petees', 'gas domes', 'oscillation cracks', or 'tufts' (see examples for MISS in Fig. 2).

The two classical study sites Mellum Island and Sabkha Bohouane demonstrate that the microbial mats establish in dependence to the hydraulic conditions that prevail in a given area. As the distribution of microbial mat types changes in characteristic way from low to high morphological situations, the term 'biofilm-catena' was introduced for such lateral successions (Noffke and Krumbein 1999). It is of importance, that not only the microbial mats, but also the MISS they form are arranged along the transect from low to high sites. Detailed studies that describe those 'biofilm-catenae' and the MISS they cause are Noffke and Krumbein 1999, as well as Noffke et al. 2001a, 2008.

4 Classification of MISS

MISS rise from the response of benthic microbiota to erosive forces, and to deposition of sediment (Noffke et al. 2003a). The microbial mats grow especially well during quiet hydraulic conditions (Gerdes et al. 1991), probably because the nutrients are readily available and the microorganisms not endangered to be ripped off their substrate by currents to be buried by a – for a photoautotrophic community – lethal drape of mud. The response to erosion is long known as bio-stabilization (term introduced by Paterson 1994; see also Paterson et al. 1994; Yallop et al. 1994), the response to fall out of sediment as baffling and trapping (classical paper by Black 1933, in detail by e.g., Reid et al. 2000). Binding is the incorporation of mineral particles into the microbial mat fabrics over time. Because biomass production (replication of cells and EPS secretion) does not play a role, binding in my opinion does not equal growth (Noffke et al. 2008; Noffke 2009). The Fig. 3 is a classification scheme that places different MISS in relation to the main microbial activity that causes this structure. Note that some MISS rise from the interference of all activities.

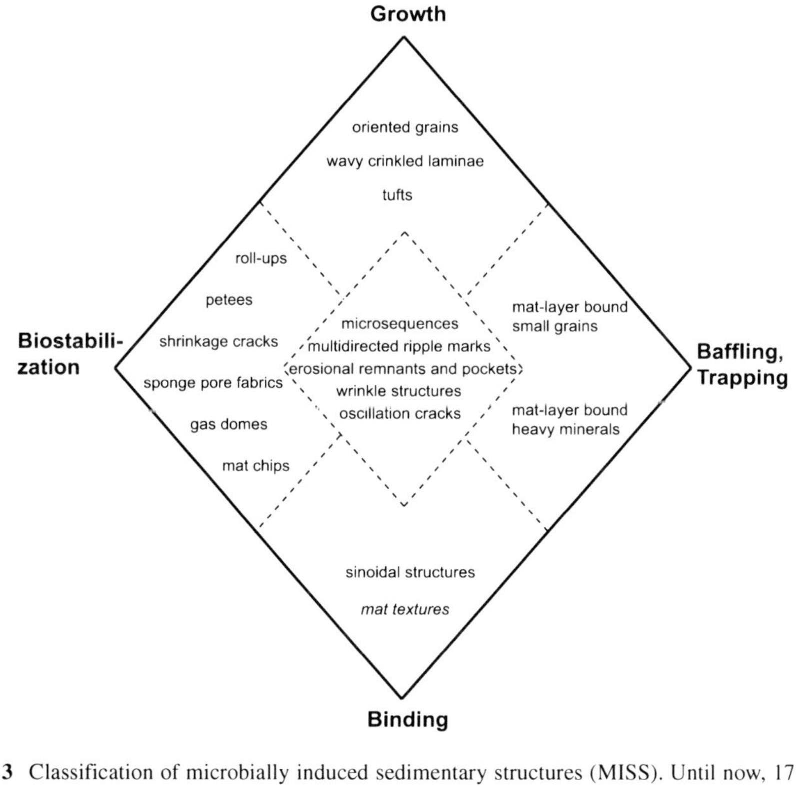

Fig. 3 Classification of microbially induced sedimentary structures (MISS). Until now, 17 main types of MISS are distinguished. With respect to their modes of formations, the structures are separated into five groups. The corners of the diagram include MISS that rise from mainly one bacterial activity such as biostabilization, or binding. In contrast, the central group of structures originates from an interference of all bacterial activities (After Noffke 2009)

5 How to Interprete Archean MISS by Using Modern Analogues of the Structures

The knowledge on the formation of the structures and on their distribution in a given area allows direct comparison of modern structures with fossil MISS (Noffke et al. 2006b). If we understand, where modern microbial mats develop and how they interact with which physical sedimentary regimen, we have the tool in hand to analyze fossil MISS in equivalent paleoenvironments, and to conclude on the potential builders of the ancient microbial mats. This application shall be outlined with some examples. Figure 2 shows microbial mat chips (Fig. 2a and b). Those chips are fragments of microbial mats that have been ripped off by water currents from their parent sites, transported and redeposited elsewhere. The chips are structures that rise from biostabilization (Fig. 3). The statistical comparison of the modern fragments with the fossil ones shows that the shape and sizes are very similar. In thin-sections, both modern and ancient mat chips are of same thickness. Because we know that the modern mat chips in Fig. 2a derive from the lower supratidal zone of Mellum Island, where *Microcoleus chthonoplastes* resides, we may conclude that the fossil mat chips might have been formed by this cyanobacterium as well. Of course a detailed study of the paleoenvironment must be conducted, and our geological survey, based on cm-scale stratigraphic profiling, revealed that indeed the ancient habitat has been a lower supratidal zone (Noffke et al. 2008).

Another example are ‘erosional remnants and pockets’ (Fig. 2c). This sedimentary structures covers sometimes square kilometers of tidal flat surfaces. It is composed of elevated, mat-covered surface portions and deeper lying, ripple-marked surface areas that are not overgrown by mats. This tidal flat morphology evolves as soon as a sediment-stabilizing microbial mat is destroyed locally so that at this spot the bare sand is exposed to the water. Starting from this location, currents now remove the sand grains, and therefore undermine the microbial mats surrounding this breakage. Over time, the small spots widen, because the microbial mat fringe is ripped off piece by piece (and mat chips form). Over time, the ‘erosional pockets’ develop with their characteristic ripple marks. The pockets easily can reach several meters extension until they connect with the neighbouring erosional depressions (Gerdes et al. 1993; Noffke 1999).

A very characteristic microscopic texture that is visible in thin-sections perpendicular through a microbial mat are ‘oriented grains’ (Noffke et al. 1997, 2006b). Figure 2e shows a modern example, and Fig. 2f a fossil equivalent one. Oriented grains are mineral particles that float independently from each other in the organic matrix of a microbial mat. Their long axes are oriented parallel to the bedding planes of the sediment layers, because the soft matrix of the organic embedding allows rotation of the grains according to the gravity force. In contrast – of course – the grains that compose the sandy substrate underneath the microbial mat cannot rotate, because the particles are fixed in their position by grain-to-grain contact. Such a microtexture rises from growth of the microbial mat. In laboratory experiment, each individual quartz grain initially is coated by a thin biofilm. If this biofilm coat grows, it grows into all directions in equal rate. After some while, the biofilm pushes its central grain upward and away from the neighbouring grains. Finally, all biofilms of all grains fuse to form a laterally continuous microbial mat layer. The original biofilm-coated grains, however, now float like individual pearls in this mat layer and witness the growth history of this microbial mat. Oriented grains are characteristic for thick, epibenthic microbial mats that contain lots of EPS. In the fossil record, we found that oriented grains indicate epibenthic microbial mats in the lower supratidal zone – again hinting on the ancient presence of *Microcoleus chthonoplastes* – like mat consortia in this ancient tidal area (Noffke et al. 2008).

Gas domes are elevations that temporarily spot microbial mat surfaces (Noffke et al. 1996, 2008). The gas domes are upheavals caused by increasing pressure of gas accumulating underneath the microbial mat (Fig. 2g and h). In the deeper sediment far below the microbial mat gases are formed by decay of organic matter, or simply by the rising groundwater pushing gas upward. Because the EPS of the microbial mats do not allow gas diffusion, the gas cannot escape the sediment. However, it may be that after some time the gas pressure becomes so high that the microbial mat is ripped off its substrate underneath and a gas dome rises. In southern Tunisia such gas domes form especially during longer periods of desiccation, when the microbial mats are exposed to dry and hot air for weeks, and a pattern of polygonal cracks disintegrates the organic carpet (Noffke et al. 2001a), see Fig. 2g. Each center of each polygon defined by the desiccation cracks rises and eventually ruptures to release intrasedimentary gases. Such a sedimentary structure can become fossil and we found equivalent structures in the same lower supratidal setting in the ancient tidal flat of the 2.9 Ga Pongola Supergroup, South Africa (Fig. 2h).

One last and well known example are 'multidirected ripple marks' (Fig. 2i and j). This surface structure is a chaotic-like pattern of ripple marks of various directions that cover the upper intertidal to lower supratidal zones. This ripple mark pattern rises from the interaction of sediment-stabilizing microbial mats with a set of subsequent storms of which each gives rise to a new a generation of ripple marks. During the time periods in between the individual storm events, each newly formed ripple mark generation is consequently overgrown and biostabilized by microbial mats, and thus cannot be reworked by following storm events. In consequence of their formation, multidirected ripple marks can be found especially well developed in the fall of a temperate climate (Noffke 1998). Fossil multidirected ripple marks are wonderfully exposed in the ancient tidal flat of the Pongola Supergroup, South Africa, witnessing a similar climate affecting the equivalent tidal zones as we know from the modern sites (Noffke et al. 2008).

6 Conclusive Remarks and the Question Which Microbiota Formed Microbial Mats in the Archean Era

MISS are sedimentary structures that rise from the interaction of benthic microbota with physical sediment dynamics in siliciclastic deposits. The physical laws of moving water affecting sandy sediments did never change in course of Earth history, that is primary sedimentary structures also never changed (Noffke et al. 2006a, b). Because of the so characteristic modification of the primary sedimentary structures by microbenthos, the MISS are highly indicative of microbial presence in such deposits. If found in the fossil record, the geometries of MISS as well as their paleoenvironmental distribution assists us to reconstruct prokaryotic evolution (Noffke 2009). For example, we think that the Pongola Supergroup includes Earth's oldest known cyanobacterial population, which resembles surprisingly precise the modern cyanobacteria we find in the modern equivalent settings. If so, the evolutionary rise of cyanobacteria must be placed much earlier than assumed until now, and must have taken place well before 3 billion years ago (Margulis 2009).

Acknowledgements The studies on modern and ancient microbial mats have been funded by the NASA Mars Exploration and Exobiology Programs, the NSF Geology-Paleontology Program, and an Old Dominion University Summer Research Grant.

References

Bland BH (1984) Arumberia Glaessner & Walter, a review of its potential for correlation in the region of the Precambrian-Camrian boundary. Geol Mag 121:625–633

Black M (1933) The algal sediments of Andros Islands, Bahamas. Philos Trans R Soc Lond Series B 222:165–192

Cameron B, Cameron D, Jones J (1985) Modern algal mats in intertidal and supratidal quartz sands, northeastern Massachusetts, USA. In: Curran A (ed) Biogenic structures: their use in interpreting depositional environments. SEPM, Tulsa, 134 p

Decho AW (1990) Microbial exopolymer secretions in ocean environments: their role(s) in food webs and marine processes. Oceanogr Mar Biol Annu Rev 28:73–154

Eriksson PG, Simpson EL, Eriksson KA, Bumby AJ, Steyn GL, Sarkar S (2000) Muddy roll-up structures in siliciclastic interdune beds of the ca. 1.8 Ga old Waterberg Group, South Africa. Palaios 15:177–183

Gehling JG (1999) Microbial mats in terminal Proterozoic siliciclastics: Ediacaran death masks. Palaios 14:40–57

Gerdes G, Krumbein W (1987) Biolaminated deposits. Springer-Verlag, Berlin, 193 p

Gehling JG (2000) Environmental interpretation and a sequence stratigraphic framework for the terminal Proterozoic Ediacara member within the Rawnsley Quarzite, South Australia. Prec Res 100:65–95

Gerdes G, Krumbein WE, Reineck HE (1991) Biolaminations – ecological versus depositional dynamics. In: Einsele G, Ricken W, Seilacher A (eds) Cycles and events in stratigraphy. Springer, Berlin

Gerdes G, Claes M, Dunajtschik-Piewak K, Riege H, Krumbein WE, Reineck HE (1993) Contribution of microbial mats to sedimentary surface structures. Facies 29:61–74

Gerdes G, Noffke N, Klenke Th, Krumbein WE (2000) Microbial signatures in peritidal sediments – a catalogue. Sedimentology 47:279–308

Golubic SG, Knoll AH (1999) Prokaryotes. In: Lipps J (ed) Fossil prokaryotes and protists. Blackwell, Boston, MA

Hagadorn JW, Bottjer DJ (1999) Unexplored microbial worlds. Palaios 14:1–93

Lowe DR, Tice MM (2007) Tectonic controls on atmospheric, climatic, and biological evolution 3.5–2.4 Ga. Prec Res 158:177–197

Margulis L (2009) Pavements along memory lane and castles of sand long before "Man". Earth Sci Rev 96:1–5

Noffke N (1998) Multidirected ripple marks rising from biological and sedimentological processes in modern lower supratidal deposits (Mellum Island, southern North Sea). Geology 26:879–882

Noffke N (1999) Erosional remnants and pockets evolving from biotic-physical interactions in a Recent lower supratidal environment. Sediment Geol 123:175–181

Noffke N (2000) Extensive microbial mats and their influences on the erosional and depositional dynamics of a siliciclastic cold water environment (lower Arenigian, Montagne Noire, France). Sediment Geol 136:207–215

Noffke N (2009) The criteria for biogeneicity of microbially induced sedimentary structures (MISS) in Archean and younger, sandy deposits. Earth Sci Rev 96:173–180

Noffke N, Krumbein WE (1999) A quantitative approach to sedimentary surface structures contoured by the interplay of microbial colonization and physical dynamics. Sedimentology 46:417–426

Noffke N, Gerdes G, Klenke T, Krumbein WE (1996) Microbially induced sedimentary structures – examples from modern sediments of siliciclastic tidal flats. Zbl Geol Palaeontol 1:307–316

Noffke N, Gerdes G, Klenke T, Krumbein WE (1997) A microscopic sedimentary succession indicating the presence of microbial mats in siliciclastic tidal flats. Sediment Geol 110:1–6

Noffke N, Gerdes G, Klenke T, Krumbein WE (2001a) Microbially induced sedimentary structures indicating climatological, hydrological and depositional conditions within Recent and Pleistocene coastal facies zones (southern Tunisia). Facies 44:23–30

Noffke N, Gerdes G, Klenke T, Krumbein WE (2001b) Microbially induced sedimentary structures–a new category within the classification of primary sedimentary structures. J Sediment Res 71:649–656

Noffke N, Knoll AH, Grotzinger J (2002) Sedimentary controls on the formation and preservation of microbial mats in siliciclastic deposits: a case study from the upper Neoproterozoic Nama Group, Namibia. Palaios 17:1–14

Noffke N, Gerdes G, Klenke T (2003a) Benthic cyanobacteria and their influence on the sedimentary dynamics of peritidal depositional systems (siliciclastic, evaporitic salty and evaporitic carbonatic). Earth Sci Rev 12:1–14

Noffke N, Hazen RM, Nhleko N (2003b) Earth's earliest microbial mats in a siliciclastic marine environment (Mozaan Group, 2.9 Ga, South Africa). Geology 31:673–676

Noffke N, Beukes NJ, Hazen RM (2006a) Microbially induced sedimentary structures in the 2.9 Ga old Brixton formation, Witwatersrand Supergroup, South Africa. Prec Res 146:35–44

Noffke N, Hazen RM, Eriksson K, Simpson E (2006b) A new window into early life: Microbial mats in a siliciclastic early Archean tidal flat (3.2 Ga Moodies Group, South Africa). Geology 34:253–256

Noffke N, Beukes N, Bower D, Hazen RM, Swift DJP (2008) An actualistic perspective into Archean worlds – (cyano-)bacterially induced sedimentary structures in the siliciclastic Nhlazatse Section, 2.9 Pongola Supergroup, South Africa. Geobiology 6:5–20

Paerl H, Pinckney JL, Steppe TF (2000) Cyanobacterial-bacterial mat consortia: examining the functional unit of microbial growth in extreme environments. Environ Microbiol 2:11–26

Paterson D (1994) Siliciclastic intertidal microbial sediments. In: Stal LJ, Caumette P (eds) Microbial mats. Springer, Berlin

Paterson DM, Yallop M, George C (1994) Stabilization. In: Krumbein WE, Paterson DM, Stal LJ (eds) Biostabilization of sediments. BIS, Oldenburg

Prave AR (2002) Life on land in the Proterozoic: evidence from the Torrodonian rocks of northwest Scotland. Geology 30:811–814

Reid RP, Visscher PT, Decho AW, Stolz JF, Bebout BM, Dupraz C, MacIntyre IG, Pearl HW, Pinckney JL, Prufert-Bebout L, Steppe TF, DesMarais DJ (2000) The role of microbes in accretion, lamination and early lithification of modern marine stromatolites. Nature 406:989–991

Runnegar BN, Fedonkin MA (1992) Proterozoic metazoan body fossils. In: Schopf JW, Klein C (eds) The Proterozoic biosphere, a multidisciplinary study. Cambridge University Press, New York

Sarkar S, Banerjee S, Eriksson PG, Catuneau O (2005) Microbial mat control on siliciclastic Precambrian sequence stratigraphic architecture: examples from India. Sediment Geol 176:191–205

Schieber J (1989) Facies and origin of shales from the mid-Proterozoic Newland Formation, Belt Basin, Montana, USA. Sedimentol 36:203–219

Schieber J, Bose, PK, Eriksson, P, Banerjee, S, Sarkar, S, Alterm, ann, W, and Catuneau, O (2007) Atlas of microbial mat features preserved within the siliciclastic rock record. Elsevier Science, Oxford, 311 p

Schopf JW, Kudryavtsev AB, Czaja AD, Tripathi AB (2007) Evidence of Archean life: stromatolites and microfossils. Prec Res 158:141–155

Stal LJ (2000) Cyanobacterial mats and stromatolites. In: Whitton BA, Potts M (eds) The ecology of cyanobacteria. Kluwer, Dordrecht, The Netherlands

Stolz J (2000) Structure of microbial mats and biofilms. In: Riding R, Awramik S (ed) Microbial Sediments. Springer-Verlag, Berlin, 256 p

Villbrandt M (1992) Interactions of nitrogen fixation and photosynthesis in marine cyanobacterial mats (Mellum, southern North Sea). Ph.D. thesis, University of Oldenburg, Oldenburg

Walter MR, Schopf JW, Rujii C (2007) Earliest evidence of life on Earth. Prec Res 158:3–4

Yallop ML, De Winder B, Paterson DM, Stal LJ (1994) Comparative study on primary production and biogenic stabilization of cohesive and non-cohesive marine sediments inhabited by microphytobenthos. Estuar Coast Shelf Sci 39:565–582

Early Life Record from Nitrogen Isotopes

Daniele L. Pinti and Ko Hashizume

Abstract Biological activity fractionates the nitrogen isotopes in a peculiar way, making them a reliable biosignature and an accurate paleoenvironmental proxy. Nitrogen has been ignored for long time, being extremely fragile compared to the more stable graphitic forms of C; however, N has an advantage over other isotopic systems such as those of C and S. The dominant source of N at the surface of the Earth, that is, the atmospheric triple-bonded N_2, is so stable that only a very limited number of metabolic processes can bridge the abiotic and biotic world. Therefore we can draw relatively simple flux models for N. In this contribution, we review the N isotopic record in the last 4 billions years. Large isotopic shifts recorded by nitrogen are related to specific metabolic changes as a direct response to major environmental stress such as the rise of oxygen in the atmosphere and the evolution of nitrifiers and denitrifiers in the ocean. These isotopic changes are not unique but well correlated with those of C and Fe, indicating that nitrogen can be successfully used for modeling the interplay of changing microbial metabolisms over Earth's history and relate them to precise environmental changes.

Keywords Nitrogen isotopes • Isotopic biomarkers • Biological fixation • Nitrification • Denitrification • Ammonium • Cherts • Great Oxygenation Event

D.L. Pinti (✉)
GEOTOP and Département des Sciences de la Terre et de l'Atmosphère, Université du Québec à Montréal, CP 8888, Centre-Ville, Montréal, Quebec H3C 3P8, Canada
e-mail: pinti.daniele@uqam.ca

K. Hashizume
Department of Earth & Space Sciences, Graduate School of Science, Osaka University, Toyonaka, Osaka 560-0043, Japan
e-mail: kohash@ess.sci.osaka-u.ac.jp

S.D. Golding and M. Glikson (eds.), *Earliest Life on Earth: Habitats, Environments and Methods of Detection*, DOI 10.1007/978-90-481-8794-2_8,

1 Introduction

The most important elements in organic chemistry – C, N, H, O and S, only except P – have multiple isotopes. The isotopic ratio (R) of these elements (normally expressed by the "delta" notation δ (‰) = $[(R_{sample}/R_{std} - 1) \times 1{,}000])$ is a potential signature of early life and a proxy of its environment. This is the case because the kinetic isotope fractionation is common in biology and can be dominant over equilibrium fractionation.

Although geological processes such as diagenesis and metamorphism may fractionate these isotopes, metabolic processes tend to produce different, and sometimes diagnostic, enrichment or depletion in certain isotopes. A classical example is the *fixation* of inorganic carbon (CO_2, HCO_3^-) by autotrophic organisms mediated by the ribulose 1,5 bisphosphate carboxylase/oxygenase enzyme, otherwise known as RubisCO. This metabolic process always produces a specific isotope fractionation favoring the lighter isotope of C. Thus, the produced C_{org} is ^{12}C-enriched with characteristic $\delta^{13}C$ values (where the ratio R is $^{13}C/^{12}C$) of −17‰ to −30‰ compared to the Dissolved Inorganic Carbon (DIC) that has a $\delta^{13}C$ value from −10‰ to 0‰. The kinetic fractionation deriving from the multi-step processes of atmospheric N_2 *fixation*, *ammonification* (products: NH_3–NH_4^+), *nitrification* (NO_3^-), *denitrification* (N_2O, N_2) produces also a distinctive isotopic signature ($\delta^{15}N$ = +5‰; where R = $^{15}N/^{14}N$) of the oceanic nitrate assimilated by primary producers such as phytoplankton (DON) (e.g., Sigman et al. 2009). In this case, the heavier isotope of N (^{15}N) is enriched in the organic product compared to the inorganic, atmospheric source. When the organic matter is buried in the sediment, it preserves the peculiar isotopic fractionation of the nitrate (Altabet and Francois 1994).

Isotopic ratios of C ($\delta^{13}C$) and N (δ ^{15}N) are normally measured in modern marine sediments. They are extremely helpful to identify the metabolic pathways regulating the fate of those elements in the ocean (Peters et al. 1978) and/or the source of organic matter that supply sediments (terrestrial against marine) in transitional environments (e.g., Thornton and McManus 1994). Changes in ocean circulation and nutrient supply, which occur in response to changes in environmental conditions, affect the relative importance and spatial extent of the major metabolic pathways of these elements, making their isotopic signatures also a sensitive paleoenvironmental proxy (Scholten 1991; Macko and Engels 1993; Altabet and Francois 1994). This is the principle of isotopic chemostratigraphy, which has been successfully applied from most recent times back to the Proterozoic.

In Precambrian sedimentary rocks, particularly those spanning in age from Paleoarchean to Mesoarchean (3.8–2.8 Ga), "isotopic fossils" such as N and C isotopes are one of the few available records of the biomass present at that time and its metabolic activity. Microfossils are indeed rare (Schopf 2006) and their interpretation is often highly debated (Schopf et al. 2002; Brasier et al. 2002; López-Gárcia et al. 2006; Pinti et al. 2009b). On the other hand, it is not well understood how metamorphic-related devolatilization processes could alter the original biological isotopic signature of C and N in the rock (Schidlowski 2001; Dauphas and Marty 2004; Pinti et al. 2001, 2009a).

In this chapter, we review the nitrogen isotopic record in Precambrian rocks. We will firstly show how N is bound in different chemical sediments and minerals. We then discuss the N isotopic signatures observed in several suites of rocks spanning from 3.8 to 0.5 Ga and how to interpret them in terms of metabolic response to major environmental changes.

2 Nitrogen in Rocks

Since the pioneering work of Stevenson (1962) and Barker (1964), nitrogen is believed to occur in rocks mainly as ammonium (NH_4^+) in two forms. The first chemical form is *exchangeable ammonium*, adsorbed on the charged surfaces of clay minerals and representing organic nitrogen that has been decomposed in sediments most recently (Williams et al. 1995). The second form is *fixed ammonium*, which is strongly bound in the lattice structure of K-bearing minerals – feldspars, plagioclase, biotite and muscovite. Indeed, ammonium has an ionic radius of 0.143 nm, slightly larger than that of K^+ (0.138 nm) and thus it can replace potassium in most silicate minerals (Honma and Itihara 1981). The best example of ammonium replacement is illustrated by buddingtonite ($NH_4AlSi_3O_8 \times 1/2H_2O$) a hydrothermal mineral containing up to 4 wt.% of ammonium (Orberger et al. 2005). Nitrogen resides mainly in the ammonium form in magmatic (Honma and Itihara 1981; Boyd et al. 1993; Hall 1999; Busigny et al. 2005), metamorphic (Duit et al. 1986; Bebout and Fogel 1992; Sadofsky and Bebout 2000), and sedimentary rocks (Williams et al. 1995; Holloway and Dahlgren 1999; Boyd 2001). Ammonium concentrations vary from several thousands ppm for unaltered mica (Hall 1999; Pitcairn et al. 2005) down to a few ppm for metamorphosed sedimentary rocks (Pinti et al. 2001).

Another important reservoir of nitrogen is the so-called organic-N (N_{org}) corresponding to N bound in biomolecules and preserved into carbonaceous matter when this latter is progressively transformed into polymeric kerogen (Ader et al. 2006; Boudou et al. 2008). Nitrogen could exist in other chemical forms (supposed but never proven NH_3 in cherts; Wlotzka 1969) and in other mineral host phases. Recent work of Gallien et al. (2004), Orberger et al. (2005), Pinti et al. (2007) and Hashizume et al. (2008) suggests that nitrogen resides in quartz, iron oxides, Fe–Ni sulfides and phosphates. Except for N_{org} chemisorbed on the intra-layer surfaces of iron oxides (Pinti et al. 2007), the other chemical forms of N in those minerals have been not yet clearly identified.

Nitrogen is also found in fluid inclusions. Research carried out in Precambrian rocks (Sano and Pillinger 1990; Pinti et al. 2001; Pinti 2002) suggested that nitrogen could be either in the form of ammonium (NH_4^+) or ammonia (NH_3). However, recent work of Nishizawa et al. (2007) in fluid inclusions from Paleoarchean cherts indicates that the main nitrogen form could be N_2 and only subordinately NH_4^+. Further research in this domain is needed to correctly identify the chemical speciation of N in fluids, its origin (inorganic versus organic) and its role in the mass and isotopic balance of N extracted from rocks.

In Precambrian rocks, nitrogen as N_{org} and NH_4^+ has been measured in kerogen and graphite (Beaumont and Robert 1999; van Zuilen et al. 2005), chert (Sano and Pillinger 1990; Pinti et al. 2001, 2009a), shales (Yamaguchi 2002; Papineau et al. 2009), iron oxides (Pinti et al. 2001, 2007; Hashizume et al. 2008) and mica (Pinti et al. 2001, 2009a; Jia and Kerrich 2004a, b; Papineau et al. 2005; Rouchon et al. 2005).

3 Nitrogen Isotopes in Modern Marine Sediments: The Cycle

Nitrogen is mainly preserved in marine sediments as NH_4^+ and this is a direct consequence of its biogeochemical cycle in the ocean. For a better understanding on how N isotopes can trace back ancient metabolic pathways, we explain briefly the N biogeochemical cycle in the modern ocean. The picture is necessarily simplified. For a complete treatment on the subject, please refer to the reviews of Galloway (2003), Brandes et al. (2007) and Sigman et al. (2009). In the following description of the N cycle, the kinetic isotopic effect "ε" of a given reaction is introduced. This corresponds to the ratio of rates with which the two N isotopes are converted from reactant to product:

$$\varepsilon\,(‰) = \left({}^{14}k/{}^{15}k - 1\right) x\ 1000 \tag{1}$$

For $|\varepsilon| << 1{,}000‰$ this expression can be simplified as:

$$\varepsilon\,(‰) = \delta^{15}N_{reactant} - \delta^{15}N_{product} \tag{2}$$

Nitrogen is mainly cycled in the ocean by microorganisms through three dominant metabolic processes, namely *fixation*, *nitrification* and *denitrification*.

Atmospheric triple-bonded N_2 enters the marine nitrogen cycle via the process of biological *fixation* (Fig. 1). N_2-fixing organisms, mainly oxygenic and anoxygenic photoautotrophs, incorporate molecular nitrogen (N_{org}) with little or no isotopic fractionation ($\varepsilon \leq 3‰$; Fogel and Cifuentes 1993). The transformation of N_{org} to nitrate (NO_3^-) is a multistep process. Organic matter is degraded and mineralized through enzymatic breakdown into short-chain organic compounds (aminoacids, proteins) and then released as NH_4^+, following the process of *ammonification*. Again, the kinetic isotopic effect related to this process is negligible (Sigman et al. 2009).

Under oxic conditions, NH_4^+ is rapidly oxidized to nitrite (NO_2^-) and subsequently to nitrate (NO_3^-) in a two-step process called *nitrification*. The kinetic isotopic effect related to this process is large, around +16‰ (Casciotti et al. 2003). In suboxic environments, where dissolved O_2 ranges between 2 and 4 μmol/L (Murray et al. 1995), nitrate is stepwise reduced to N_2O and/or N_2 by heterotrophic bacteria that can use NO_3^- as an alternative electron acceptor when O_2 is not readily available. This process, called *denitrification*, is the most important sink of fixed nitrogen in the ocean, through the production of gaseous N_2 escaping back to the

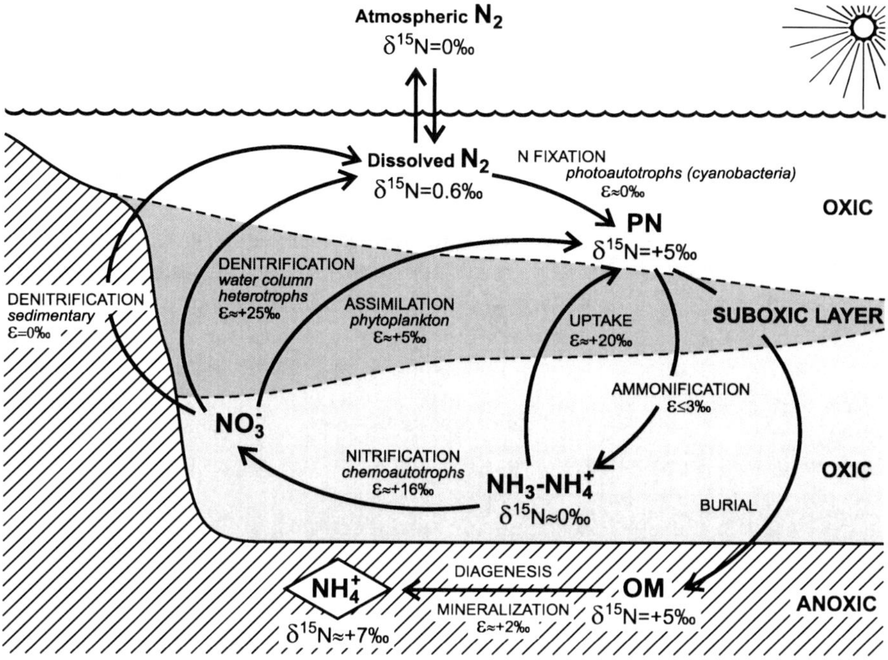

Fig. 1 Schematic illustration showing a simplified nitrogen cycle in the modern ocean. Atmospheric N_2 ($\delta^{15}N$ = 0‰ by definition) is the primary input in the ocean. Atmospheric N_2 is fixed into biomolecules, mediated by the nitrogenase enzyme. Degradation of organic matter brings newly formed N_{org} to be reduced to NH_3 (ammonia) with little or no fractionation. At the pH = 8 of the ocean, NH_3 is quickly protonated to NH_4^+ (ammonium). Heterotrophic microorganisms use free O_2 in the ocean to oxidize ammonium to NO_3^- (nitrate) with large fractionation (ε = +16‰). In suboxic areas (O_2 from 2 to 4 μmol/L) often corresponding to the oceanic oxygen minimum layer, nitrates are reduced back to N_2O and then to N_2 (*denitrification*) with large fractionation (ε = +25‰). This is the major output of N back to the atmosphere. The residual nitrates are the largest nutrient pool assimilated by phytoplankton to feed the sinking particulate N_{org}; kinetic isotopic fractionation and mass balance produce a characteristic $\delta^{15}N$ of +5‰. This isotopic signature is little fractionated during organic matter burial and diagenesis in anoxic sediments. Produced ammonium fixed in minerals or N_{org} trapped in carbonaceous matter (kerogen) has an average $\delta^{15}N$ of +7‰

atmosphere. There are two *denitrification* processes that take place in oceanic environments: water-column and sedimentary *denitrification* (Sigman et al. 2009). The kinetic isotopic fractionation related to sedimentary *denitrification* is mostly nil (~0‰) while water-column *denitrification* produces a large isotopic shift of +25–30‰, due to the preferential use of $^{14}NO_3^-$ during this process (Cline and Kaplan 1975). The consequence is that the gaseous N_2 and N_2O products are thus depleted in ^{15}N, whereas the residual nitrate pool in the ocean is ^{15}N-enriched, with a global oceanic value of $\delta^{15}N$ of +5‰ (Sigman et al. 2009). Recently, another mechanism of N loss has been identified: the *anaerobic ammonium oxidation*, or "*anammox*" (e.g., Strous et al. 2002; Kuypers et al. 2003), in which nitrite (NO_2^-)

is used to reduce ammonium to N_2 in a totally anaerobic environment. This process has yet unknown isotopic effects on isotope distribution in the oceans (Sigman et al. 2009).

It is worth to note that when the N oceanic budget is at steady state, the $\delta^{15}N$ of the N removed by *denitrification* will equal the $\delta^{15}N$ of the fixed N added. If *denitrification* with an isotopic effect of 25‰ would occur homogeneously, the residual nitrate should have a $\delta^{15}N$ value of +25‰ and not +5‰ as measured. This difference can be explained assuming a relative importance of sedimentary *denitrification*, which has no or little isotopic effect. Indeed, from isotopic and mass balance calculation it turns out that water-column *denitrification* accounts for only 30% of the N that flows back to the atmosphere (Sigman et al. 2003).

Primary producers, such as phytoplankton, assimilate the residual oceanic nitrate (NO_3^-) with an isotopic fractionation of +5‰ through the process of *assimilation* (Fig. 1). This is an important step in the N cycle because this nitrate is assimilated with no or little isotopic fractionation with the consequence that the $\delta^{15}N$ of the sinking flux (and of sedimentary organic matter prior to diagenesis) will be close to +5‰ (Fig. 1). Indeed, modern marine sediments exhibit an average $\delta^{15}N$ value of +7‰ (Peters et al. 1978; Shen et al. 2006) close to the value for seawater nitrate.

The fate of N_{org} in sediments depends primarily on the redox conditions (Lehmann et al. 2004). In aerobic surface sediments, NH_4^+ produced from degradation of N_{org} is rapidly converted to nitrate and then back to N_2. In anaerobic sediments, NH_4^+ is partially adsorbed on, or fixed to, clay minerals (Williams et al. 1995) or replaces K^+ in the lattice structure of K-bearing silicate minerals (Honma and Itihara 1981). The isotopic effects of diagenesis during these reactions are still poorly known. The fact that bulk marine sediments have a $\delta^{15}N$ value close to that of seawater nitrate indicates that diagenesis does not modify significantly the original isotopic ratio, which is at the base of the use of N as a paleoenvironmental proxy and biosignature of life. Experimental studies have produced contrasting results. Lehmann et al. (2002) showed that a slight negative isotope shift of 3‰ occurs during microbial degradation of organic matter during suboxic early sedimentary diagenesis. However, N isotopic shifts towards more positive values have also been reported (Freudenthal et al. 2001). Most recently, Prokopenko et al. (2005, 2006) reported isotopic shifts of less than 1‰ in deep-sea sediments, indicating again that diagenesis could play a minor role in altering the N isotopic signal.

4 The Preservation of the Isotopic Signature of N in the Rock Record: The Role of Metamorphism and Alteration

The preservation of the pristine biological-fractionated isotopic ratio for nitrogen is a matter of debate. If diagenesis could be responsible for only a few permil variations compared to the pristine biological signature, contrasting results exist for the role of metamorphism and hydrothermal alteration. The pioneering work of Haendel et al. (1986) showed that nitrogen isotopes measured in rocks from contact metamorphism aureoles are fractionated towards higher $\delta^{15}N$ moving from the external zone ($\delta^{15}N = +5‰$) to the internal one ($\delta^{15}N = +16‰$). Concomitant to this isotopic fractionation (^{14}N is kinetically lost more rapidly than heavier ^{15}N), the bulk N content in the rock decreases. The effect of N loss on isotopic composition can be modeled as: (a) a batch volatilization (one-step process), where fluids that are released equilibrate with the rock in a closed system; or (b) Rayleigh distillation, where each small aliquot of volatile that is released is immediately removed from the rock in an open system. The equations that govern these two processes are, respectively:

$$\delta^{15}N_{final} = \delta^{15}N_0 + (1 - f)\cdot 1000\, ln\, \alpha \tag{3}$$

$$\delta^{15}N_{final} = \delta^{15}N_0 - 1000\cdot[f^{(\alpha-1)} - 1] \tag{4}$$

where $\delta^{15}N_0$ and $\delta^{15}N_{final}$ are the isotopic compositions before and after the metamorphic/hydrothermal alteration, and α is the fractionation factor. The term f is the residual fraction of nitrogen in the rock. Haendel et al. (1986) and Hanschamn (1981) calculated fractionation factors α for isotopic exchanges between N species (N_2–NH_4^+–NH_3) contained in rocks and the metamorphic/hydrothermal fluids interacting with them.

This phenomenon has been clearly documented by other authors in metamorphic and hydrothermal systems (Bebout and Fogel 1992; Sadofsky and Bebout 2000; Mingram and Brauer 2001; Pitcairn et al. 2005; Jia 2006; Svensen et al. 2008) indicating various degrees of fractionation with $\Delta^{15}N$ from 6‰ to 15‰. On the other hand, works from Busigny et al. (2003) and Ader et al. (2006) have showed that during high pressure, high temperature metamorphism, the $\delta^{15}N_{NH4+}$ shift is limited to +2–3‰, while the $\delta^{15}N_{kerogen}$ shift is null.

Contrasting results on the preservation of the N isotopic signal dominate also the Precambrian record. Boyd and Philippot (1998) showed the effect of devolatilization on the $\delta^{15}N_{NH4+}$ of muscovite and biotite from Proterozoic metasediments of the Moine Succession, Scotland. These sediments were affected by amphibolite facies metamorphism. Boyd and Philippot (1998) measured $\delta^{15}N_{NH4+}$ values from +8 to +12‰ and NH_4^+ concentrations of 140–420 ppm in the biotite/muscovite host phases. Using a qualitative approach based on Haendel et al. (1986) work, Boyd and Philippot (1998) estimated the pristine $\delta^{15}N_{NH4+}$ values, prior of metamorphism, to be +5‰ to +10‰. This suggests that amphibolite facies metamorphism could have produced an isotopic shift from +2‰ to +8‰. Pinti et al. (2001) suggested possible metamorphic overprinting of the $\delta^{15}N$ values in Early Archean cherts up to +20‰, particularly in metachert and quartzite from the 3.8 Ga Isua Supracrustal Belt, West Greenland. These cherts are metamorphosed up to the amphibolite facies. Dauphas and Marty (2004) suggested an isotopic shift up to +15‰, produced by devolatilization processes, to explain the high $\delta^{15}N$ values, up to +23‰, measured in mica of Late Archean age by Jia and Kerrich (2004a, b).

Most recently, Pinti et al. (2009a) showed that metamorphic/hydrothermal fluids could strongly alter both the $\delta^{15}N$ and the $\delta^{13}C$ values in Archean cherts. Figure 2a shows a rough correlation (except one sample) between bulk $\delta^{15}N$ and $\delta^{13}C$ measured in kerogeneous black cherts from eastern and western terranes of Pilbara Craton, Western Australia. This co-variation could be explained by the progressive C and N loss (Fig. 2b) from a unique source having an initial $\delta^{15}N$ of −6‰ and $\delta^{13}C$ of −38‰.

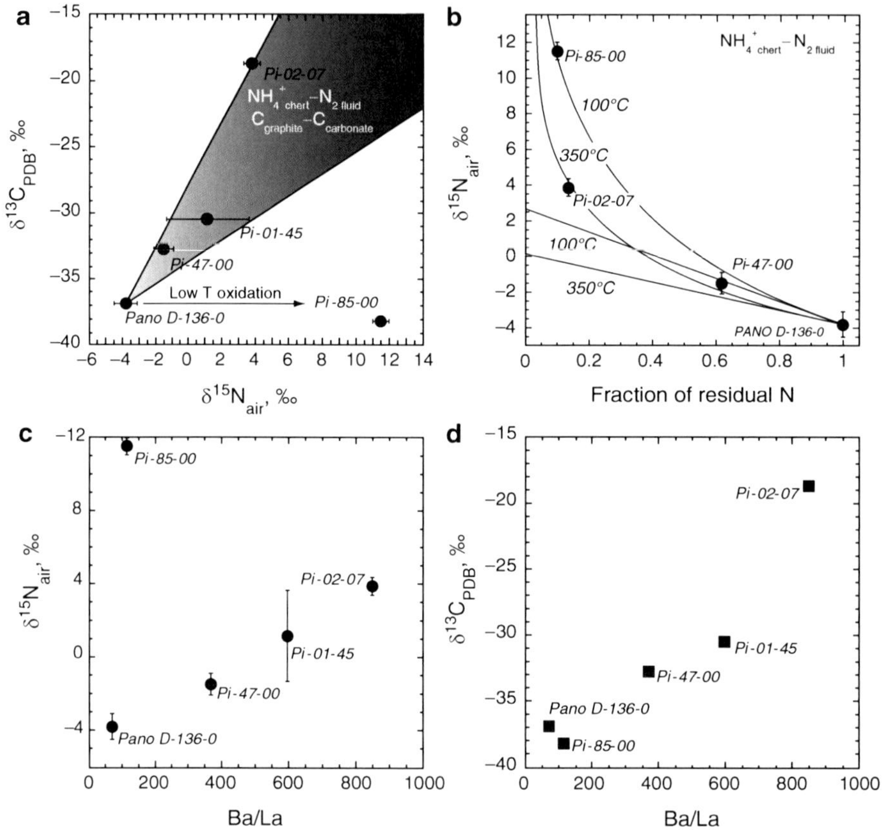

Fig. 2 (**a**) A rough correlation between the bulk $\delta^{15}N$ and $\delta^{13}C$ measured in 3.5–3.2 Ga kerogeneous cherts from Pilbara Craton, Western Australia (From Pinti et al. 2009a, redrawn). *Shaded area* is the isotopic variations expected for devolatilization of cherts and isotopic exchange between N_2–CO_2 rich metasomatic fluids and chert at 100–350°C. *Arrow* indicates the isotopic variation of C and N by low-temperature oxidation of organic matter (Ueno et al. 2004; Pinti et al. 2009a). (**b**) The fraction of residual N in cherts (F) against the N isotopic composition, after devolatilization of a pristine source represented here by sample PANO D-136-0 (Data from Pinti et al. 2001). *Straight lines* represent results of a batch volatilization and isotopic exchange between NH_4^+ (rock) and N_2 (fluid) at 100°C and 350°C, respectively. *Curves* represent results of a Rayleigh volatilization and isotopic exchange between NH_4^+ (rock) and N_2 (fluid) at 100°C and 350°C, respectively. Fractionation factors are calculated from data of Hanschamn (1981). (**c**, **d**) The N and C isotopic composition of studied cherts plotted against the Ba/La ratio. The correlation suggests progressive alteration of the pristine signatures by hydrothermal or metamorphic fluids

The residual N and C pool is consequently enriched in 15N and 13C, with the highest $\delta^{15}N$ and $\delta^{13}C$ values measured around +4 ‰ and −19‰, respectively (Fig. 2a).

This process could be related to, hydrothermal alteration as observed in the North Pole silica veins (Ueno et al. 2004); or metamorphic devolatilization (Fig. 2b; Pinti et al. 2001). Correlation between C and N isotopic composition and trace element ratios Co/As (not shown here) and Ba/La (Fig. 2c) suggests hydrothermal alteration as the main process of modification of the N isotopic composition in cherts. Indeed, there is a direct correlation between the Ba/La ratio and the C and N isotopic composition (Fig. 2c) that could reveal the progressive enrichment of Ba in the hydrothermal altered residue (Humphris and Thompson 1978). The nitrogen isotopic shift can be produced by isotopic exchange at 350°C between the chert and the hydrothermal/metamorphic fluids following the reaction (Hanschamn 1981):

$$^{15}N^{14}N_{fluid} + {}^{14}NH_4^+{}_{rock} \rightleftharpoons {}^{15}NH_4^+{}_{rock} + {}^{14}N_{2fluid} \tag{5}$$

Glikson et al. (2008) measured $\delta^{13}C$ values from −36.5‰ to −32.1‰ of the Dresser Formation carbonaceous matter. The C isotopic signature was negatively correlated with TOC (0.13–0.75%) and positively correlated with the C/N ratio (134–569), which was interpreted to reflect the relative abundance of high reflectance/oxidized/recycled carbonaceous matter and preferential loss of ^{12}C and N during thermal maturation. Though devolatilization could explain part of the observed C isotopic variations, a more plausible mechanism to make carbonaceous matter enriched in ^{13}C to such a large extent, with $\delta^{13}C$ values up to −20‰ (Fig. 2a) is by isotopic exchange with carbonate, in presence of metamorphic CO_2 fluids (Valley and O'Neil 1981; Ueno et al. 2002). This enrichment is favored by the relatively high diffusivity of the carbonate phase, originally enriched in ^{13}C, compared to the graphitic phase, by which the latter phase becomes progressively enriched in ^{13}C (Hoefs 2004). This process could take place at temperatures as low as 270°C (Morikiyo 1984) corresponding to the temperatures reached by the greenschist facies metamorphism of the Pilbara Craton.

It is worth to note that Schidlowski et al. (1983) already suggested isotopic exchange between graphite and carbonate to explain the C isotopic variations in early Archean rocks. Thus, isotopic exchange between N_2 and NH_4^+ and between carbonate C and graphite C, promoted by pervasive CO_2–N_2 hydrothermal or metamorphic fluids at temperatures ≤ 350°C, could strongly alter the pristine isotopic composition of N and C in Precambrian rocks.

5 The Significance of ^{15}N-Depleted Nitrogen in Early Archean Organic Matter

Beaumont and Robert (1999) showed that nitrogen extracted from Early Archean kerogens has a peculiar isotopic composition with $\delta^{15}N$ values from −6‰ to 0‰, while Proterozoic and Phanerozoic kerogens have $\delta^{15}N$ shifted towards ^{15}N-enriched values (from +6‰ to +20‰). Pinti et al. (2001, 2009a) and Ueno et al. (2004)

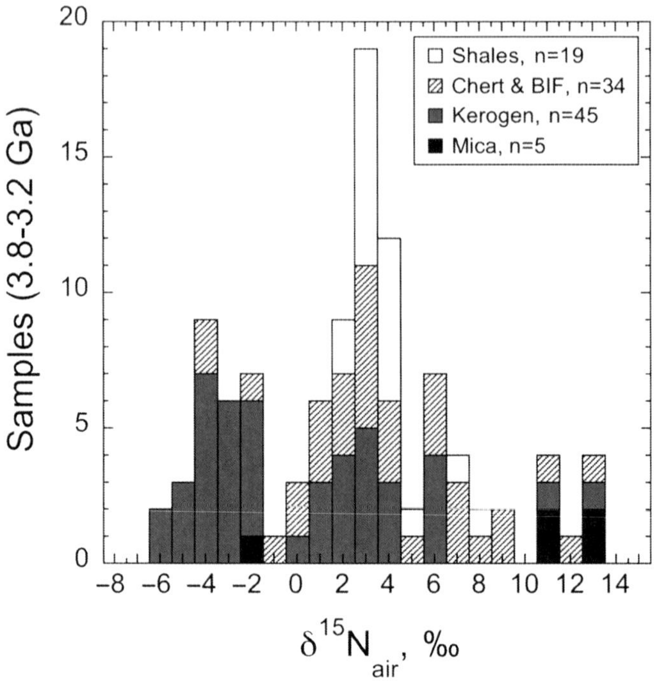

Fig. 3 Histogram of the N isotopic composition measured in the Paleoarchean bulk rock record. Data are: kerogen (Beaumont and Robert 1999; van Zuilen et al. 2005); cherts and BIF (Sano and Pillinger 1990; Pinti et al. 2001, 2007, 2009a, b; unpublished data); mica (Pinti et al. 2001; Papineau et al. 2005); shales (Yamaguchi 2002). Data have been compiled and are available in Thomazo et al. (2009)

confirmed the existence of nitrogen with $\delta^{15}N$ values ranging from −7‰ to −4‰ in the Archean hydrothermal silica veins of the 3.49 Ga Chert-Barite Unit at North Pole, Pilbara Craton, Western Australia (Fig. 3). Papineau et al. (2005) and van Zuilen et al. (2005) measured $\delta^{15}N$ values from −3‰ to +6‰ from mica and graphite recovered from the 3.8 Ga metasediments of the Isua Supracrustal Belt, West Greenland. Finally, Yamaguchi (2002) measured $\delta^{15}N$ values from +2.5‰ to +8.0‰ in Archean shales (Fig. 3).

The occurrence of lighter $\delta^{15}N$ values in Archean rocks and organic matter compared to the Proterozoic equivalents was supposed to represent a shift in the metabolic pathways of nitrogen, from a biosphere dominated by N-fixation in an anoxic Archean ocean, to an oxic ocean where nitrifiers and denitrifiers flourished, cycling the nitrogen (Beaumont and Robert 1999). Pinti and Hashizume (2001) challenged this interpretation by suggesting that the cherts studied by Beaumont and Robert (1999) are biased samples of Archean biota because most of them are deposition of silica from hydrothermal fluids. Pinti and Hashizume (2001) and Pinti et al. (2001) related the ^{15}N-depleted values measured in Archean kerogens to inorganic sources of N (N_2, NH_3), possibly mantle N, metabolized by chemosynthetic bacteria living in proximity to the hydrothermal vents. These communities are indeed able to fix inorganic N, producing large negative isotopic shifts in the biological residue ($\delta^{15}N$ from −4‰ to −10‰; Conway et al. 1994).

Nishizawa et al. (2007) measured the isotopic signature of N contained in Archean fluid inclusions. Isotopic variations were interpreted as a mixing between a paleoseawater component with $\delta^{15}N_{N2}$ from −0.7‰ to −0.2‰ and a hydrothermal component containing nitrogen depleted in ^{15}N ($\delta^{15}N_{N2}$ = −3.0 and $\delta^{15}N_{NH4+}$ = −12.6‰). The biological assimilation of this hydrothermal NH_4^+ component might explain the ^{15}N-depleted values found in Early Archean kerogens.

The occurrence of carbonaceous matter with $\delta^{13}C_{org}$ of less than −38‰ (Ueno et al. 2004) and associated fluids with $\delta^{13}C_{CH4}$ of less than −56‰ (Ueno et al. 2006) have been interpreted as the result of the metabolic activity of methanogens in a seafloor hydrothermal environment, supporting the hypothesis of Pinti and Hashizume (2001). But lighter $\delta^{13}C_{CH4}$ values have been also interpreted as resulting from abiological processes (Sherwood-Lollar and McCollom 2006), although the same processes would be unable to mimic the lighter $\delta^{15}N$.

Jia and Kerrich (2004a, b) reported $\delta^{15}N$ values up to +24‰ in ca. 2.7 Ga old kerogens, higher and more varied than those measured in contemporary cherts of any other ages. Kerrich et al. (2006) interpreted the high $\delta^{15}N$ values as a residual signature of a veneer atmosphere having an initial chondritic composition, with $\delta^{15}N$ values from +30‰ to +42‰ (Gilmour 2003). The isotopic shift from +24‰ to 0‰ in the present atmosphere was interpreted as the combination of three processes: (1) degassing of ^{15}N-depleted mantle N_2 ($\delta^{15}N$= −5‰ ± 2‰); (2) progressive sequestration of atmospheric chondrite-like N_2 in sedimentary rocks by fixing organisms; and (3) return flux of ^{15}N-depleted nitrogen to the atmosphere as a byproduct of some sort of metabolism.

Although an interesting alternative hypothesis, there are several lines of evidence against an N extraterrestrial input:

1. Early Archean paleoseawater N isotopic values suggest that the $\delta^{15}N$ of the atmosphere was between −1‰ to +2‰ since 3.5 Ga (Sano and Pillinger 1990; Pinti et al. 2001, 2009a; Nishizawa et al. 2007).
2. If the $\delta^{15}N$ and associated $\delta^{13}C$ values measured in Late Archean kerogens are similar to those of molecular organics from chondrites (e.g., Gilmour 2003), the elemental ratio C/N is not (Pinti and Hashizume 2008).
3. If we assume the validity of the model of Javoy (1998) of an Enstatite Chondrite source for the mantle N ($\delta^{15}N$ = −25‰) mixed with a Late Veneer chondritic source ($\delta^{15}N$ = +42‰), then the occurrence of $\delta^{15}N$ values of −5‰ in Early Archean diamonds clearly indicate that the surface's reservoirs of the Earth has been isotopically homogenized very early in the Earth's history (Cartigny et al. 2001), excluding a strong isotopic unbalance among the surface reservoirs such the atmosphere.

Both positive and negative $\delta^{15}N$ values have been recorded in organic matter (N_{org}), minerals and metasedimentary rocks (NH_4^+) of ages from 3.8 to 2.9 Ga (Sano and

Pillinger 1990; Beaumont and Robert 1999; Pinti et al. 2001, 2007, 2009a; Yamaguchi 2002; Ueno et al. 2004; Papineau et al. 2005; Van Zuilen et al. 2005) (Fig. 3). If some of the positive values can be attributed to devolatilization and isotopic fractionation (Fig. 2 and Pinti et al. 2001, 2009a) others seem to be pristine. Pinti et al. (2007) measured a $\delta^{15}N$ value of +6‰ and a C/N molar ratio of 50 in iron hydroxides possibly precipitated from very hot hydrothermal fluids. These values are similar to those measured in organic matter preserved in modern marine sediments, but it is not a modern organic contamination. The N component is indeed strongly bound by chemisorption on the iron surface and then embedded in the lattice during iron oxidation and crystal growth.

Figure 3 shows the frequency distribution of the measured $\delta^{15}N$ values in the Paleoarchean rock record (3.8 – 3.2 Ga). From 3.2 to 2.7 Ga there is a lack of data. Most of samples are centered at 2‰ with a second population centered at −4‰ and a few sparse data with $\delta^{15}N$ values higher than +8‰, mostly being mica from the 3.8 Ga terranes of Isua. Shen et al. (2006) already noticed this bimodal distribution that could be explained by a diversified biosphere with anoxygenic photosynthetic mats, stromatolites and cyanobacteria at the surface of the ocean, and methanogens and chemolithoautotrophs at depth, close to hydrothermal habitats (Nisbet and Sleep 2001; Shen et al. 2006). Both ecosystems use inorganic sources of N, cycled through biological *fixation*: the first microorganisms used energy from sunlight (268 mW/m^2) and atmospheric N (Rosing et al. 2006); the seconds used energy derived from chemical gradients (87 mW/m^2) and mantle N (Pinti et al. 2001, 2009a, b). Biological reduction of N mediated by nitrogenase – i.e. the N_2 *fixation* – could be indeed one of the oldest metabolic reactions (Raymond 2003) and recently Mehta and Baross (2006) showed that *fixation* might be responsible for the transformation of mantle N into N_{org} in chemosynthetic Archaea communities.

Recently, Zerkle et al. (2008) examined the effects of varying Fe, Mo, and P concentrations on $\delta^{15}N$ fractionation during N_2 *fixation* in the cyanobacterium *Anabaena variabilis*. They showed that when it is cultured in Fe-enriched media ([Fe] = 50 nmol/L), this organism produces biomass up to 3‰ lower in $\delta^{15}N$ than when it is cultured in Fe-limited media ([Fe] < 50 nmol/L). Zerkle et al. (2008) concluded that enhanced Fe availability during periods of widespread ocean anoxia (as during Early Archean) could have further stimulate the production of ^{15}N-depleted biomass by N-fixing organisms, contributing to the isotopic record.

Whether the N isotopic record during the Paleo- and Mesoarchean reflects N biological *fixation* by different microorganisms in different paleoenvironments (photosynthetizers against chemosynthetizers; Pinti and Hashizume 2001) or just reflect variations in the supply of nutrients to the ocean (Zerkle et al. 2008), the isotopic record cannot give conclusive arguments on the metabolic processes regulating the N outputs in the Archean ocean. N *fixation* is indeed the N input in the modern (and possibly ancient) biological cycle. In the modern ocean, the N output is regulated by *denitrification* of nitrates to N_2, that returns back to the atmosphere. In absence of significant oxygen in the Early Archean, seawater nitrate should be scarce. Thus we should expect alternative metabolic pathways to *nitrification-denitrification*.

The easiest sink for N would be the hydrothermal alteration of N_{org} or NH_4^+ buried in sediments (Summers 1999). Indeed most of the Archean terranes show widespread hydrothermal alteration and hot fluids could have rapidly degraded N. Up to 90% of the N_{org} trapped in Archean metasedimentary rocks could have been removed by this process, as indicated in Fig. 2b (Pinti et al. 2009a). An appealing metabolic process that has been advocated as the major N sink in the Archean (e.g., Papineau et al. 2005) is the *anaerobic ammonium oxidation*, or "*anammox*". Yet, we need to explain how half of the nitrogen involved in the reaction could have been partially oxidized to nitrite (NO_2^-) to participate to the reaction, in a completely anoxic Archean ocean. Further, the isotopic fractionation induced by this metabolic process is not yet known (Sigman et al. 2009). This hypothesis is yet at the stage of pure speculation.

6 Isotopic Shifts Recorded by N, C and Fe at Late Archean: A Reaction to the Progressive Oxygenation of the Earth

Most of the measured $\delta^{15}N_{org}$ in the Late Archean range between +6‰ and +24‰ (e.g., Jia and Kerrich 2004a, b), while a few banded iron formations and shales show $\delta^{15}N_{org}$ values up to +40‰ (Beaumont and Robert 1999; Yamaguchi 2002; Fig. 4). These are the highest isotopic values found in the rock record.

The large isotopic shift at around 2.7 Ga is observed also in the isotopic records of carbon ($\delta^{13}C$) and of iron ($\delta^{56}Fe$) (Fig. 4). The carbon isotopic record of carbonaceous matter is restricted to $\delta^{13}C$ values between −20‰ to −40‰ since 3.5 Ga (Fig. 4) except for data from the 3.8 Ga Isua supracrustal rocks. These latter have $\delta^{13}C$ of the graphitic C from −25‰ to −10‰ indicating likely a mixing between organic sources (the lighter; Mojzsis et al. 1996) and inorganic C produced by metamorphic reactions (the heavier; van Zuilen et al. 2002). At around 2.7 Ga, the lowest $\delta^{13}C_{org}$ values ever recorded, down to −60‰, have been measured (Fig. 4).

Iron is a nutrient and often a catalyst in metabolic reactions. For this reason, the isotopic fractionation of iron ($\delta^{56}Fe$ with $R = {}^{56}Fe/{}^{54}Fe$) is related to biological activity (Beard et al. 2003; Johnson et al. 2008). However, the isotopic variability of iron is relatively restricted compared to that of C and N, generally to ±1‰ around the standard value (Fig. 4). At around 2.8–2.7 Ga, iron shows the largest isotopic shift ever recorded, with measured $\delta^{56}Fe$ values down to −3‰ (Fig. 4).

The concomitant large isotopic shifts observed for these three isotopic biosignatures could indicate a simultaneous change in the metabolic pathways of these three elements, as a direct response to a major environmental stress: the progressive oxygenation of the ocean. Johnson et al. (2008) already noticed the isotopic co-variation of C and Fe between 3.0 and 2.4 Ga, suggesting that both isotopic fractionations could be related to a change in metabolisms as a response to the slow rising of free O_2 in the ocean. This does not necessarily mean a full-oxygenated ocean, as proposed by others (Ohmoto 1997), but instead local condition of enhanced oxygenation (oxygen oasis; Johnson et al. 2008).

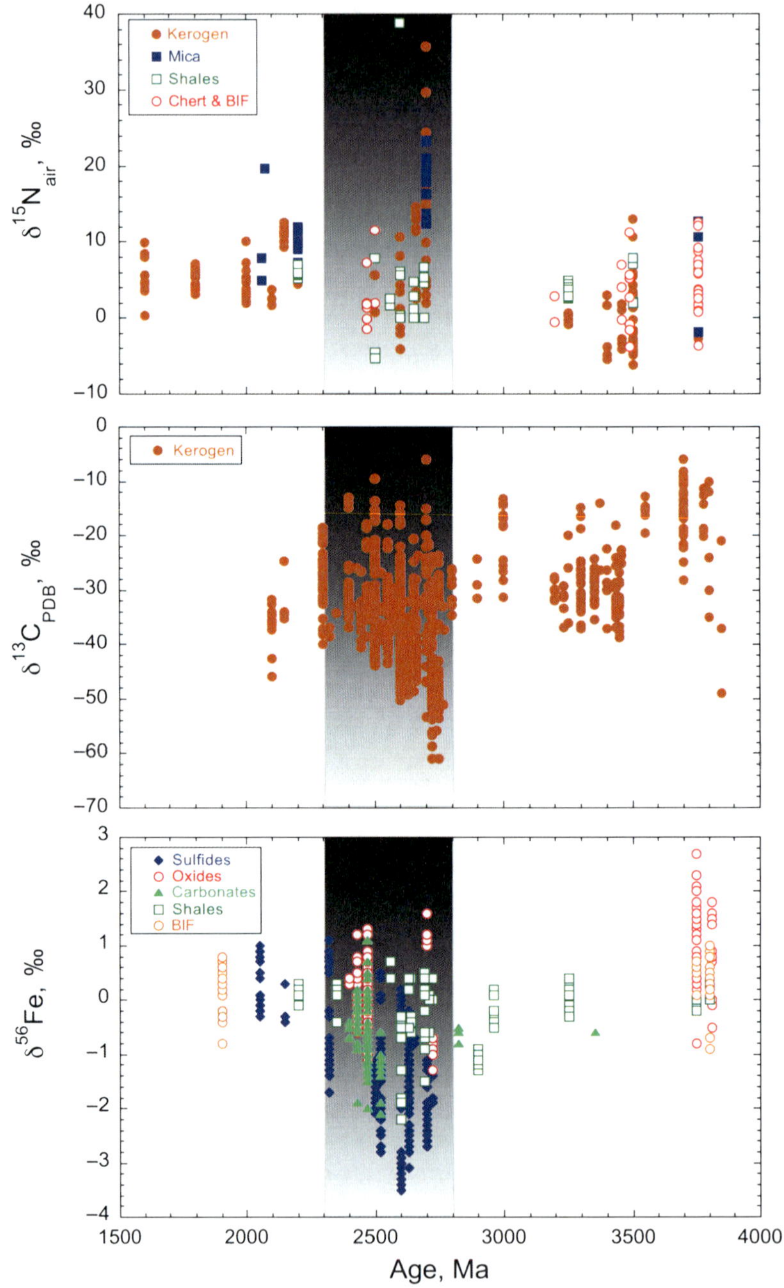

Fig. 4 N, C and Fe isotopic record between 3.8 and 1.5 Ga. Refer to Thomazo et al. (2009) for compilation of literature data. The N database has been updated with unpublished data of Pinti. The shaded area between 2.8 and 2.3 Ga highlights the large and contemporaneous isotopic shifts measured for the three isotopic systems

Indeed, the strong decrease in $\delta^{13}C$ values for organic carbon at around 2.7 Ga (Fig. 4) could reflect *methanotrophy* in the presence of free O_2 (Hayes et al. 1983; Johnson et al. 2008), which occurs via the reaction:

$$7CH_4 + 8O_2 \rightarrow 6CH_2O + CO_2 + 8H_2O \quad (6)$$

But alternative explanations have been also suggested, such as the *anaerobic oxidation* of methane (Hayes et al. 1987) or changes in the burial rate of organic carbon (Bjerrum and Canfield 2004).

Among the abiologic and biologic processes that involve redox or bonding changes, microbial Fe^{3+} reduction to Fe^{2+} or DIR (Dissimulatory Iron Reduction) is the one that produce the largest quantities of isotopically distinct Fe (lowest $\delta^{56}Fe$). This reaction can be expressed as:

$$4Fe(OH)_3 + CH_2O + 8H^+ \rightarrow 4Fe^{2+}_{aq} + CO_2 + 11H_2O \quad (7)$$

DIR is enhanced by the increased availability of Fe^{3+} through the oxidation of Fe^{2+} by O_2 and larger sources of organic carbon, possibly supplied by oxygenic photosynthesis (Johnson et al. 2008). Thus, the C and Fe isotopic variations could be related in some extent to the oxygen rise in the Late Archean.

7 Enhanced *Denitrification* or N Imbalanced Fluxes in the Ocean as a Response to the Oxygenation of the Earth

The large N positive isotopic shift recorded around 2.7 Ga (Fig. 4) could be also caused by the slow rise of oxygen in the ocean.

Simplifying the N biogeochemical cycle (Fig. 1) and assuming steady-state conditions, the $\delta^{15}N_{biosphere}$ can be explained by three fluxes in the ocean:

1. An input that is N_2 *fixation* ($\delta^{15}N_{input} = \delta^{15}N_{N2} - \Delta_{input} \approx 0‰$)
2. A sink (output) that is denitrification ($\delta^{15}N_{output\text{-}A} = \delta^{15}NO_3^- - \Delta_{output\text{-}A}$)
3. Another sink that is nitrate *assimilation* and particulate nitrogen (PN) burial ($\delta^{15}N_{output\text{-}B} = \delta^{15}NO_3^- - \Delta_{outut\text{-}B}$)

The proportions of these three fluxes are 1: α : 1 − α. Given that nitrate is the major N form in the ocean, we can assume that:

$$\delta^{15}N_{biosphere} \approx \delta^{15}NO_3^- = \Delta_{output\text{-}A} \times \alpha + \Delta_{output\text{-}B} \times (1-\alpha) \quad (8)$$

Depending on the relative importance of the *water-column denitrification* or of *assimilation*, we can assume that:

$$0 \le \Delta_{output\text{ - }A} \le \varepsilon_{water\text{ - }column\ denitrification} \approx +25‰ \quad (9)$$

$$0 \le \Delta_{\text{output - B}} \le \varepsilon_{\text{assimilation}} \approx +5‰ \tag{10}$$

and consequently,

$$0 \le \delta^{15}N_{\text{biosphere}} \le \max(\varepsilon_{\text{denitrification}}, \varepsilon_{\text{assimilation}}) \tag{11}$$

In the present ocean, *denitrification* is limited (and thus its isotopic effect on residual nitrate) because denitrifiers need suboxic conditions to operate ($O_2 = 10^{-2}$ Present Atmospheric Level or 3 μmol/L). These environmental conditions are reached in localized areas of the ocean of high primary productivity, where minimum oxygen layers can develop extensively during C oxidation (such as the Arabian Sea; Altabet et al. 2002). The large isotopic shift recorded around 2.7 Ga (up to +24‰; Fig. 4) could indicate that *denitrification* (and its isotopic effect on nitrate) was dominant over that of *assimilation*. This is not surprising if free O_2 were slowly rising in the atmosphere and the ocean. Indeed this process could have enhanced the formation of widespread suboxic layers where the denitrifiers could have consumed the few nitrates formed, leaving largely fractionated residual nitrate pool to be assimilated by primary producers and buried with organic matter (Fig. 1).

A dynamic imbalance between the input and output fluxes of N in the ocean could also explain the large isotopic shift observed in Fig. 4. The degree of isotopic fractionation of the residual nitrate which is assimilated by microorganisms and then buried within sediments as PN ($\delta^{15}N_{\text{biosphere}} \approx \delta\ ^{15}NO_3^-$) depends on the availability of the input nitrate pool, which constitutes the reactant in the reaction of *denitrification*. If the reactant N (NO_3^-) is continuously supplied and consumed, then the residual N reactant is exported at a constant rate. The gross supply of reactant N equals the sum of the product N and the residual reactant N exported (e.g., Sigman et al. 2009). This is the steady-state condition and the isotopic ratio of the residual nitrate, after *denitrification*, will follow the expression (Mariotti et al 1981; Sigman et al. 2009):

$$\delta^{15}NO^-_{3\ \text{residual}} = \delta^{15}NO^-_{3\ \text{initial}} + \varepsilon(1 - f) \tag{12}$$

where $\delta^{15}NO^-_{3\ \text{residual}}$ is the isotopic ratio of the residual nitrate; $\delta^{15}NO^-_{3\ \text{initial}}$ is the isotopic ratio of the initial nitrate pool; ε is the kinetic isotope effect; *f* is the fraction of residual nitrate $[(NO_3^-)/(NO^-_{3\ \text{initial}}]$, i.e. not consumed by the *denitrification* reaction. In Fig. 5, we reported the result of this calculation assuming a steady-state conditions between an input N pool with $\delta^{15}N = 0‰$ and the consumed NO_3^-. It can be observed (straight line) that when *denitrification* is nil, the $\delta^{15}NO^-_{3\ \text{residual}} = \delta^{15}N_{\text{initial}} = 0‰$. Approaching the maximum rate of *denitrification*, the $\delta^{15}NO^-_{3\ \text{residual}}$ approaches the value of +25 ($\varepsilon_{\text{denitrification}}$) (Eq. 11).

If the nitrate pool is limited, which means that the reactant is consumed more rapidly than refueled, then the isotopic effect of the reaction will be described in terms of Rayleigh fractionation, following the expression (Sigman et al. 2009):

$$\delta^{15}NO^-_{3\ \text{residual}} = \delta^{15}NO^-_{3\ \text{initial}} - \varepsilon[\ln(f)] \tag{13}$$

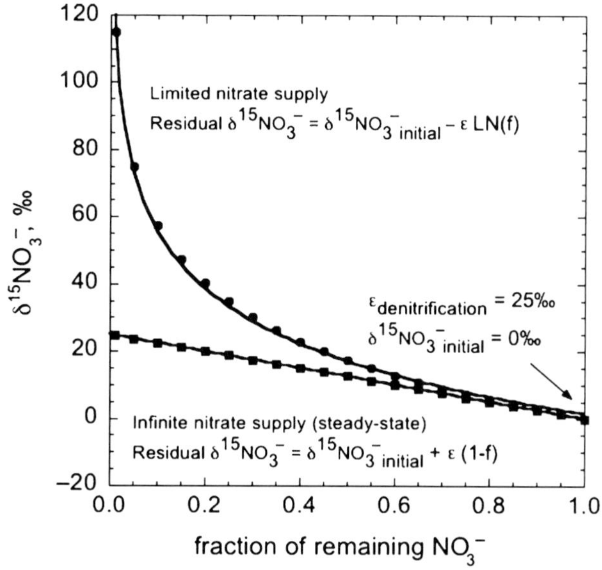

Fig. 5 Isotopic fractionation of residual nitrates (product) after *denitrification* of a limited nitrate pool (reactant), modeled following a Rayleigh distillation (exponential curve). The same process, but with a continuous supply of reactant (steady-state) is represented by the *straight line*

The isotopic effect on the residual seawater nitrate after *denitrification* on a limited nitrate pool (exponential curve) is much larger than at the steady-state (straight line; Fig. 5). The residual nitrate would be entirely consumed by primary producers, recording a larger isotopic fractionation, following the expression (Altabet 1988; Altabet and Francois 1994):

$$\delta^{15}N^{-}_{PN} = \delta^{15}NO_3^{-}{}_{\text{residual}} - \varepsilon_{\text{assimilation}} \tag{14}$$

A dynamic imbalance between input and output N fluxes could have thus created a significant diversity in the isotopic composition among the buried organic matter of same age, sometimes exhibiting positive $\delta^{15}N$ values of +25‰ and higher, as observed in the geological record (Fig. 4). The presence of a limited nitrate pool ($N_{input} < N_{output}$) could be explained by the fact that slowly rising free O_2, was not yet sufficient to produce a large (and theoretically infinite) nitrate pool to be consumed during *denitrification* reactions.

8 Conclusions

Nitrogen isotopes have been neglected as a record of the early metabolic pathways, compared to the carbon isotopes. Recent advances in the understanding of the main metabolic pathways regulating the modern cycle in the ocean (e.g., Brandes et al. 2007;

Sigman et al. 2009) and the effects of geological time in the preservation of biological isotopic signals (Busigny et al. 2005; Ader et al. 2006; Jia 2006; Pinti et al. 2009a) have certainly boosted N isotopes as a reliable biosignature in Precambrian rocks. Yet, research in this domain need to advance further. Recent organic contamination should be excluded for the correct interpretation of pristine isotopic signatures. Analysis of separate mineral phases can be useful to avoid inter-grain organic contamination (Papineau et al. 2005; Pinti et al. 2007). Yet, we need to perform step-heating or step-combustion isotopic analyses for separating N isotopic signatures coming from different sources and preserved in different mineral host phases (Pinti et al. 2001, 2009a). Coupling multiple techniques is also useful for tracing back and isolating these sources (Pinti et al. 2007). Nitrogen, as NH_4^+ or N_{org}, resides in several minerals and its position in their lattice structure and bonding is not yet completely understood (Orberger et al. 2005; Pinti et al. 2007). Most recent results indicated that the mineral structure and the type of chemical bonding of ammonium could influence the isotopic fractionation of N in mica (Pinti et al. 2009a). Finally, the still poor N temporal database needs to be enlarged (Fig. 4; Thomazo et al. 2009). If the present amount of data is helpful for tracing the main environmental changes affecting the metabolic pathways of N (Fig. 4), yet we lack sufficient temporal resolution for modeling the interplay of changing microbial metabolisms over Earth's history and relating them to the environmental changes of the surface's reservoirs (Garvin et al. 2009).

Acknowledgments This review is the result of a long-standing and fruitful collaboration between DLP and KH, which was enriched during years with precious expertise and friendship from C. Cloquet, J.P. Gallien, M. Massault, B. Orberger, V. Rouchon, C. Wagner, R. Wirth and many others. Research of DLP was funded by JSPS (Japan), European Community, CNRS (France) and NSERC (Canada). Research of KH was mainly supported by JSPS and MEXT. We wish to thank an anonymous reviewer for the useful comments and the editors of this book for inviting to write this review. Michelle Laithier drew Fig. 1. This is GEOTOP contribution 2009–0016. In loving memory of my mother, Magda (DLP).

References

Ader M, Cartigny P, Boudou JP et al (2006) Nitrogen isotopic evolution of carbonaceous matter during metamorphism: methodology and preliminary results. Chem Geol 232:152–169

Altabet MA (1988) Variations in nitrogen isotopic composition between sinking and suspended particles: implications for nitrogen cycling and particle transformation in the open-ocean. Deep-Sea Res 35:535–554

Altabet MA, Francois R (1994) Sedimentary nitrogen isotopic ratio as a recorder for surface ocean nitrate utilization. Global Biogeochem Cycles 8:103–116

Altabet MA, Higginson MJ, Murray DW (2002) The effect of millennial-scale changes in Arabian Sea denitrification on atmospheric CO_2. Nature 415:159–162

Barker DS (1964) Ammonium in alkali feldspars. Am Mineral 49:851–858

Beard BL, Johnson CM, Skulan JL et al (2003) Application of Fe isotopes to tracing the geochemical and biological cycling of Fe. Chem Geol 196:43–56

Beaumont V, Robert F (1999) Nitrogen isotope ratios of kerogens in Precambrian cherts: a record of the evolution of atmosphere chemistry? Precambr Res 96:63–82

Bebout GE, Fogel ML (1992) Nitrogen-isotope compositions of metasedimentary rocks in the Catalina Schist, California: implications for metamorphic devolatilization history. Geochim Cosmochim Acta 56:2839–2849

Bjerrum CJ, Canfield DE (2004) New insights into the burial history of organic carbon on the early Earth. Geochem Geophys Geosyst. doi:10.1029/2004GC000713

Boudou JP, Schimmelmann A, Ader M, Mastalerz M, Sebilo M, Gengembre L (2008) Organic nitrogen chemistry during low-grade metamorphism. Geochim Cosmochim Acta 72:1199–1221

Boyd SR (2001) Ammonium as a biomarker in Precambrian metasediments. Precambr Res 108:159–173

Boyd SR, Philippot P (1998) Precambrian ammonium biogeochemistry: a study of the Moine metasediments, Scotland. Chem Geol 144:257–268

Boyd SR, Hall A, Pillinger CT (1993) The measurement of $\delta^{15}N$ in crustal rocks by static vacuum mass-spectrometry – application to the origin of the ammonium in the Cornubian Batholith, Southwest England. Geochim Cosmochim Acta 57:1339–1347

Brandes JA, Devol AH, Deutsch C (2007) New developments in the marine nitrogen cycle. Chem Rev 107:577–589

Brasier MD, Green OR, Jephcoat AP, Kleppe AK, Van Kranendonk MJ, Lindsay JF, Steele A, Grassineau NV (2002) Questioning the evidence for Earth's oldest fossils. Nature 416:76–81

Busigny V, Cartigny P, Philippot P et al (2003) Massive recycling of nitrogen and other fluid-mobile elements (K, Rb, Cs, H) in a cold slab environment: evidence from HP to UHP oceanic metasediments of the Schistes Lustres nappe (western Alps, Europe). Earth Planet Sci Lett 215:27–42

Busigny V, Laverne C, Bonifacie M (2005) Nitrogen content and isotopic composition of oceanic crust at a superfast spreading ridge: a profile in altered basalts from ODP Site 1256, Leg 206. Geochem Geophys Geosyst. doi:10.1029/2005GC001020

Cartigny P, Harris JW, Javoy M (2001) Diamond genesis, mantle fractionations and mantle nitrogen content: a study of $\delta^{13}C$-N concentrations in diamonds. Earth Planet Sci Lett 185:85–98

Casciotti KL, Sigman DM, Ward BB (2003) Linking diversity and stable isotope fractionation in ammonia-oxidizing bacteria. Geomicrobiol J 20:335–353

Cline JD, Kaplan IR (1975) Isotopic fractionation of dissolved nitrate during denitrification in the eastern tropical North Pacific Ocean. Mar Chem 3:271–299

Conway NM, Kennicutt MC, Van Dover CL (1994) Stable isotopes in the study of marine chemosynthetic-based ecosystems. In: Lajtha K, Michener RH (eds) Stable isotopes in ecology and environmental science. Blackwell, Oxford

David P. Summers (1999) Sources and sinks for ammonia and nitrite on the early Earth and the reaction of nitrite with ammonia. Origins of Life and Evolution of the Biosphere 29:33–46

Dauphas N, Marty B (2004) "A large secular variation in the nitrogen isotopic composition of the atmosphere since the Archaean?": response to a comment on "The nitrogen record of crust-mantle interaction and mantle convection from Archaean to Present" by R Kerrich and Y Jia. Earth Planet Sci Lett 225:441–450

Duit W, Jansen JBH, Van Breemen A et al (1986) Ammonium micas in metamorphic rocks as exemplified by Dome de l'Agout (France). Am J Sci 286:702–732

Fogel ML, Cifuentes LA (1993) Isotope fractionation during primary production. In: Engels MH, Macko SA (eds) Organic geochemistry. Plenum, New York

Freudenthal T, Wagner T, Wenzhofer F et al (2001) Early diagenesis of organic matter from sediments of the eastern subtropical Atlantic: evidence from stable nitrogen and carbon isotopes. Geochim Cosmochim Acta 65:1795–1808

Gallien J-P, Orberger B, Daudin L et al (2004) Nitrogen in biogenic and abiogenic minerals from Paleozoic black shales: an NRA study. Nucl Instrum Meth Phys Res B 217:113–122

Galloway J (2003) The global nitrogen cycle. Treat Geochem 8:557–583

Garvin J, Buick R, Anbar A et al (2009) Isotopic evidence for an aerobic nitrogen cycle in the latest Archean. Science 323:1045–1048

Gilmour I (2003) Structural and isotopic analysis of organic matter in carbonaceous chondrites. Treat Geochem 1:269–290

Glikson M, Duck LJ, Golding SD et al (2008) Microbial remains in some earliest Earth rocks: comparison with a potential modern analogue. Precambr Res 164:187–200
Haendel D, Mühle K, Nitzsche H-M, Stiehl G, Wand U (1986) Isotopic variations of the fixed nitrogen in metamorphic rocks. Geochim Cosmochim Acta 50:749–758
Hall A (1999) Ammonium in granites and its petrogenetic significance. Earth Sci Rev 45:145–165
Hanschamn G (1981) Berechnung von Isotopieeffekten auf quantenchemischer Grundlage am Beispiel stick-stoffhaltiger Moleküle. Zfi-Mitt 41:19–39
Hashizume K, Soyama H, Cloquet C et al (2008) Covariation of nitrogen and iron isotopic ratios in a banded iron formation. Geochim Cosmochim Acta 72(suppl 1):A356
Hayes JM, Kaplan IR, Wedeking KW (1983) Precambrian organic geochemistries, preservation of the record. In: Schopf WJ (ed) Earth's earliest biosphere. Cambridge University Press, Cambridge
Hayes JM, Takigiku R, Ocampo R, Callot HJ, Albrecht P. (1987). Isotopic compositions and probable origin of organic molecules in the Eocene Messle shale. Nature 329:48–51
Hoefs J (2004) Stable isotope geochemistry. Springer, Berlin
Holloway JM, Dahlgren RA (1999) Geologic nitrogen in terrestrial biogeochemical cycling. Geology 27:567–570
Honma H, Itihara Y (1981) Distribution of ammonium in minerals of metamorphic and granitic rocks. Geochim Cosmochim Acta 45:983–988
Humphris SE, Thompson G (1978) Trace element mobility during hydrothermal alteration of oceanic basalts. Geochim Cosmochim Acta 42:127–136
Javoy M (1998) The birth of the Earth's atmosphere: the behaviour and fate of its major elements. Chem Geol 147:11–25
Jia Y (2006) Nitrogen isotope fractionations during progressive metamorphism: a case study from the Paleozoic Cooma metasedimentary complex, southeastern Australia. Geochim Cosmochim Acta 70:5201–5214
Jia Y, Kerrich R (2004a) Nitrogen 15–enriched Precambrian kerogen and hydrothermal systems. Geochem Geophys Geosyst. doi:07010.01029/02004GC000716
Jia Y, Kerrich R (2004b) A reinterpretation of the crustal N-isotope record: evidence for a 15N-enriched Archean atmosphere? Terra Nova 16:102–108
Johnson C, Beard B, Roden E (2008) The iron isotope fingerprints of redox and biogeochemical cycling in the modern and ancient Earth. Annu Rev Earth Planet Sci 36:457–493
Kerrich R, Jia Y, Manikyamba C et al (2006) Secular variations of N-isotopes in terrestrial reservoirs and ore deposits. In: Kesler SE, Ohmoto H (eds) Evolution of early Earth's atmosphere, hydrosphere, and biosphere–constraints from ore deposits. Geological Society of America, Boulder, CO, Memoir 198
Kuypers MMM, Sliekers AO, Lavik G et al (2003) Anaerobic ammonium oxidation by anammox bacteria in the Black Sea. Nature 422:608–611
Lehmann MF, Bernasconi SM, Barbieri A et al (2002) Preservation of organic matter and alteration of its carbon and nitrogen isotope composition during simulated and in situ early sedimentary diagenesis. Geochim Cosmochim Acta 66:3573–3584
Lehmann MF, Sigman DM, Berelson WM (2004) Coupling the $^{15}N/^{14}N$ and $^{18}O/^{16}O$ of nitrate as a constraint on benthic nitrogen cycling. Mar Chem 88:1–20
López-Garcia P, Moreira D, Douzery E et al (2006) Ancient fossil record and early evolution (ca. 3.8 to 0.5 Ga). Earth Moon Planet 98:247–290
Macko SA, Engel MH (1993) Organic geochemistry: principles and applications. Plenum, New York
Mariotti A, Germon JC, Hubert P et al (1981) Experimental determination of nitrogen kinetic isotope fractionation: some principles; illustration for the denitrification and nitrification processes. Plant Soil 62:413–430
Mehta MP, Baross JA (2006) Nitrogen fixation at 92°C by a hydrothermal vent archaeon. Science 314:1783–1786
Mingram B, Brauer K (2001) Ammonium concentration and nitrogen isotope composition in metasedimentary rocks from different tectonometamorphic units of the European Variscan Belt. Geochim Cosmochim Acta 65:273–287

Mojzsis SL, Arrhenius G, Friend CRL (1996) Evidence for life on Earth before 3,800 million years ago. Nature 384:55–57

Morikiyo T (1984) Carbon isotopic study on coexisting calcite and graphite in the Ryoke metamorphic rocks, northern Kiso district, central Japan. Contrib Mineral Petrol 87: 251–259

Murray JW, Codispoti LA, Friederich GE (1995) Oxidation-reduction environments: the suboxic zone in the Black Sea. In: Huang CP, O'Melia CR, Morgan JJ (eds) Aquatic chemistry: interfacial and interspecies processes. American Chemical Society, Washington, DC

Nisbet EG, Sleep NH (2001) The habitat and nature of early life. Nature 409:1083–1091

Nishizawa M, Sano Y, Ueno Y, Maruyama S (2007) Speciation and isotope ratios in fluid inclusions from seafloor hydrothermal deposits at ca. 3.5 Ga. Earth Planet Sci Lett 254:332–344

Ohmoto H (1997) When did the Earth's atmosphere become oxic? Geochem News 93(12–13):26–27

Orberger B, Gallien J-P, Pinti DL et al (2005) Nitrogen and carbon partitioning in diagenetic and hydrothermal minerals from Paleozoic Black Shales (Selwyn Basin, Yukon Territories, Canada). Chem Geol 218:249–264

Papineau D, Mojzsis SJ, Karhu JA et al (2005) Nitrogen isotopic composition of ammoniated phyllosilicates: case studies from Precambrian metamorphosed sedimentary rocks. Chem Geol 216:37–58

Papineau D, Purohit R, Goldberg T et al (2009) High primary productivity and nitrogen cycling after the Paleoproterozoic phosphogenic event in the Aravalli Supergroup, India. Precambr Res 171:37–56

Peters KE, Sweeney RE, Kaplan IR (1978) Correlation of carbon and nitrogen stable isotope ratios in sedimentary organic matter. Limnol Oceanogr 23:598–604

Pinti DL (2002) The isotopic record of Archean nitrogen and the evolution of the early earth. Trends Geochem 2:1–17

Pinti DL, Hashizume K (2001) N-15-depleted nitrogen in Early Archean kerogens: clues on ancient marine chemosynthetic-based ecosystems? Precambr Res 105:85–88

Pinti DL, Hashizume K (2008) $\delta^{15}N$-$\delta^{13}C$ Covariations in organic matter through eons: tracing the evolution of metabolic pathways. Geochim Cosmochim Acta 72(Suppl 1):A751

Pinti DL, Hashizume K, Matsuda J (2001) Nitrogen and argon signatures in 3.8 to 2.8 Ga metasediments: clues on the chemical state of the Archean ocean and the deep biosphere. Geochim Cosmochim Acta 65:2301–2315

Pinti DL, Hashizume K, Orberger B et al (2007) Biogenic nitrogen and carbon in Fe-Mn-oxyhydroxides from an Archean chert, Marble Bar, Western Australia. Geochem Geophys Geosyst. doi:10.1029/2006GC001394

Pinti DL, Hashizume K, Sugihara A et al (2009a) Isotopic fractionation of nitrogen and carbon in Paleoarchean cherts from Pilbara Craton, Western Australia: origin of ^{15}N-depleted nitrogen. Geochim Cosmochim Acta 73:3819–3848

Pinti DL, Mineau R, Clement V (2009b) Hydrothermal alteration and microfossils artefacts of the 3,465-million-year-old Apex chert. Nat Geosci 2, 640–643

Pitcairn IK, Teagle DAH, Kerrich R et al (2005) The behavior of nitrogen and nitrogen isotopes during metamorphism and mineralization: evidence from the Otago and Alpine Schists, New Zealand. Earth Planet Sci Lett 233:229–246

Prokopenko MG, Hammond DE, Spivack A et al (2005) Impact of long-term diagenesis on ^{15}N of organic matter in marine sediments: sites 1227 and 1230. In: BB Jørgensen, SL D'Hondt, DJ Miller (eds) Proceedings of the ocean drilling program, scientific results, vol 201. College Station, TX, pp 1–30

Prokopenko M, Hammond DE, Berelson W et al (2006) Nitrogen cycling in the sediments of Santa Barbara basin and Eastern Subtropical North Pacific: nitrogen isotopes, diagenesis and possible chemosymbiosis between two lithotrophs (Thioploca and Anammox)–"riding on a glider". Earth Planet Sci Lett 242:186–204

Raymond J (2003) The natural history of nitrogen fixation. Mol Biol Evol 21:541–554

Rosing M, Bird D, Sleep N et al (2006) The rise of continents–an essay on the geologic consequences of photosynthesis. Palaeogeogr Palaeoclimatol Palaeoecol 232:99–113
Rouchon V, Pinti DL, Gallien JP et al (2005) NRA analyses of N and C in hydromuscovite aggregates from a 3.5 Ga chert from Kittys Gap, Pilbara, Australia. Nucl Instrum Methods Phys Res Sec B 231:536–540
Sadofsky SJ, Bebout GE (2000) Ammonium partitioning and nitrogen-isotope fractionation among coexisting micas during high-temperature fluid-rock interactions: examples from the New England Appalachians. Geochim Cosmochim Acta 64:2835–2849
Sano Y, Pillinger CT (1990) Nitrogen isotopes and N_2/Ar ratios in cherts: an attempt to measure time evolution of atmospheric δ^{15}N value. Geochem J 24:315–325
Schidlowski M (2001) Carbon isotopes as biogeochemical recorders of life over 3.8 Ga of Earth history: evolution of a concept. Precambr Res 106:117–134
Schidlowski M, Hayes JM, Kaplan IR (1983) Isotopic inferences of ancient biochemistries: carbon, sulfur, hydrogen, and nitrogen. In: Schopf WJ (ed) Earth's earliest biosphere. Princeton University Press, Princeton, NJ
Scholten SO (1991) The distribution of nitrogen isotopes in sediments. Geol Ultraiectina 81:101, pp
Schopf WJ (2006) Fossil evidence of Archaean life. Phil Trans R Soc B 361:869–885
Schopf WJ, Kudryavtsev AB, Agresti DG, Wdowiak TJ, Czaja AD (2002) Laser-Raman imagery of Earth's earliest fossils. Nature 416:73–76
Shen Y, Pinti DL, Hashizume K (2006) Biogeochemical cycles of sulfur and nitrogen in the Archean ocean and atmosphere. In: K Benn, JC Mareschal, KC Condie (eds) Archean geodynamics and environments. AGU Geophys Monogr 164
Sherwood-Lollar B, McCollom TM (2006) Biosignatures and abiotic constraints on early life. Nature 444:E18. doi:10.1038/nature05499
Sigman DM, Robinson R, Knapp AN et al (2003) Distinguishing between water column and sedimentary denitrification in the Santa Barbara Basin using the stable isotopes of nitrate. Geochem Geophys Geosyst 4:1040. doi:10.1029/2002GC000384
Sigman DM, Kash KL, Casciotti KL (2009) Ocean process tracers: nitrogen isotopes in the ocean. In: Steele JH, Turekian KK, Thorpe SA (eds) Encyclopedia of ocean science, 2nd edn. Elsevier, Amsterdam
Stevenson FJ (1962) Chemical state of the nitrogen in rocks. Geochim Cosmochim Acta 26:797–809
Strous M, Kuenen JG, Fuerst JA, Wagner M (2002) The anammox case–a new experimental manifesto for microbiological eco-physiology. Antonie Leeuwenhoek 81:693–702
Svensen H, Bebout GE, Kronz A et al (2008) Nitrogen geochemistry as a tracer of fluid flow in a hydrothermal vent complex in the Karoo Basin, South Africa. Geochim Cosmochim Acta 72:4929–4947
Thomazo C, Pinti DL, Busigny V, Ader M, Hashizume K, Philippot P (2009) Biological activity and Earth's surface evolutions: insights from carbon, sulfur, nitrogen and iron stable isotopes in the rock record. Comptes Rendus Palevol. doi:10.1016/j.crpv.2009.02.003
Thornton SF, McManus J (1994) Application of organic carbon and nitrogen stable isotope and C/N ratios as source indicators of organic matter provenance in estuarine systems: evidence from the Tay estuary, Scotland. Estuar Coast Shelf Sci 38:219–233
Ueno Y, Yurimoto H, Yoshioka H, Komiya T, Maruyama S (2002) Ion microprobe analysis of graphite from ca. 3.8 Ga metasediments, Isua supracrustal belt, West Greenland: relationship between metamorphism and carbon isotopic composition. Geochim Cosmochim Acta 66:1257–1268
Ueno Y, Yoshioka H, Maruyama S et al (2004) Carbon isotopes and petrography of kerogens in similar to 3.5-Ga hydrothermal silica dikes in the North Pole area, Western Australia. Geochim Cosmochim Acta 68:573–589
Ueno Y, Yamada K, Yoshida N et al (2006) Evidence from fluid inclusions for microbial methanogenesis in the early Archaean era. Nature 440:516–519
Valley JW, O'Neil JR (1981) ^{13}C/^{12}C exchange between calcite and graphite: a possible thermometer in Grenville marbles. Geochim Cosmochim Acta 45:411–419

van Zuilen MA, Lepland A, Arrhenius G (2002) Reassessing the evidence for the earliest traces of life. Nature 418:627–630

van Zuilen MA, Mathew K, Wopenka B et al (2005) Nitrogen and argon isotopic signatures in graphite from the 3.8-Ga-old Isua Supracrustal Belt, Southern West Greenland. Geochim Cosmochim Acta 69:1241–1252

Williams LB, Ferrell RE Jr, Hutcheon I et al (1995) Nitrogen isotope geochemistry of organic matter and mineral during diagenesis and hydrocarbon migration. Geochim Cosmochim Acta 59:765–779

Wlotzka F (1969) Nitrogen. In: Wedepohl KH (ed) Handbook of geochemistry. Springer, Berlin

Yamaguchi K (2002) Geochemistry of Archean-Paleoproterozoic black shales: the early evolution of atmosphere, oceans and biosphere. PhD thesis, Pennsylvania State University, Pennsylvania

Zerkle A, Junium C, Canfield D et al (2008) Production of ^{15}N-depleted biomass during cyanobacterial N_2-fixation at high Fe concentrations. J Geophys Res 113:1–9

Part III
Distinguishing Biological From Abiotically Synthesized Organic Matter in the Early Archean

Integration of Observational and Analytical Methodologies to Characterize Organic Matter in Early Archaean Rocks: Distinguishing Biological from Abiotically Synthesized Carbonaceous Matter Structures

Miryam Glikson, Arthur H. Hickman, Lawrence J. Duck, Suzanne D. Golding, and Robyn E. Webb

Abstract Transmission Electron Microscopy (TEM) was applied to observe and characterize carbonaceous materials (CM) extracted from black cherts and argillite in drill core from the Warrawoona Group of the Pilbara Craton, Western Australia. The black chert came from a 'white smoker type' seafloor deposit in the ca. 3.49 Ga Dresser Formation, whereas the black argillites were obtained from the 3.46 Ga Apex Basalt. The samples were observed and analyzed in TEM combined with electron dispersive spectral analysis (EDS), high resolution TEM (HRTEM) to determine molecular ordering, and C-isotope geochemistry. The TEM and HRTEM observations revealed significant morphological and structural differences between the carbonaceous materials of the Dresser and Apex samples enabling interpretations in terms of primary and secondary origins, as well as metamorphic history.

Organic petrology using reflected light microscopy was used on whole rock samples to observe mineral – organic matter relationship and CM structure relative to host rock texture. The results support an in situ, syn-depositional origin for the Dresser Formation CM. Reflectance % (Ro) of CM determined on polished whole rock samples and polished resin-embedded CM-concentrates enabled the reconstruction of thermal history. Several Ro populations were identified in the Dresser Formation samples: probable microbial cells preserved in fluid inclusions within quartz crystals, thermally degraded CM originally belonging to microbial cells,

M. Glikson (✉)
P.O. Box 553, Palmwoods, Queensland 4555, Australia
e-mail: miryam7@bigpond.com

L.J. Duck and S.D. Golding
School of Earth Sciences, University of Queensland, Brisbane, Queensland 4072, Australia

A.H. Hickman
Geological Survey of Western Australia, 100 Plain Street, East Perth, Western Australia 6004, Australia

R.E. Webb
Centre for Microscopy and Microanalysis, University of Queensland, Brisbane, Queensland 4072, Australia

S.D. Golding and M. Glikson (eds.), *Earliest Life on Earth: Habitats, Environments and Methods of Detection*, DOI 10.1007/978-90-481-8794-2_9,

CM coating mineral grains and reworked CM particles. In contrast, the Apex Basalt samples yielded consistently very high Ro values corresponding to graphite stage organic metamorphism. The weak optical anisotropy of the Apex graphite is inconsistent with formation during regional metamorphism. Two main graphite forms were identified, namely platy and tubular varieties. In HRTEM the tubular form showed nano-tubes and fullerenes within mono-layered spheres. Furthermore, TEM and HRTEM show that the void enclosed mono-layered carbon nano-spheres are more often detached from tubular graphite, forming clusters outside the nanotubes. These forms are a key to the distinction between biologically and abiotically synthesized CM bodies, both by their small size, perfect outline and especially their resistance to thermal degradation.

Dresser Formation samples are isotopically light in the range of −32.1‰ to −38.2‰ consistent with a biological source. Although TEM indicated four distinct types of CM, C-isotope analysis was undertaken on mixed CM concentrates. Nevertheless, the isotopically lighter samples contain a notable input from less thermally degraded low Ro material. On the other hand the isotopically heavier samples contain predominantly thermally degraded high Ro CM. C-isotope compositions of the Apex CM are generally heavier than Dresser samples, between −22.5‰ and −28.6‰ consistent with high thermal stress. The samples show a C-isotope trend in which CM at 143 m depth is isotopically lighter, whereas above and below this level CM becomes increasingly and consistently isotopically heavier. The upper part of the section is dominated by platy graphite with rare nano-tubes. Predominantly tubular graphite and fullerenes characterize sample SAL-13 at 142 m depth coinciding with the isotopically lightest values. Below 143 m CM is less well preserved and predominantly of the fragmented platy graphite type, becoming increasingly isotopically heavier signifying close association with peridotite intrusion.

Keywords Carbonaceous materials (CM) • Transmission electron microscopy (TEM) • Fullerenes • Carbon isotopes • Dresser formation • Apex basalt

1 Introduction

Some of Earth's best preserved rocks containing carbonaceous matter (CM) are found in the 3.53–3.43 Ga Warrawoona Group of Western Australia's Pilbara Craton (Fig. 1) considered a potential host for earliest life on Earth. For the best part of the last 30 years, the Warrawoona Group has been the source of contentious microbial remains (e.g. Awramik et al. 1983; Schopf and Walter 1983; Schopf and Packer 1987; Schopf 1993; Furnes et al. 2004). The search for demonstrable 'microfossils' from such early Archaean rocks has so far focused primarily upon hydrothermal chert dykes that fed into the depositional system. More recently the use of geochemical analytical methodologies (Ueno et al. 2001, 2002, 2004) has thrown light on the possible source of the CM. However, only relatively minor attention has been directed toward an examination of the bedded units of the

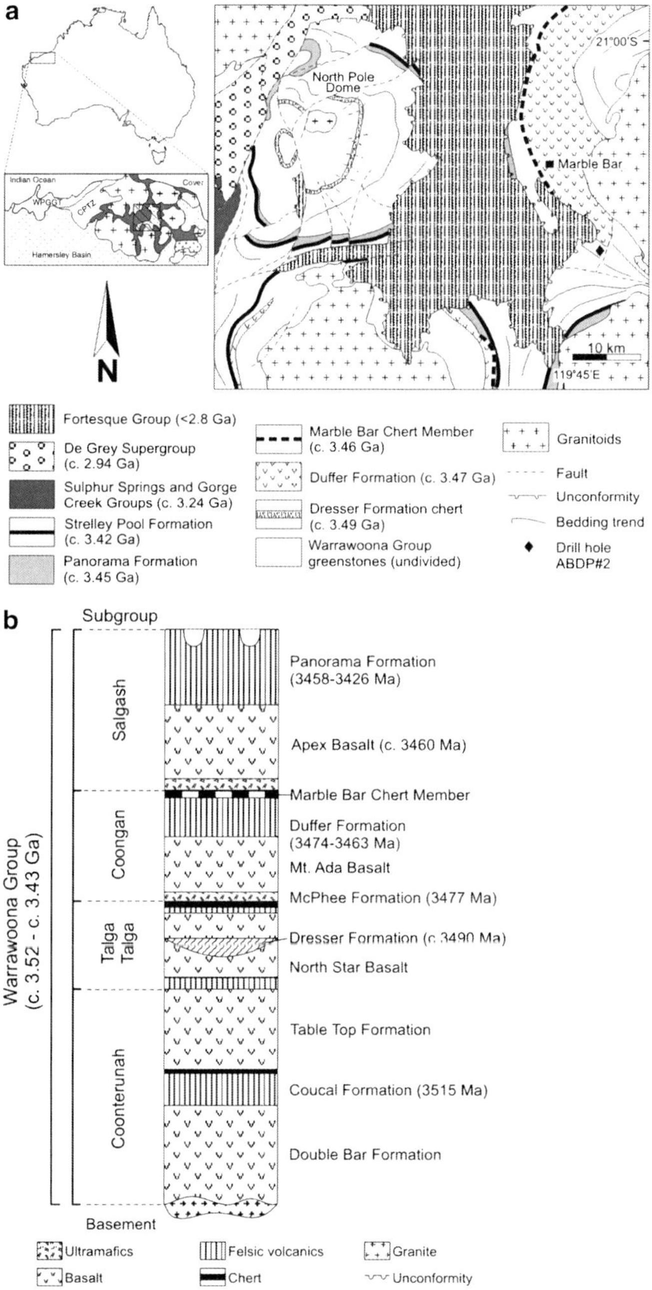

Fig. 1 (**a**) Location map of Dresser and Salgash (Apex Basalt) drillhole and other samples. (**b**) Stratigraphic section of formations dealt with in the study

Table 1 Carbon isotope and major element composition for Dresser Formation CM samples; data for samples DF-7, DF-10, DF-16, DF-17, EZ2-180.5 and EZ2-199.5 from Glikson et al. (2008)

Sample number	$\delta^{13}C$‰	C%	S%	H%	N%	H/C atomic	N/C atomic
DF-1	−37.7	64.63	17.00	0.66	0.23	0.122	0.003
DF-2	−38.2	43.32	35.52	0.53	0.18	0.146	0.003
DF-3	−36.2	40.95	38.05	0.47	0.19	0.138	0.004
DF-5	−36.9	11.15	63.85	0.17	0.07	0.182	0.005
DF-6	−34.8	32.90	44.51	0.16	0.07	0.058	0.002
DF-7	−33.3	14.63	51.07	0 .24	0.03	0.196	0.002
DF-9	−34.9	25.22	50.93	0.04	0.13	0.019	0.004
DF-10	−32.1	68.02	16.41	0.73	0.19	0.128	0.003
DF-16	−34.4	19.77	54.19	0.18	0.12	0.109	0.005
DF-17	−36.5	27.92	52.48	0.16	0.14	0.068	0.004
EZ2-180.5	−34.0	29.90	34.55	1.17	0.26	0.466	0.007
EZ2-199.5	−34.4	8.11	51.79	0.17	0.07	0.250	0.007

Warrawoona Group (e.g. Dunlop et al. 1978; Awramik et al. 1983; Marshall et al. 2007), the seemingly rich CM content of which suggests a most promising source of possible microbial remains (Van Kranendonk 2000, 2006).

The rock samples used in the present study include drill core obtained as part of the Archaean Biosphere drilling project (ABDP, Hickman 2004, 2005) and the Pilbara drilling project (PDP, Van Kranendonk et al. 2006). Drill core obtained from >50 m depth is generally free of near-surface alteration effects, and is unlikely to be contaminated by 'modern' organic matter. Two rock packages within the lower Warrawoona Group (Fig. 1b), were analysed for CM residues; the ~3.49 Ga Dresser Formation of the Talga Talga Subgroup, and the 3.46 Ga Apex Basalt member of the Salgash Subgroup (Warrawoona Group stratigraphy as revised by Van Kranendonk et al. 2006). The Dresser Formation samples reported in the present study were dated using lead isotopes of pyrite (model ages = 3.48–3.52 Ga; Golding et al. this volume). The Apex Basalt was not directly dated, but its age is closely constrained to ~3.46 Ga by the underlying 3.474–3.463 Ga Duffer Formation and the overlying 3.458–3.427 Ga Panorama Formation (Van Kranendonk et al. 2002, 2006). An objective of this project has been the isolation and study of CM sourced from these bedded units, utilising a series of analytical and observational techniques with the aim of determining its nature, thermal history, and possible origin/s. The Dresser Formation CM was compared with the extant hyperthermophilic archaeon *Methanocaldococcus jannaschii* (Glikson et al. 2008), providing further constraints on the origin of the morphologically and structurally distinct entities found in the CM in these very old lithologies (Table 1).

2 Geological Setting of the Warrawoona Group

The Pilbara Craton is a 250,000 km^2 ovoid segment of 3.80–2.83 Ga Archean crust that underlies the north-western part of Western Australia. Most of the southern part of the craton is concealed by overlying 2.77–2.45 Ga rocks of the Fortescue and

Hamersley Basins, but in the north the craton is well exposed within a 60,000 km^2 inlier. Here, the craton comprises six major components, in order of decreasing age: 3.80–3.55 Ga early crust (only very rarely exposed); the 3.53–3.17 Ga East Pilbara Terrane; the 3.27–3.06 Ma West Pilbara Superterrane (three granite-greenstone terranes); the 3.20–2.89 Ga Kurrana Terrane; the 3.02–2.93 Ma De Grey Superbasin (five predominantly sedimentary basins, and three contemporaneous, mainly granitic, supersuites); and the 2.89–2.83 Ga Split Rock Supersuite (late- to post-tectonic granitic intrusions).

The Warrawoona Group is the oldest lithostratigraphic group of the Pilbara Supergroup that comprises the supracrustal succession of the East Pilbara Terrane (Van Kranendonk et al. 2006). The group comprises four predominantly volcanic subgroups (i.e. 3.52–3.49 Ga Coonterunah Subgroup, 3.49–3.48 Ga Talga Talga Subgroup, 3.47–3.46 Ga Coongan Subgroup and 3.46–3.43 Ga Salgash Subgroup). Each of these subgroups is made up of one or more volcanic cycles, commencing with the eruption of komatiite, followed by much thicker (generally 2–4 km) accumulations of pillow basalt flows. The basalts are overlain by dacite and rhyolite formations that typically underlie thin units of clastic sedimentary rocks and chert. Granitic intrusion was closely contemporaneous with felsic volcanism. The volcanic cycles indicate repeated episodes of heating and melting of the middle to lower crust and upper mantle, probably in response to mantle plumes (Van Kranendonk et al. 2002, 2006). Tectonic activity included uplift and extension above granitic intrusions, followed by collapse of the upper parts of the volcanic piles to establish shallow, faulted, hydrothermally active basins. These caldera-like environments lasted for several million years, before being inundated by lavas of the next volcanic cycle. Such post-collapse hydrothermal deposition is recorded at 3.48 Ga (Van Kranendonk 2006), 3.46 Ga (Dimarco and Lowe 1989), and 3.44 Ga (Nijman and de Vries 2004). Volcanic activity ceased at 3.43 Ga, and was followed by a 75 million year period of extensive subaerial erosion, leading to the development of the 30,000 km^2 shallow-water marine to fluviatile basin of the Strelley Pool Formation (Hickman 2008). Overlying basal conglomerate and sandstone, shallow-water carbonate rocks of this formation (now commonly replaced by secondary chert) contain structures widely interpreted as stromatolites (Hofmann et al. 1999; Grey et al. 2002; Allwood et al. 2006, 2007).

2.1 Dresser Formation

The stratigraphic units sampled for the present study are hosted within the East Pilbara Terrane of the east Pilbara Craton (Fig. 1). The oldest of these is the 3,490 ± 10 Ma (Thorpe et al. 1992; Hickman and Van Kranendonk 2008; Van Kranendonk et al. 2008) Dresser Formation that outcrops only in the North Pole Dome locality of the Panorama greenstone belt, approximately 160 km south of Port Hedland and 40 km west of Marble Bar in Western Australia. An upper formation of the Talga Talga Subgroup of the Warrawoona Group (Hickman and

Van Kranendonk 2008), the Dresser Formation is a 1,000 m thick sequence of interbedded black chert, massive barite, minor felsic volcaniclastic arenites and pillowed to massive basalts (Buick and Dunlop 1990; Nijman et al. 1999; Van Kranendonk et al. 2002; Van Kranendonk 2006). This seafloor exhalative hydrothermal system has a mineralogy and architecture consistent with deposition from a white-smoker style vent (Nijman et al. 1999; Huston and Logan 2004; Van Kranendonk and Pirajno 2004), and is interpreted to have formed in water depths of <100 m (Nijman et al. 1999). Along its strike length of 50 km it comprises up to five lenticular members of bedded chert ± barite ± carbonate ± jasper, interbedded with pillow basalts. The dominant lithology is extensively carbonate altered and silicified volcanics (Barley 1993), which is reported as of upper greenschist-lower amphibolite metamorphic facies (Nijman et al. 1999; Van Kranendonk and Pirajno 2004). The Dresser Formation has been recently interpreted as deposited in a tectonically active environment within the caldera of a felsic volcano (Van Kranendonk 2006).

Dresser Formation samples for this study (n = 12) were collected during field trips in 2003 and 2005 from diamond drill cores of the North Pole Dome massive barite deposit, with a focus on the bedded chert-barite units (Glikson et al. 2008).

2.2 *Apex Basalt*

A suite of samples was collected from the ca. 3.46 Apex Basalt of the Salgash Subgroup in the Marble Bar greenstone belt of the East Pilbara Terrane (Fig. 1). This greenstone belt outcrops around the margins of the Mount Edgar Dome, the total area of which is approximately 5,000 km^2. Regional variations of strain in the greenstones, combined with variations in the degree of both contact and burial metamorphism result in metamorphic grades from lower greenschist facies to lower amphibolite facies (Van Kranendonk et al. 2002).

The Apex Basalt overlies the volcanoclastic sedimentary rocks of the Duffer Formation, and consists largely of pillowed Mg-rich and tholeitic basalt containing numerous thin (1–10 m thick) chert units (Hickman 1983; Zegers and van Keken 2001).

Samples (n = 20) were collected from diamond drill core of hole ABDP#2 (Fig. 1a), drilled as part of the ABDP in 2003 (Hickman 2004, 2005), which intersected the clastic sedimentary rocks and ultramafic units in the basal section of the formation. Specifically, samples were drawn from the black slate/argillite horizons of the basal part of the last volcanic cycle of the Warrawoona Group (Hickman 2004). These lower units consist of a series of chert beds intercalated with sulphidic (Cu–Zn–Fe), carbonaceous argillites and sandstones separated by volcanic peridotite flows (Hickman and Van Kranendonk 2008). Recent analysis of the carbonaceous slates showed that they contain from 0.7–5.2% organic carbon (Nedachi 2004).

3 Materials and Methodology

In the laboratory all samples were individually washed with distilled water, before the cores were halved lengthwise on a water-cooled diamond saw and separated into analytical and reference portions. Those portions destined for analysis were halved again, with one half being block embedded in casting resin and polished for organic petrology and reflectance (Ro) measurements. These same polished blocks were later used for SEM-EDS analysis of areas marked during observations in reflected light microscopy. The remaining sample portion was subsequently powdered, as outlined below, for use in elemental and stable isotope determinations, TEM and HRTEM.

The preparation of the CM concentrates was conducted in a clean rock preparation laboratory, where each sample was hydraulically crushed then powdered in a Retsch MM200 stainless steel ball-mill. Powders were subsequently demineralised by HF/HCl acid dissolution to concentrate the carbonaceous material for elemental analysis, C-isotope analysis and TEM. This was achieved by placing a portion (~5 g) of powdered sample in separate 50 ml Falcon centrifuge tubes and dampening with distilled water prior to half filling with warm (50°C) 38% HCl. Sealed and agitated, the tubes were left to heat and settle at 90°C for 2 h in a water bath, prior to the fluid fraction being decanted. Concentrated (50%) hydrofluoric acid (HF) was then added to again half-fill each tube, which were resealed and agitated before placement in a 160°C water bath for 12–15 h in order to increase reaction rate. Samples were subsequently removed from the bath, the fluid phase decanted and warm (50°C) 38% HCl added prior to sample re-agitation. Following settling (1–2 h), the fluid fraction was again decanted and the samples rinsed with distilled water until a near neutral pH was obtained. The remaining concentrate slurry was transferred into ceramic crucibles and oven dried overnight at 60°C. In the final step, the dried residue was powdered in a methanol-rinsed agate mortar before being sealed in labelled Alltech screw-cap 12 × 32 mm glass vials.

Two different techniques were used in the preparation of these samples for TEM analyses. Firstly, sample powders were embedded directly into EPON resin, cut into ultrathin sections 60–70 nm thick using a diamond knife and placed onto form-var coated copper grids or perforated carbon-coated grids for TEM or HRTEM respectively. Alternatively, CM concentrates were ground in 100% ethanol using a mortar and pestle then pipetted onto the grid and left to air dry. This variance in methodology was used in an effort to identify and therefore avoid the classification of any potential artifacts introduced as a part of the resin embedding, and also to allow the three dimensional structure of the CM to be observed.

3.1 Organic Petrology

Samples for organic petrology and Ro determination were prepared as polished whole rock blocks and resin-embedded CM concentrates polished for reflected light microscopy using established techniques (Taylor et al. 1998). Petrological analysis of the mineral matter in association with resident CM was further undertaken in both transmitted and reflected light on polished thin sections. CM reflectance (%Ro) measurements were obtained using a MPV-2 photomicroscope equipped with photomultiplier operating at 546-nm wavelength, with glass standards being used for Ro calibration. Subsequent conversion of %Ro to a temperature scale is based on the Aizawa (1990) correlations of vitrinite Ro and fluid inclusion temperatures, whereby Ro of lipidic organic matter (microbial/algal) conversion to a vitrinite Ro scale has been documented in earlier studies (cf. Saxby et al. 1986; Glikson et al. 1992). However, all organic matter converges at 1.0%Ro (Taylor et al. 1998), and no immature CM is present in the samples of this study. Therefore conversion to vitrinite Ro is considered obsolete in this case and generally in CM from Archaean rocks.

3.2 *Electron Microscopy*

Transmission electron microscopy (TEM) was applied to ultra-thin sections of resin-embedded organic matter concentrates, following the method outlined by Glikson and Taylor (1985). Further TEM investigations were also carried out on grid-adhered loose carbonaceous concentrate powders to eliminate the presence of resin artefacts. Ultrathin sections of freeze-dried *M. jannaschii* cultures were also observed in TEM and HRTEM (Glikson et al. 2008).

Minerals and CM features were analysed by electron dispersive spectrometry (EDS).

Observations and analyses were typically carried out at 80 kV on Jeol 1010 and 2010 transmission electron microscopes. High Resolution TEM (HRTEM) investigations were conducted on both a JEOL 2010 and FEI TECNAI F30 TEM at 200 kV utilising the same sample grids used for conventional TEM observations. A Jeol 6400 field emission SEM equipped with EDS was utilised in back-scatter mode for mineral identification, a process conducted on the same polished blocks used in reflected light observations.

3.3 *C-Isotope Analysis*

$\delta^{13}C$ values were determined with a stable isotope ratio mass spectrometer (GV Isoprime), coupled in continuous flow mode with an elemental analyser (Carlo Erba NA1500). Calibration was performed by use of two standards, USGS24 (−16.1‰ $\delta^{13}C$ V-PDB) and NAT76H (−29.26‰ $\delta^{13}C$ V-PDB), which were interspersed throughout analytical runs, with a third standard, ANU-SUCROSE (−10.5‰ $\delta^{13}C$ V-PDB), being run as an unknown. Each sample was analysed in duplicate, using 50–200 μg of concentrate, which was combusted at 1,020°C in 3.5 × 5 mm tin capsules. Any sample with a beam size outside the working range of 1

$\times 10^{-9}$ to 9×10^{-9} A, or with a result variation between samples of >0.4‰ was reanalyzed, in accordance with laboratory quality control practices.

3.4 *Total Organic Carbon (TOC) and Elemental Analysis*

TOC was determined on a CM5014 Coulometer. The determination involved a two-step process whereby the first of two portions of an individual sample was analysed for total C content by furnace combustion at 910°C (UIC 1995b). Total inorganic C was determined on the other portion of the sample by dissolving it in hot 50% H_2SO_4 (UIC 1995a). TOC was determined from the difference. Carbon (C), sulfur (S), hydrogen (H) and nitrogen (N) contents of both whole rock powders and carbonaceous concentrates were determined on a Carlo Erba Model 1112 EA Elemental Analyser.

4 Results and Discussion

Carbonaceous matter (CM) recovered from 3.49 Ga Dresser Formation black cherts and 3.46 Ga Apex Basalt argillites were examined using organic petrology, TEM, microanalysis and geochemical techniques. The CM when observed in light microscopy displays a marked variation between formations, with that from the Dresser occurring as distinct morphological entities within the cherts. CM predominantly forms bedded strands and bands in the siliceous host rock indicating an in situ syn-depositional origin (Fig. 2). CM also constitutes dispersed particles of

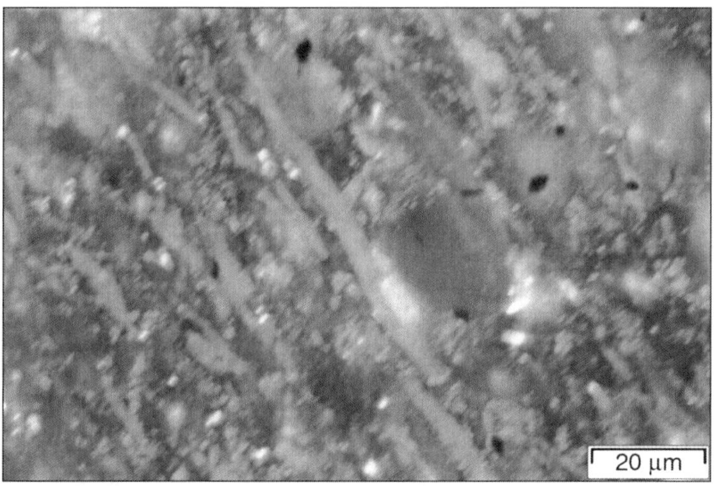

Fig. 2 Carbonaceous matter (CM) in Dresser chert is parallel to bedding. Reflected light microscopy of polished whole rock surface cut perpendicular to bedding

re-mobilized material within host rock and crosscutting veins. The layered CM, which can be up to millimetres in length, is typically composed of clumps of aggregates that vary in size from 1 to >20 μm. The Apex Basalt CM is markedly different and forms platy and rod-shaped bodies in reflected light microscopy. However, the location of the Apex CM within the crenulated bedding planes of the host rock also suggests a syn- depositional origin. The Apex Basalt CM is closely associated with sulfides (pyrite, pyrrhotite, chalcopyrite, sphalerite), and reveals only faint surficial morphological features in light microscopy observations. Additional distinctions and detailed features are revealed using electron microscopical techniques, mainly TEM and HRTEM (cf. Glikson et al. 2008).

4.1 Dresser Formation Carbonaceous Matter

Several morphologically and physically distinct CM types from the Dresser Formation cherts have been identified and characterized (Table 2). The first type described here will be referred to as CM-D. Type D CM is clearly visible in TEM mode and forms coating of mineral grains, or fill of spaces between mineral grains (Fig. 3a, b). Voids in the CM concentrate as seen in TEM represent minerals removed from the rock during processing. This CM is amorphous in nature and lacks any distinguishable internal or specific surface structure apart from occasional

Table 2 Types and characteristics in TEM of CM in Dresser Formation

CM type	Structural characteristics	Ro%	Comments
Type A	Porous loosely adhering aggregates. Single bodies characterized by very small central cavity.	2.5–2.8	Predominant CM in all Dresser samples. In light microscopy appears as silica-impregnated more or less rounded bodies. Thermally degraded microbial colonies suggested as source.
Type A1	Differs from type A by higher density and absence of distinct porosity.	3.1–3.8	Often partially enclosed by multi-septate 'cell' wall. Interpreted microbial origin.
Type B	Irregular-shaped shards/particles randomly distributed through Dresser samples.	3.2–4.8	Extreme hardness of these particles expressed as 'chatter' lines. Type B is the fragmented, re-worked type D.
Type C	Spherical to elliptical, thin-walled vacuous bodies. Remnant of a surrounding membrane occasionally present.	1.2–1.8	Interpreted as the primary cells of microbial remains possibly preserved in fluid inclusions in quartz crystals.
Type D	CM coating mineral grains.	2.8–3.1	CM that has softened following thermal effect and re-solidified around mineral grains. This type of CM may be abiotically-formed by FTT reactions in the seafloor hydrothermal system.

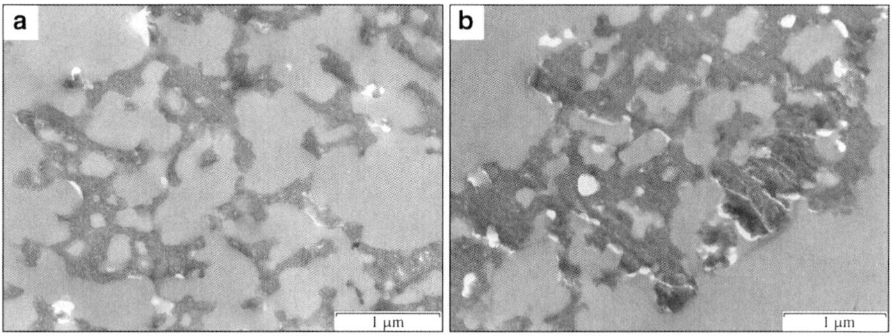

Fig. 3 (**a**, **b**). TEM observation of CM coating mineral grains. The latter are represented by voids after dissolution following CM concentration. CM is amorphous as no primary structure has been preserved during softening/liquefaction

'chatter' lines (artifacts formed during sectioning of very hard material) as in type B. The observed homogenous nature and mode of coating of mineral grains of this type is consistent with re-solidification from CM that softened or fluidized as a result of being subjected to temperatures >200°C, following the destruction of any primary structures (Glikson 2001; Glikson et al. 2006; Duck et al. 2007). Softening of carbonaceous matter has been demonstrated experimentally to occur at temperatures around 400°C (Hower 1989). The degree of softening and fluidization of organic matter depends on its original composition (Khavari-Khorasani and Michelsen 1993), which would control molecular fluidity. This CM yields some of the higher Ro values recorded from the Dresser Formation samples, and may represent material of microbial origin subjected to extensive thermal/hydrothermal effects (Table 2). This may have occurred when the sediments were affected by renewed hydrothermal activity. A different primary composition of this material, which in younger sediments may be responsible for the different response to thermal stress and subsequently to a different mode of preservation, is unlikely when dealing with Archaean CM. On the other hand, there is a distinct resemblance of this CM to that reported and illustrated by De Gregorio and Sharp (2003) and De Gregorio et al. (this volume) coating quartz grains in Apex chert. The same study interprets this CM as having precipitated from suspension in hydrothermal fluids syngenetically with the precipitation of quartz. An abiotic source is attributed to this CM by De Gregorio and Sharp (2003, 2006) and De Gregorio et al. (2005). The distinction of 'chatter' lines in TEM observations of this CM type at times points to high temperature thermal oxidation as discussed below. It has been suggested by De Gregorio (personal communication) that this type of CM may have been formed abiotically in close proximity of venting by FTT reactions documented from these environments. It has been suggested previously that both biologically-formed as well as abiotically synthesized CM may be present in the Dresser Formation rocks (e.g. Philippot et al. 2006).

Another CM type observed as minute dispersed particles in reflected light microscopy, and of high Ro, shows distinct 'chatter' marks when observed in TEM

Fig. 4 Oxidized, reworked and fragmented CM shards/particles of type B, showing characteristic '*chatter*' lines. The latter are parallel to the cutting direction, and occur when hard CM is ultra-sectioned

mode (Fig. 4, 3b). Chatter marks are a typical hardness indicator (O'Brien and McCully 1981). A close resemblance to the coal maceral 'inertinite' is evident, the latter being an oxidation product. This material, termed type 'B' in an earlier paper (Glikson et al. 2008) is a typical thermal oxidation product of CM recycled from higher temperature zones within the depositional environment, and as such is inert and does not lend itself to softening during subsequent thermal/hydrothermal stress (Taylor et al. 1998). The fragmented often shard-like nature of this CM with no distinct morphology may suggest it had its origin in a shattered type D. The co-occurrence of such material with in situ parallel-to-bedding CM in the same samples is usually explained by transportation from surrounding areas (Dow 1977). This is supported by its dispersed mode within the host rock matrix in contrast to the parallel -to-bedding form of the predominant CM type A reported in a previous study (Glikson et al. 2008). The combination of higher Ro values of CM and lower Ro value CM in the same samples has been attributed to inclusion of transported/recycled CM (Dow 1977), which was brought to the depositional site to be mixed with in situ CM. HRTEM of type B highest Ro group CM shows poorly graphitized molecular ordering such as is expected from oxidized/re-worked CM (Oberlin et al. 1980).

A third type of CM which has been designated in an earlier study 'A' and 'A1' when observed in reflected or transmitted light microscopy is seen in top view as blobs of rounded to sub-rounded particles ranging in size from <1 to >20 μm (Fig. 5a, b). This CM is a characteristic feature in all Dresser Formation, and renders very low Ro values in whole rock samples as a result of Si impregnation of the bulk

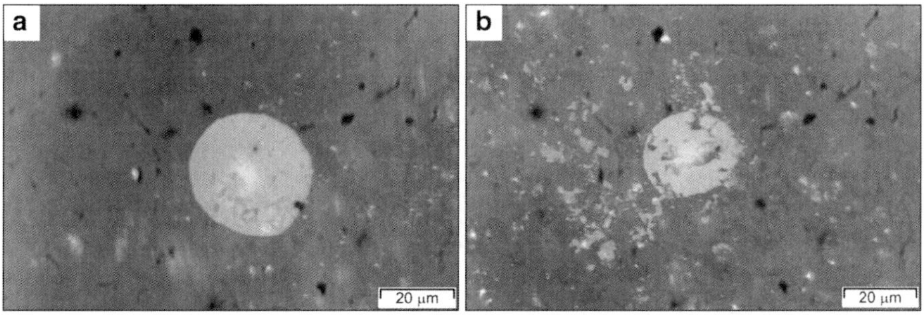

Fig. 5 (**a**, **b**) Top view of CM in Dresser samples as observed in reflected light microscopy of polished whole rock cut parallel to bedding, (**c**) EDS of the above highlighting the intimate association of silica with CM in Dresser samples

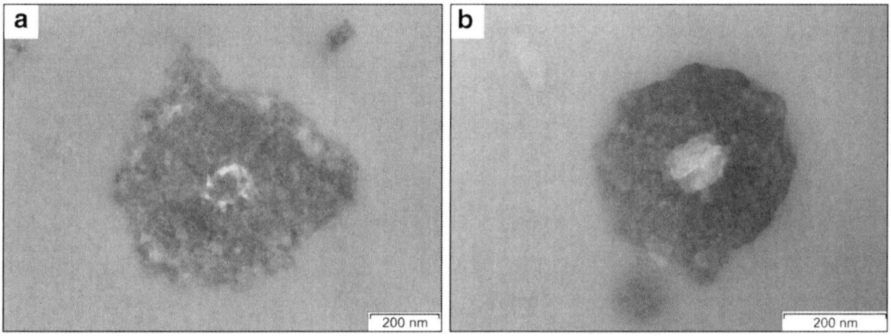

Fig. 6 (**a**, **b**) Porous CM entities (type A) observed in TEM of CM concentrate where silica has been dissolved leaving micro-pores

CM blobs (Glikson et al. 2008). Closer examination and EDS analysis confirm that minute pore spaces within these bodies are filled with microcrystalline silica (Fig. 5c). CM concentrates of type A in resin-embedded pellets give Ro values significantly higher than those obtained from silica-impregnated CM in whole rock samples. When observed as a CM concentrate in TEM, the bodies appear as porous, loosely packed carbonaceous material (Glikson et al. 2008). The porous nature of this CM, the result of post-thermal degradation provided conduits and voids for infusion and infiltration of silica – saturated fluids in the hydrothermal environment, filling voids and spaces that formed following the death of a microbial colony (Fig. 6a, b). The process of silica infiltration would have enhanced CM preservation. This interpretation is reinforced through TEM observations of cultured *M. jannaschii*, autoclaved at 132°C, in which the mounting resin was seen to fill spaces formed by the breakdown and ejection of cellular material (Glikson et al. 2008). Often a small central cavity observed in TEM (Fig. 6a, b) is seen to be filled with a single silica grain when observed as a whole rock sample in reflected light. The silica grain having been dissolved in the CM concentration process has left a void. The latter is interpreted to be the remains of the cell cavity as evidenced from thermal degradation simulation of live cultures of *M. jannaschii* (Glikson et al. 2008). The intimate association of CM and silica in the Dresser samples is reminiscent of silicified microbial colonies, such as have been documented from active hydrothermal systems by Al-Hanbali et al. (2001) who suggested these as analogues of Archaean hydrothermal sites.

Close similarities appear between the type A material, and a model of protoarchaea illustrated by Martin and Russell (2003). Some of the spherical or sub-spherical entities of type A CM, appear as densely spaced aggregates devoid of the siliceous component (Fig. 7a, b), in an earlier study designated as type 'A1' (Glikson et al. 2008). Type A1 forms occasionally have a defined rim in cross section and display highly developed molecular ordering in HRTEM (Fig. 7c). The rims are suggestive of outer walls of cells that have become partially or entirely detached from the cells during thermal degradation, and display a marked resemblance to TEM-observed specimens of *M. jannaschii* subjected to thermal degradation (Glikson et al. 2008). This mode of detachment is replicated after subjecting cells of *M. jannaschii* grown

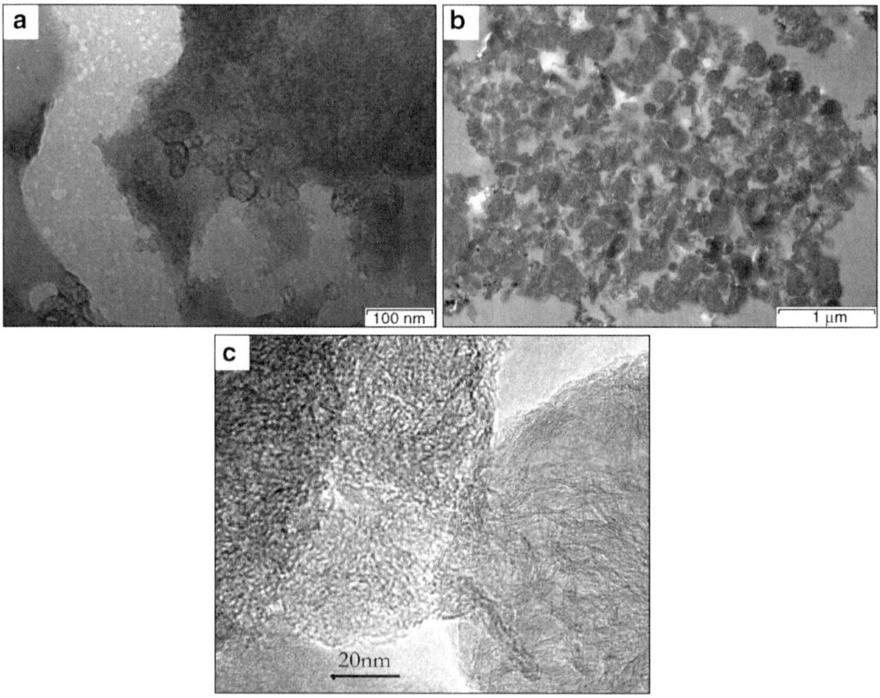

Fig. 7 (**a**, **b**) Spherical CM lacking in distinctive porosity (type A1), showing an inner body surrounded by a faintly outlined wall of higher electron density (7A) than the inner body. (**c**) HRTEM of type A and A1. The difference in molecular ordering is evident. The body on the left hand side of the micrograph shows a lower molecular order of type A, whereas the body on the right hand side shows high molecular order of sub-graphite, characterizing the denser type A1

at 79°C to autoclaving at 100°C (1 atm) and 132°C (2 atm), in an effort to simulate in situ degradation by hydrothermal stress (Glikson et al. 2008). Type A1 appears to be a more cohesive, non porous form of type A displaying a better defined cell shape. The lack of porosity prevented infiltration of SiO_2-rich fluids and silica precipitation in micro-pores within the cell. Porosity was a result of severe breakdown following thermal degradation. It is not clear why some cells escaped such severe disintegration. The answer may lie in positioning of certain cells in relation to hydrothermal fluid flow, which would have been an enhancing agent in tearing apart the CM residue. Variations in temperature within the medium may also be a contributing factor to differential preservation. Furthermore, it has been suggested (De Gregorio, 2009) that microbial cells will likely be clumping together, and those in the centre of clumps will be somewhat shielded from invading fluids. Similar variations in degradation were observed in the simulation of thermal maturation of *M. jannaschii* (Glikson et al. 2008).

Type 'C' (Glikson et al. 2008) is assigned to the relatively rare occurrence of spherical to egg-shaped electron transparent globules that appear in clusters among fragments of type B carbonaceous material and grainy porous type A (Fig. 8). A fuzzy material, the remains of an outer membrane -like structure, is occasionally seen to partially surround the globular cells. The relatively low electron density of this material as well as the clearly defined outline and hollow interior point to material that has not been subjected to high thermal stress and thermal degradation. In an earlier study (Glikson et al. 2008) we have interpreted this material to be primary microbial cells preserved within fluid inclusions in quartz crystals; the latter having been dissolved during CM concentration proce-

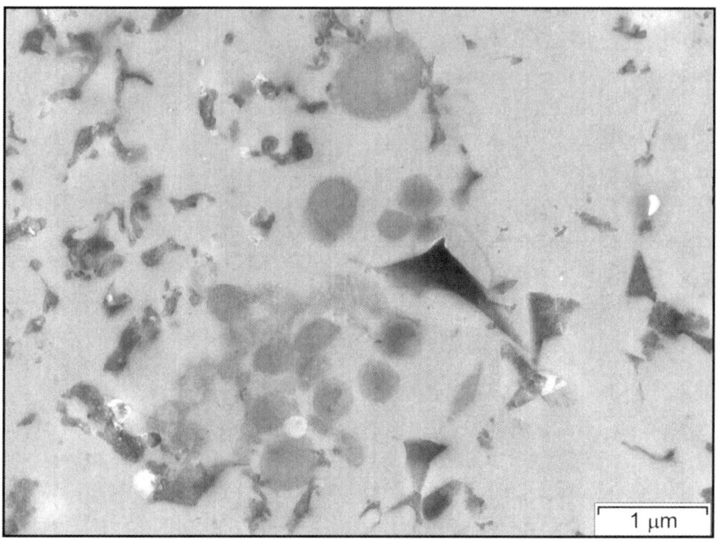

Fig. 8 A cluster of type C CM, representing the primary thermally unaffected cells preserved in fluid inclusions within quartz mineral grains

dure, thus liberating the cells within. The lower Ro of this material confirms relatively low thermal stress at the time of deposition, as also suggested by previous studies (Ueno et al. 2004).

4.2 Apex Basalt Carbonaceous Matter

The younger Apex Basalt, represented by samples from ABDP#2 (Salgash) drillhole (Table 2), is significantly different from the Dresser Formation samples in Ro values, organic petrology, TEM/HRTEM and carbon isotopic composition. Organic petrology has shown the CM to be graphitic material, with the measured Ro on carbonaceous concentrates in the range of 10–13%, outside the scale of what is considered 'normal' for common thermal or hydrothermal effects on CM (Taylor et al. 1998).

The noteworthy difference between the Apex Basalt 'graphite' and the standard graphite mineral is the absence of pronounced bi-reflectance in the Apex material, suggestive of different processes responsible for its formation. Standard graphite is the product of regional metamorphism coupled with deformation, with strain expressed as optical anisotropy. On the other hand it has been shown experimentally that very rapid heating of CM may lead to high Ro without optical anisotropy (Khavari-Khorasani and Michelsen 1993). Therefore the anomalous optical properties of Apex graphite may have been the result of rapid, very high temperature thermal effect in the absence of deformation. Such a high temperature process (>1,000°C) would have brought about the volatilization of hydrocarbons, CO_2 and other gaseous compounds from heating the initial CM, resulting in graphitization of the residual carbonaceous material. Processes responsible for rapid high temperature effects may be magma intrusion, intense volcanism or bolide impact. A bolide impact event has been documented from the Mount Ada Basalt of the North Pole Dome (Byerly et al. 2002) that occurs about 2 km below the Duffer Formation and Marble Bar Chert Member, and thus cannot have effected the younger Apex Basalt. The rapid injection of very high temperature (around 1,500°C) peridotite magmas may explain the effect on the CM and its unusual optical properties. The presence of fullerenes within nano-tubes and on a platy kind of graphite form in these samples further supports very high thermal effect, a mechanism leading to graphitization of in situ CM. Fullerenes in abundance were found in the Murchison and Allende meteorites (Becker et al. 2000). Becker et al. (2000) and Heymann et al. (1994) have also reported fullerenes in sedimentary rocks at the K-T boundary and the P-T boundary, both of which are associated with major meteorite impacts. In addition, Buseck et al. (1992) reported fullerenes from Shungite. Shungite is an unusual carbonacous deposit the formation of which is uncertain, although it is associated with extensive volcanic flows which may have been triggered by a bolide impact. The CM of the Shungite consists of two distinct kinds that greatly differ from each other, namely a bright glassy type with conchoidal fracturing and a dull friable CM.

The bright glassy type is the only one containing fullerenes (Buseck et al. 1992; Parthasarathy et al. 1998). This type displays high Ro equivalent to temperatures >400°C and lacks pronounced optical anisotropy as reported by the present author (in Mastalerz et al. 2000), in contrast to CM underlying the massive Bushveld Complex that displays extreme bi-reflectance, the result of deformation (Glikson and Taylor 2000).

Temperatures exceeding 1,000°C experienced by the Apex Basalt rocks in the lead up to graphitization and the resulting high thermal maturation state (10–13% Ro) support Melker et al. (2000) computer simulation of fullerene formation at similarly high temperatures where perfect fullerene structure and stability can be achieved. The same study found that at lower temperatures incomplete fullerene structures are formed.

Observations using TEM reveal two main forms of graphite: a common platy form (Fig. 9a, b), and a tubular form (Fig. 10a–d). Two different structural entities of graphite have also been reported by Buseck et al. (1985) in a sample from a black slate in a chlorite metamorphic zone. Observations at higher TEM magnification of the platy graphite reveal a spongy texture with dark electron dense spots (Fig. 9b), which in HRTEM appear as detached nano-size particles and occasionally show fullerene molecular organization.

A closer look at higher TEM magnification of the tubular graphite reveals occasional fullerene texture, which is also suggested from cross sections of individual spheres within the nano-tubes. Separate spheres, detached from the nano-tubes, are seen as void-encapsulating carbonaceous clusters with rounded to sub-rounded encasing monolayer (Fig. 11a) closely resembling those observed and illustrated using TEM and HRTEM in CM from the Allende meteorite (Harris and Vis 2003) and Lake Tagish meteorite (Garvie and Buseck 2004). HRTEM reveals very rare fullerene structure present within some of these detached spheres observed in our study (Fig. 11b). Most spheres appear to have a hollow interior suggesting either loss of their fullerenes while retaining only the outer mono-layered envelope, or alternately fullerenes have not been formed

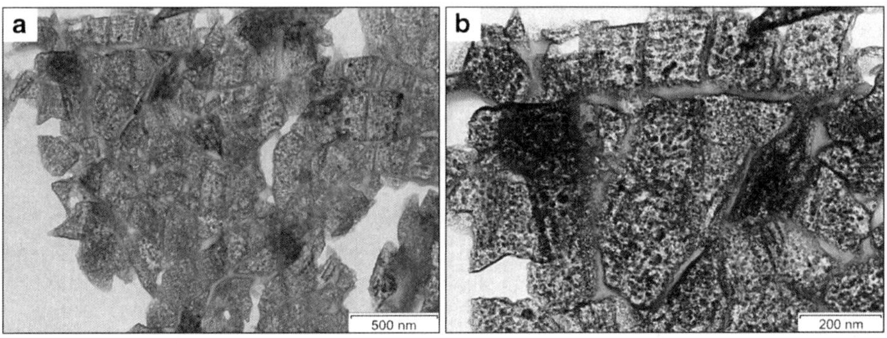

Fig. 9 (**a, b**) Platy graphite. At higher magnification TEM in B, the 'spongy' look of the plates is observed with dark electron dense spots scattered within the plates

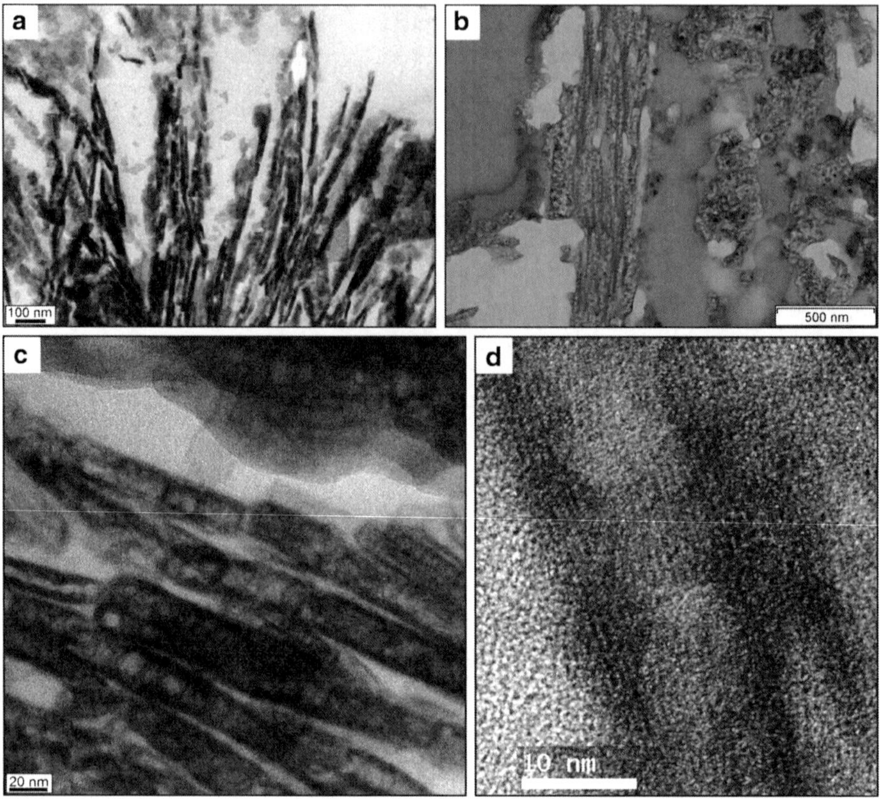

Fig. 10 (**a**) Tubular graphite at relatively low magnification TEM. (**b**) Tubular graphite at higher magnification TEM, showing some internal structure. (**c**) Tubular graphite in HRTEM showing detail of individual nano-tubes within tubes. The latter are distinguished by partitions marking compartments or vacuous chambers within the tubes. (**d**) HRTEM highlights the string-like round structure of the vacuous chambers within the nano-tubes, which contained the mono-layered hollow spheres that may contain fullerenes. Nano-tube walls (outer monolayer) exhibit perfect crystalline graphitic molecular order

within most of these carbon spheres. The second alternative is adopted here as most likely due to a probable varied temperature and pressure regime within the carbonaceous muds during peridotite injection into the sediments. Thus conditions for fullerene formation have not been achieved throughout the entire carbonaceous sediment package.

The tubular graphite in HRTEM reveals the distinct highest level of molecular ordering characteristic of graphite with long, closely spaced, continuous parallel lattice fringes (Fig. 10d). This form of graphite structure has been documented in earlier studies (Buseck and Bo-Jun 1985; Buseck et al. 1988). However, in the Apex CM, this typical graphitic molecular structure makes up the external mono-layer of the nano-tubes. Within the nano-tubes TEM and HRTEM reveal closely-spaced vacuous sphere spaces <5 nm in diameter, in an almost continuous string-like

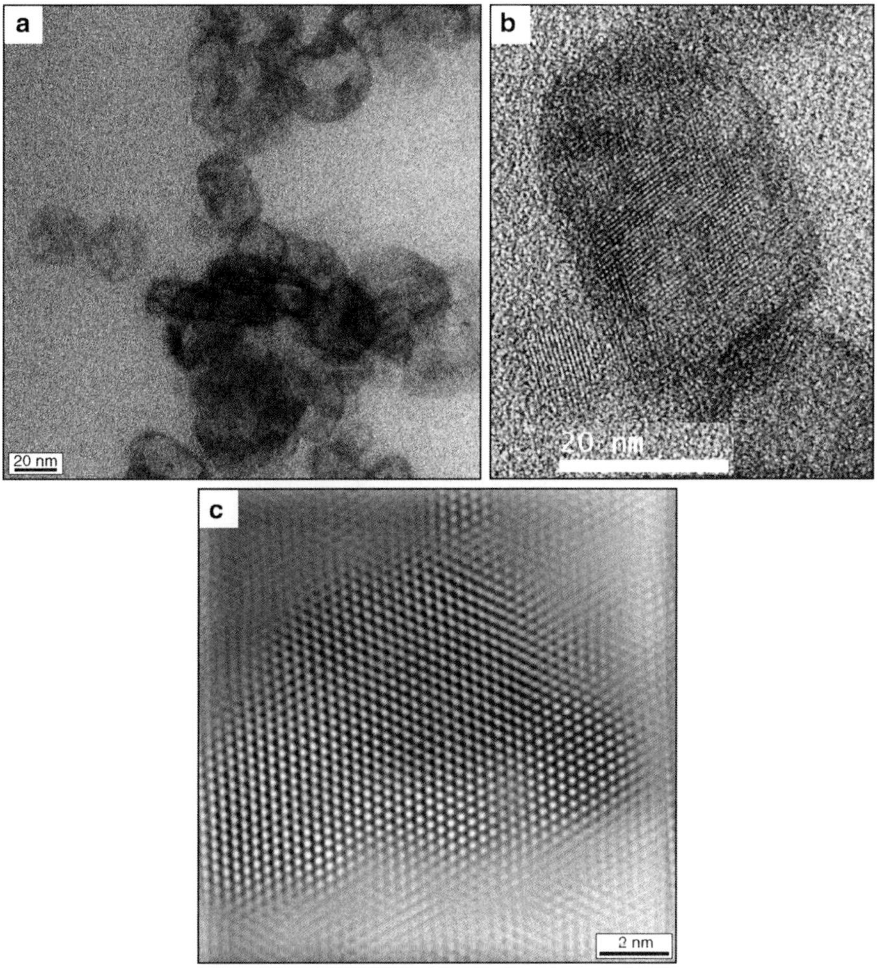

Fig. 11 (**a**) HRTEM of spheres which have been detached from the nano-tubes, closely resembling forms observed in the Allende meteorite, (**b**, **c**) Fullerene structures in detached spheres

formation (Fig. 10c) suggesting the fallout of nano-spheres that likely occupied these spherical spaces. Further HRTEM observations focusing on these areas suggest indeed loss of a sphere-like entity (Fig. 11). A nano-tube as in the Apex CM is identical to type D of Akbulut (2008), who demonstrated experimentally the formation of fullerenes within mono to multilayered nano-tubes. According to the same study, fullerenes are an intermediate stage between conventional graphite and nano-tubes. Another study (Iijima 1991) of graphitic carbon synthesized using arc evaporation discharge noted that this form of graphite, as well as soot and fullerenes, may form directly from vapor phase deposition of carbon. Although graphite may form at temperatures >500°C, nano-tubes are known to form in the laboratory at 900°C (Akbulut 2008), and the intermediate fullerenes at 600–1,000°C. The presence of all forms in the same samples representing different and varied temperatures suggests mixing of CM during the injection pulses of ultra basic peridotite magma/volcanism under pressure into carbonaceous muds on the sea floor.

4.3 Elemental and Carbon Isotope Compositions of CM

Dresser Formation percent elemental H (Table 1) and N in most samples, show a strong positive correlation with C (Fig. 12a). The strong positive correlation trends between carbon and hydrogen, and carbon and nitrogen, indicate that these three elements are intimately associated with the CM, which is supportive of a biogenic origin for this material (Durand and Monin 1980). Conversely, the strong negative

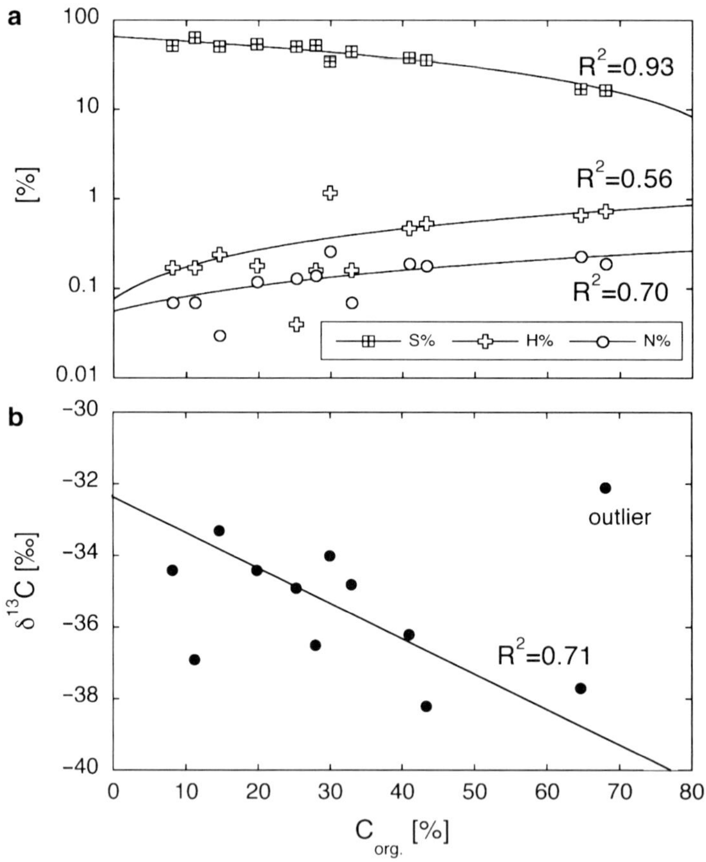

Fig. 12 (**a**) Relationship between elemental compositions in Dresser Formation CM samples. (**b**) Relationship between C-isotope composition and total organic carbon in Dresser concentrates. High sulfide content in organic concentrates is a good proxy for thermal/hydrothermal alteration of samples

correlation that exists between carbon and sulfur indicates inorganic association and a non-biogenic source for the sulfur. Generally, both H/C and N/C ratios are extremely low, as expected in Archaean sediments (Brocks et al. 2003). Sample DF-10 is dominated by high Ro material and also displays very low N/C. As the bulk of the samples were analyzed, the differences in the elemental composition may be attributed to predominance of different Ro populations in different samples. Thus extremely low H/C ratios indicate predominance of previously oxidized/recycled material such as type B. However, as all these samples have been subjected to high thermal stress, the H/C ratios are generally low in all samples as expected (Clayton and Bostick 1986).

Carbon isotopic compositions of the Dresser Formation samples (Table 1, Fig. 12b), range from −32.1‰ to −38.2‰, and are considered typical of isotopic fractionations produced by biogenic processes (Schidlowski et al. 1983; Schidlowski 1988, 2001; House et al. 2003). The values obtained in our study compare favorably with results previously reported from Pilbara samples of the same age (Glikson and Taylor 2000; Ueno et al. 2001, 2004). Ueno et al. (2004) suggested that the ^{13}C-depleted compositions of CM in silica dykes cutting the metabasalt footwall of the Dresser Formation were likely the product of anaerobic chemoautotrophs such as methanogens rather than aerobic autotrophy or abiogenic synthesis. Ueno et al. (2004) also reported carbon isotope compositions ranging from −29.4‰ to −31.2‰ for four bedded chert samples of the Dresser Formation. However, C-isotopic compositions alone may not provide an undisputed indicator of biogenic activity according to a study by McCollom and Seewald (2006). The same study obtained isotopic values in the range of −45‰ to −50‰ for specific organic compounds synthesized abiotically under hydrothermal conditions. The range of isotopic composition is thus quite different from that in the Dresser Formation CM. This may indicate different primary carbonaceous compounds from those obtained in McCollom and Seewald's study, or as also suggested by the same authors, the Dresser isotopic compositional range may be the result of high temperature thermal effects that is supported by Ro values in the present study. As all samples analyzed contain a mix of several carbonaceous matter types equivalent to several Ro populations, it is not possible to differentiate between them on the basis of isotopic compositions. Nevertheless, it may be true to say that most of the CM in the Dresser samples has not been subjected to temperatures >250°C (Ro equivalent of 3.0%; Aizawa 1990) or temperatures >300°C (Ro equivalent according to Price 1983) such as are expected to rupture ^{13}C–^{12}C and ^{13}C–^{13}C bonds and render the CM significantly isotopically heavier. Significant isotopic temperature effects are known to be noticeable at Ro = 3.0, with a more pronounced effect in aromatics than saturates (Clayton and Bostick 1986). Thermal maturation thus invariably enriches the residual OM in ^{13}C. The mixed Ro populations provide evidence for temperature varied thermal/hydrothermal effect on the CM, reflected in the $\delta^{13}C$ range from −32.1‰ to −38.2‰ in the Dresser data set.

In contrast to the Dresser, the Apex samples (Table 3) are enriched in $\delta^{13}C$ (values predominantly between −22.5‰ and −28.4‰), which is a direct effect of the extremely high temperatures that were also responsible for the graphitization of the

Table 3 Carbon isotope composition of Apex Basalt CM

Sample number	Depth (m)	$\delta^{13}C‰$
SAL-1	55.50	−25.2
SAL-2	56.95	−27.4
SAL-3	57.60	−27.0
SAL-4	58.00	−27.6
SAL-5	58.70	−28.2
SAL-6	60.30	−28.6
SAL-7	67.55	−26.3
SAL-8	67.90	−26.7
SAL-10	85.40	−28.4
SAL-11	102.00	−27.6
SAL-12	108.00	−29.8
SAL-13	142.30	−30.4
SAL-14	158.40	−26.9
SAL-15	173.30	−24.1
SAL-16	185.25	−24.6
SAL-17	186.15	−24.4
SAL-18	192.10	−23.8
SAL-19	203.50	−23.6
SAL-20	219.30	−22.5

CM. The range of $\delta^{13}C$ values of some 8‰ is larger than expected for normal thermal maturation (Des Marais 1997), which further supports unusual thermal processes responsible for the organic metamorphism encountered in these samples.

A C-isotope trend in the Apex suite (Fig. 13) shows the samples becoming progressively isotopically lighter from the top of the borehole at 55.5 m downward to a sample taken at 142 m depth, ranging thus from −25.2‰ at the top to −30.4‰ at 142 m depth. An inverse trend is then encountered below the above section, where the CM becomes progressively and consistently isotopically heavier towards the bottom of the sampled section, with a range from around −30.0‰ at the top of that segment to −22.5‰ at the bottom. The 'heaviest' values at the bottom of the section must have experienced highest temperatures. The overlying inverted trend whereby the isotopically heaviest values are at the top are difficult to explain without involving an overlying heat source (e.g. Maricko Basin, Mount Isa; Glikson et al. 2000, 2006). However, differences in molecular ordering of CM as a result of its response to thermal stress should also be considered following the observations of Buseck et al. (1988) who have discussed the isotopic variation among graphites, in which two texturally different graphites were also distinguishable by differences in their carbon isotope composition. A distinction between two graphites, namely tubular and platy was also observed in our samples although the samples in the borehole generally yielded a mixture of the graphite types. Nevertheless, platy graphite appeared to be more common in the top samples and predominant in the lowest drill-hole samples. Tubular graphite with nano-tubes and fullerenes was the major CM component in sample at 142 m depth with isotopically lightest values, and

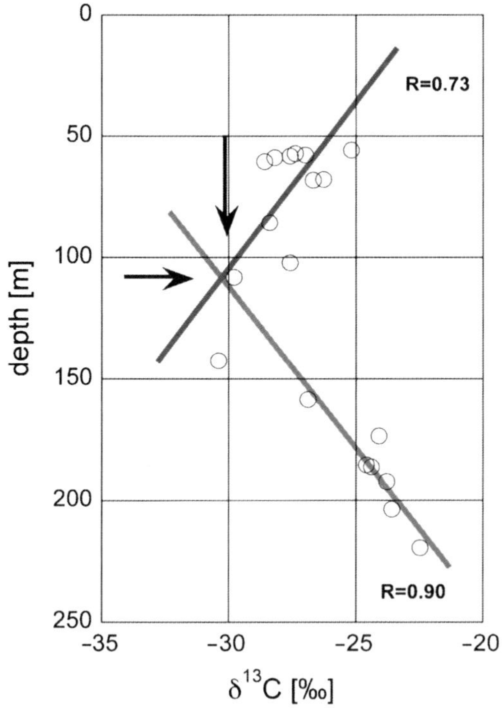

Fig. 13 C-isotope composition of Apex samples with depth showing an inverse trend down to about 115 m depth, and a more normal trend from there on downwards. The lower correlation coefficient of the upper part of the section is also indicative of the mixed nature of the graphite compared to the significantly more homogeneous graphite in the lower part of the section

became scarcer away from that point both upwards, and down the hole, with diminishing presence of tubular graphite type.

A study by Binu-Lal et al. (2003) reported preferential enrichment in $\delta^{13}C$ as a result of partial melting and re-precipitation of earlier graphite. Pulses of high temperature peridotite injected into the Apex sequence may have remobilized and partially melted the initially formed tubular graphite formed during preceding peridotite injection. The latter re-solidified as platy graphite and experienced further loss of the light isotope of carbon.

5 Summary and Conclusions

Various organic compounds, including graphitic carbon, can be formed abiotically in hydrothermal systems, in addition to possible extraterrestrial source (e.g. Ehrenfreund et al. 2002; Horita 2005), such that evidence for early life must necessarily combine geological, structural, ultra-structural, visual and geochemical data to be compelling. In our current study, carbonaceous materials have been thoroughly observed by applying microscopy and microanalysis. A reconstruction of some of the CM based on comparison with a living microbial analogue was reported earlier (Glikson et al. 2008).

TEM has been shown as a powerful tool in resolving fine structure in carbonaceous materials in sediments, and its integration with analytical methodologies is most rewarding in defining and characterizing organic matter that underwent severe thermal degradation. HRTEM further enhances the understanding of CM by observing its molecular ordering.

Carbonaceous materials obtained from the 3.49 Ga Dresser Formation and 3.46 Ga Apex Basalt of the Pilbara's Warrawoona Group, have been identified and characterized based on organic petrology, structural details observed in TEM, and molecular ordering determined in HRTEM. These were matched with elemental and C-isotope compositions. It is noteworthy that the average C-isotopic composition of Apex CM is −26.0‰ compared to −17‰ to −20‰ of standard graphite (Tissot and Welte 1984).

Several types of CM have been identified in the Dresser Formation samples. These have been classified based on their Ro values, structure as revealed in TEM and molecular ordering highlighted by HRTEM.

Reflectance determination (Ro%) of the CM demonstrate thermal/hydrothermal effects, which were responsible for the thermal degradation of the primary CM in the Dresser Formation samples. Variations in structure and texture aid in classification and distinction between recycled and in situ CM. It is possible that biogenic and abiotic (type 'D') CM are present in the same samples.

Carbonaceous material (CM) occurs in both dispersed and layered forms, interlayered with fine-grained silica. The intimate association of the CM and formerly amorphous silica strongly resembles silicified microbial colonies from active hydrothermal systems, which have been previously proposed as analogues of Archaean hydrothermal sites. Furthermore, the geological setting and mineralogy of the 3.49 Ga Dresser Formation support its formational environment as a silica-barite dominated seafloor hydrothermal deposit, most likely analogous to modern "white smokers".

The younger Apex Basalt Formation of the Salgash Subgroup yielded consistently very high Ro values. The carbonaceous material from the Apex Basalt samples has the appearance of graphite in reflected light microscopy. However, the Apex graphite differs optically from the typical mineral graphite, a metamorphic carbonaceous rock, with distinct optical anisotropy that is the result of the deformation commonly accompanying regional metamorphism. The 'graphite' in the Apex samples lacks distinct anisotropy suggesting a different mode of formation. Rapid heating at very high temperature in an isotropic pressure regime such as during injection of magma into seafloor carbonaceous muds is suggested for the Apex Basalt graphite.

Relatively heavy $\delta^{13}C$ of the Apex material is in contrast with the markedly lighter isotopic compositions of the Dresser Formation CM supporting extremely high temperatures in the formation of the Apex CM. Within the Apex suite, isotopically lighter values coincide with fullerene and nano-tube abundance, whereas isotopically heavier values are typically associated with predominance of platy graphite. Thus two types of graphite observed in TEM are also distinguished by their isotopic composition.

TEM and HRTEM of the Apex CM revealed fullerenes occasionally encased in spheres residing within nano-tubes. The hollow mono-layered minute spheres are often observed as clusters released from nano-tubes. As these entities formed at temperatures of >1,000°C, it may demonstrate that structures resembling microbial cells can form abiotically. In contrast, biologically produced cells undergo thermal degradation leading to their breakdown and disintegration at temperatures of 100°C, marking the limits of life. Thermal degradation of CM may be used as an indicator of biogenic source.

Acknowledgments Thanks are due to the Western Australian Geological Survey for access to ABDP#2 drill cores. Graham Rowbottom (OIC Vibrational Spectroscopy Unit, University of Tasmania) is gratefully acknowledged for elemental analyses. Thanks also to Kim Baublys (UQ Stable Isotope Laboratory) for assistance with isotope analytical methodologies and to Rick Webb, Graeme Auchterlonie and Jamie Riches (Centre for Electron Microscopy and Microanalyses) for technical advice. Special thanks to Robert Bolhar (Earth Sciences, UQ) for graphical data presentation and to Bradley De Gregorio for constructive and valuable comments. Arthur Hickman publishes with permission of the Executive Director of the Geological Survey of Western Australia. The study has been funded by an Australian Research Council grant (ARC- Discovery).

References

Aizawa J (1990) Paleotemperatures from fluid inclusions and coal rank of carbonaceous material of the tertiary formations in northwestern Kyushu, Japan. J Jpn Assoc Miner Petrol Econ Geol 85:145–154

Akbulut A (2008) Nanotube structure. Am Sci 1–64

Al-Hanbali HS, Sowerby SJ, Holm NG (2001) Biogenicity of silicified microbes from a hydrothermal system: relevance to the search for evidence of life on earth and other planets. Earth Planet Sci Lett 191:213–218

Allwood AC, Walter MR, Kamber BS et al (2006) Stromatolite reef from the Early Archaean era of Australia. Nature 441:714–718

Allwood AC, Walter MR, Burch IW et al (2007) 3.43 billion-year-old stromatolite reef from the Pilbara Craton of Western Australia: ecosystem-scale insights to early life on Earth. Precambr Res 158:198–227

Awramik SM, Schopf JW, Walter MR (1983) Filamentous fossil bacteria from the Archean of Western Australia. Precambr Res 20:357–374

Barley ME (1993) Volcanic, sedimentary and tectonostratigraphic environments of the 3.46 Ga Warrawoona Megasequence: a review. Precambr Res 60:47–67

Becker L, Poreda RJ, Bunch TE (2000) Advances in space research: stability of interstellar fullerenes. Proc Natl Acad Sci USA 97:2979–2983

Binu-Lal SS, Wilbert Kehelpannala KV, Satish-Kumar M et al (2003) Multistage graphite precipitation through protracted fluid flow in sheared metagranitoid, Digana, Sri Lanka: evidence from stable isotopes. Chem Geol 197:253–270

Brocks JJ, Love GD, Snape CE et al (2003) Release of bound aromatic hydrocarbons from late Archaean and Mesoproterozoic kerogens via hydropyrolysis. Geochim Cosmochim Acta 67:1521–1530

Buick R, Dunlop JSR (1990) Evaporitic sediments of Early Archaean age from the Warrawoona Group, North Pole, Western Australia. Sedimentology 37:47–277

Buseck PR, Bo-Jun H (1985) Conversion of carbonaceous material to graphite during metamorphism. Geochim Cosmochim Acta 49:2003–2016

Buseck PR, Bo-Jun H, Miner B (1988) Structural order and disorder in Precambrian kerogens. Org Geochem 12:221–234

Buseck PR, Tsipursky ST, Hettich R (1992) Fullerenes from the geological environment. Science 257:215–217

Byerly GR, Lowe DR, Wooden JL et al (2002) An Archaean impact layer from the Pilbara and Kaapvaal Cratons. Science 297:1325–1327

Clayton JL, Bostick NH (1986) Temperature effects on kerogen and on molecular and isotopic composition of organic matter in Pierre Shale near an igneous dike. Org Geochem 10:135–143

De Gregorio BT, Sharp TG (2003) Determining the biogenicity of microfossils in the Apex Chert, Western Australia. Lunar Planet Sci XXXIV:Abstract 1267

De Gregorio BT, Sharp TG (2006) Possible abiotic formation of kerogen-like carbon in the Strelley Pool chert. Lunar Planet Sci XXXVII:1–3

De Gregorio BT, Sharp TG, Flynn GJ (2005) A comparison of the structure and bonding of carbon in Apex Chert kerogenous material and Fischer-Tropsch-type carbons. Lunar Planet Sci XXXVI:2 pp

Des Marais DJ (1997) Isotopic evolution of the biogeochemical carbon cycle during the Proterozoic Eon. Org Geochem 27:185–193

DiMarco MJ, Lowe DR (1989) Stratigraphy and sedimentology of an Early Archaean felsic volcaniclastic sequence, eastern Pilbara Block, Western Australia, with special reference to the Duffer formation and implications for crustal evolution. Precambr Res 44:147–169

Dow WG (1977) Kerogen studies and geological interpretations. J Geochem Explor 7:79–99

Duck LJ, Glikson M, Golding SD et al (2007) Microbial remains and other carbonaceous forms from the 3.24 Ga sulphur springs black smoker deposit, Western Australia. Precambr Res 154:205–220

Dunlop JSR, Muir MD, Milne VA et al (1978) A new microfossil assemblage from the Archaean of Western Australia. Nature 274:676–678

Durand B, Monin JC (1980) Elemental analysis of kerogens. In: Durand B (ed) Kerogen: insoluble organic matter from sedimentary rocks. Éditions Technip, Paris

Ehrenfreund P, Irvine W, Becker L et al (2002) Astrophysical and astrochemical insights into the origin of life. Rep Prog Phys 65:1427–1487

Furnes H, Banerjee NR, Muehlenbachs K et al (2004) Early life recorded in Arcahaean pillow lavas. Science 304:578–581

Garvie LAJ, Buseck PR (2004) Nanosized carbon-rich grains in carbonaceous chondrite meteorites. Earth Planet Sci Lett 224:431–439

Glikson M (2001) The application of electron miscroscopy and microanalysis in conjunction with organic petrology to further the understanding of organic-mineral association: examples from Mount Isa and McArthur Basins, Australia. Int J Coal Geol 47:139–159

Glikson M, Taylor GH (1985) Cyanobacterial mats: major contributors to the organic matter in Toolebuc Formation oil shales. Geol Soc Aust Spec Publ 12:273–286

Glikson M, Taylor D (2000) Nature of organic matter in the early Proterozoic, earliest life forms and metal associations. In: Glikson M, Mastalerz M (eds) Organic matter and mineralisation: thermal alteration, hydrocarbon generation and role in metallogenesis. Kluwer, London

Glikson M, Taylor D, Morris DG (1992) Lower Paleozoic and Precambrian petroleum source rocks and the coalification path of Alginite. Org Geochem 18:881–897

Glikson M, Golding SD, Southgate P (2006) Thermal evolution of the ore-hosting Isa superbasin: central and northern Lawn Hill Platform. Econ Geol 101:1211–1229

Glikson M, Duck LJ, Golding SD et al (2008) Microbial remains in some Earth oldest rocks: comparison with *Methanocaldococcus.jannaschii*, a potential analogue. Precambr Res 164:187–200

Grey K, Hickman AH, Van Kranendonk MJ et al (2002) 3.45 billion year-old stromatolites in the Pilbara region of Western Australia: proposals for site protection and public access: Western Australia Geological Survey, Record 2002/17

Harris PJF, Vis RD (2003) High-resolution electron microscopy of carbon and nanocrystals in the Allende meteorite. Proc R Soc Lond A 459:2069–2076

Heymann D, Chibarite LPF, Brock RR et al (1994) Are biogenic PAHs precursors of fullerenes on Earth? Science 265:645–647

Hickman AH (1983) Geology of the Pilbara Block and its environs. Geol Surv West Aust Bull 127

Hickman A (2004) Geological setting of diamond drilling for the Archean Biosphere Drilling Project, Pilbara Craton, Western Australia. Eos Trans AGU 85(47), Fall Meet Suppl #B32B-01

Hickman AH (2005) Evidence of early life from international collaborative drilling in the Pilbara Craton. In: Western Australia Geological Survey Annual Report 2004–05

Hickman AH (2008) Regional review of the 3426–3350 Ma Strelley Pool Formation, Pilbara Craton, Western Australia. In: Western Australia Geological Survey Report 2008/15

Hickman AH, Van Kranendonk MJ (2008) Marble Bar, W.A. Sheet 2855. Western Australia Geological Survey, 1:100,000 Geological Series

Hofmann HJ, Grey K, Hickman AH, Thorpe R et al (1999) Origin of 3.45 Ga coniform stromatolites in Warrawoona Group, Western Australia. Geol Soc Am Bull 111:1256–1262

Horita J (2005) Some perspectives on isotope biosignatures for early life. Chem Geol 218:171–186

House CH, Schopf JW, Stetter KO (2003) Carbon isotopic fractionation by Archaeans and other thermophilic prokaryotes. Org Geochem 34:345–356

Hower JC (1989) Petrology of liquefaction residues from the Breckinridge cannel, Western Kentucky. Org Geochem 14:299–305

Huston DL, Logan GA (2004) Barite, BIFs and bugs: evidence for the evolution of the Earth's early hydrosphere. Earth Planet Sci Lett 220:41–55

Iijima S (1991) Helical microtubules of graphitic carbon. Nature 354:56–58

Khavari-Khorasani G, Michelsen JK (1993) Thermal evolution of solid bitumens, bitumen reflectance, and kinetic modeling of reflectance: application in petroleum and ore prospecting. Energy Sour 15:181–204

Marshall CP, Love GD, Snape CE et al (2007) Structural characterization of kerogen in 3.4Ga Archean cherts from the Pilbara Craton, Western Australia. Precambr Res 155:1–23

Martin W, Russell MJ (2003) On the origins of cells: a hypothesis for the evolutionary transitions from abiotic geochemistry to chemoautotrophic prokaryotes, and from prokaryotes to nucleated cells. Phil Trans R Soc Lond B 358:59–85

Mastalerz M, Glikson M, Stankiewicz BA et al (2000) Organic and mineral matter in a Precambrian shungite deposit from Karelia, Russia. In: Glikson M, Mastalerz M (eds) Organic matter and mineralisation: thermal alteration, hydrocarbon generation and role in metallogenesis. Kluwer, London

McCollom TM, Seewald JS (2006) Carbon isotope composition of organic compounds produced by abiotic synthesis under hydrothermal conditions. Earth Planet Sci Lett 243:74–84

Melker AI, Romanov SN, Korilov DA (2000) Computer simulation of formation of carbon fullerenes. Mater Phys Mech 2:42–50

Nedachi M (2004) New discoveries from the Archean Biosphere Drilling Project (ABDP). Eos Trans AGU 85(47), Fall Meet Suppl #B32B-03

Nijman W, De Vries ST (2004) Early Archaean crustal collapse structures and sedimentary basin dynamics. In: Eriksson PG, Altermann W, Nelson DR et al (eds) The Precambrian Earth: tempos and events. Elsevier, Amsterdam

Nijman W, de Bruijne KCH, Valkering ME (1999) Growth fault control of Early Archaean cherts, barite mounds and chert-barite veins, North Pole Dome, Eastern Pilbara, Western Australia (Part 2). Precambr Res 95:247–274

O'Brien TP, McCully ME (1981) The study of plant structure: principles and selected methods. Termarcarphi Pty Ltd, Melbourne

Oberlin A, Boulmier JL, Villey M (1980) Electron microscopic study of kerogen microtexture. Selected criteria for determining the evolution path and evolution stage of kerogen. In: Durand B (ed) Kerogen: insoluble organic matter from sedimentary rocks. Éditions Technip, Paris

Parthasarathy G, Srinivasan R, Vaivamari M et al (1998) Occurrence of natural fullerenes in low-grade metamorphosed Proterozoic shungite from Karelia, Russia. Geochim Cosmochim Acta 62:3541–3544

Philippot P, Lepot K, Thomas C et al (2006) Abiotic versus biotic carbonaceous matter in 3525 – Myr-old hydrothermally altered sub-seafloor sediments. Pilbara Craton, Western Australia. Abstract EGU 6, Vienna

Price LC (1983) Geologic time as a parameter in organic metamorphism and vitrinite reflectance as an absolute palaeothermometer. J Petrol Geol 6:5–38

Saxby JD, Bennett AJR, Corcoran JF et al (1986) Petroleum generation: simulation over six years of hydrocarbon formation from torbanite and brown coal in a subsiding basin. Org Geochem 9:69–81

Schidlowski M (1988) A 3,800-million-year isotopic record of life from carbon in sedimentary rocks. Nature 333:313–318

Schidlowski M (2001) Carbon isotopes as biogeochemical recorders of life over 3.8 Ga of Earth history: evolution of a concept. Precambr Res 106:117–134

Schidlowski M, Hayes JM, Kaplan IR (1983) Isotopic inferences of ancient biochemistries: carbon, sulfur, hydrogen, and nitrogen. In: Schopf JW (ed) Earth's earliest biosphere: its origin and evolution. Princeton University Press, Princeton, NJ

Schopf JW (1993) Microfossils of the Early Archaean Apex Chert: new evidence of the antiquity of life. Science 260:640–646

Schopf JW, Packer BM (1987) Early Archaean (3.3-billion to 3.5-billion-year-old) microfossils from Warrawoona Group, Australia. Science 237:70–73

Schopf JW, Walter MR (1983) Archean microfossils: new evidence of ancient microbes. In: Schopf JW (ed) Earth's earliest biosphere: its origin and evolution. Princeton University Press, Princeton, NJ

Taylor GH, Teichmüller M, Davis A et al (1998) Organic petrology. Gebrüder Borntraeger, Berlin

Thorpe RI, Hickman AH, Davis DW et al (1992) Constraints to models for Archaean lead evolution from precise zircon U-Pb geochronology for the Marble Bar region, Pilbara Craton, Western Australia. In: Glover JE, Ho SE (eds) The Archaean: terrains, processes and metallogeny. Publication 22, Geology Department and University Extension, The University of Western Australia, Western Australia

Tissot BP and Welte DH (1984) Petroleum formation and occurrence 699pp, Springer publication

Ueno Y, Isozaki Y, Yurimoto H et al (2001) Carbon isotopic signatures of individual Archean microfossils (?) from Western Australia. Int Geol Rev 43:196–212

Ueno Y, Yurimoto H, Yoshioka H et al (2002) Ion microprobe analysis of graphite from ca. 3.8 Ga metasediments, Isua supracrustal belt, West Greenland: relationship between metamorphism and carbon isotopic composition. Geochim Cosmochim Acta 66:1257–1268

Ueno Y, Yoshioka H, Maruyama S et al (2004) Carbon isotopes and petrography of kerogens in 3.5-Ga hydrothermal silica dikes in the North Pole area, Western Australia. Geochim Cosmochim Acta 68:573–589

UIC (1995a) CAN#3: determination of carbonate carbon and dissolved CO_2 in solid and liquid samples. In: Coulometer Application Notes, UIC Incorporated Coulometrics, Joliet, Illinois, USA, 13–14

UIC (1995b) CAN#10: determination of organic and total carbon in solids and sediments and on substrates, including catalysts. In: Coulometer Application Notes, UIC Incorporated Coulometrics, Joliet, Illinois, USA, 33–34

Van Kranendonk MJ (2000) Geology of the North Shaw 1:100,000 sheet. Western Australian Geological Survey 1:100,000 Geological Series Explanatory Notes

Van Kranendonk MJ (2006) Volcanic degassing, hydrothermal circulation and the flourishing of early life on Earth; a review of the evidence from c.3490–3240 Ma rocks of the Pilbara Subgroup, Pilbara Craton, Western Australia. Earth Sci Rev 74:197–240

Van Kranendonk MJ, Pirajno F (2004) Geochemistry of metabasalts and hydrothermal alteration zones associated with c. 3.45 Ga chert and barite deposits: implications for the geological setting of the Warrawoona Group, Pilbara Craton, Australia. Geochem Explor Environ Anal 4:253–278

Van Kranendonk MJ, Hickman AH, Smithies RH et al (2002) Geology and tectonic evolution of the Archaean North Pilbara Terrain, Pilbara Craton, Western Australia. Econ Geol 97:695–732

Van Kranendonk MJ, Philippot P, Lepot K (2006) The Psilbara Drilling P. c.2.72 Ga Tumbiana Formation and 3.49 Ga Dresser Formation, Pilbara Craton, Western Australia. Western Australian Geological Survey, Record 2006/14

Van Kranendonk MJ, Philippot P, Lepot K et al (2008) Geological setting of Earth's oldest fossils in the ca. 3.5 Ga Dresser formation, Pilbara Craton, Western Australia. Precambr Res 167:93–124

Zegers TE, van Keken PE (2001) Middle Archean continent formation by crustal delamination. Geology 29:1083–1086

Bugs or Gunk? Nanoscale Methods for Assessing the Biogenicity of Ancient Microfossils and Organic Matter

Bradley T. De Gregorio, Thomas G. Sharp, Ahmed I. Rushdi, and Bernd R.T. Simoneit

Abstract Ancient microfossils and stromatolites represent some of our best evidence of early living organisms on Earth. However, due to increasingly critical evaluation of such biomarkers in the last decade, assessing the biogenicity of preserved microbe-like features and stromatolitic structures is far from trivial. Carbonaceous matter associated with *bona fide* microfossils and stromatolites should contain a general microstructure consistent with that of kerogen, the biogenic organic matter common in oil, gas, and coal source rocks. Although kerogen structure and composition can vary, there are consistent characteristics that may be used to identify "kerogen-like" carbonaceous material. Fischer-Trospch-type (FTT) reactions and other organic synthesis mechanisms may be a significant source of abiotic organic matter on the early Earth. Many minerals associated with hydrothermal activity, which was prevalent on the early Earth, have been shown to catalyze FTT synthesis. However, significant research into the maturation and preservation of FTT-derived carbonaceous matter is necessary before a robust distinction can be made from biogenic kerogen.

Controversial microfossils from the 3.458–3.465 Ga Apex Basalt and stromatolites from the 3.350–3.458 Ga Strelley Pool Chert are associated with hydrothermal activity, leading to the suggestion that the carbonaceous matter comprising these features is derived from abiotic FTT synthesis rather than biological processes. Carbon-rich black chert dike samples were obtained from the Apex and Strelley Pool

---

B.T. De Gregorio (✉)
NASA Johnson Space Center, Mail Code KT,
2101 NASA Parkway, Houston, TX 77058, USA
e-mail: bradley.t.degregorio@nasa.gov

T.G. Sharp
School of Earth and Space Exploration, Arizona State University, P.O. Box 871404, Tempe, Arizona 85287-1404, USA

A.I. Rushdi and B.R.T. Simoneit
COGER, College of Agriculture and Food Science, King Saud University, P.O. Box 2460, Riyadh 11452, Saudi Arabia

B.R.T. Simoneit
Oregon State University, Corvallis, OR, USA

S.D. Golding and M. Glikson (eds.), *Earliest Life on Earth: Habitats, Environments and Methods of Detection*, DOI 10.1007/978-90-481-8794-2_10,

localities for comparison with kerogen from the 1.9 Ga Gunflint Formation and laboratory-derived FTT carbonaceous matter. In situ scanning-transmission X-ray microscopy (STXM) and transmission electron microscopy (TEM) reveal the presence of ~100 nm carbonaceous films along quartz grain boundaries in the chert samples, which may be separated from the quartz grains in ultramicrotomed sections. Electron energy-loss near-edge structure spectroscopy (ELNES) and X-ray absorption near-edge structure spectroscopy (XANES) indicate Apex and Strelley carbonaceous matter contains a complex microstructure with abundant aromatic domains and oxygenated organic functional groups similar to that observed in Gunflint kerogen. FTT samples were chemically distinct, containing abundant carboxyl functional groups and lacking polyaromatic domains. Although it is possible that FTT material could mature into an abiotic kerogen-like material with similar characteristics, it is more likely that the Apex and Strelley Pool carbonaceous matter formed from biological processes. Biogenic material could be preserved within an ancient hydrothermal system either through gravitational influx of microbial material from the surface or transport of carbonaceous matter from a deeper sediment.

Keywords Microfossil • Kerogen • Fischer-Tropsch • Apex chert • Strelley pool chert • TEM • XANES

1 Introduction

Bacterial fossils (microfossils) and other microbially-mediated sedimentary structures (stromatolites) are some of the best evidence for the presence of Archean living organisms on the early Earth (Schopf 2006; Schopf et al. 2007). However, many of the oldest carbon-rich sedimentary rocks have been exposed to billions of years of thermal stress. Hydrothermal and metamorphic fluids may degrade and transport biogenic organic matter, destroying fine-scale cellular features in microfossils and laminated biofilms (Kaźmierczak and Kremer 2002). In addition, thermal processes change the organic chemistry and structure of macromolecular carbon compounds, leading to the generation of oil and natural gas in massive sedimentary deposits of biogenic organic matter (Tissot and Welte 1978). In many Archean metasedimentary rocks it is unclear whether the apparent microfossils and stromatolite structures are *bona fide* biogenic features. In this chapter, we describe analytical methods for the characterization of potentially biogenic carbonaceous matter in the context of Archean micropaleontology. These analytical methods primarily probe the macromolecular structure and organic chemistry of the carbonaceous matter and in some cases may clearly distinguish biogenic and abiotic carbonaceous matter.

The dimensions of most extant microorganisms and well-preserved microfossils are on the order of 1–10 µm. However, thermally-stressed carbonaceous films lining mineral grain boundaries are commonly only ~100 nm thick (De Gregorio and Sharp 2006; Moreau and Sharp 2004). Therefore, sub-µm scale analytical techniques are required to adequately resolve much of this carbonaceous matter. In addition, many of these Archean carbon-rich sediments do not contain enough carbon for typical bulk analytical techniques used to characterize coal deposits, and acid extraction methods for concentrating the carbonaceous matter may potentially destroy some of this precious material. A few modern analytical techniques allow in situ nanoscale analysis of ancient carbonaceous matter, including laser-Raman spectroscopy, synchrotron-based Fourier-transform infrared spectroscopy (FTIR), scanning-transmission X-ray microscopy (STXM), and transmission electron microscopy (TEM). These analytical techniques are also nominally non-destructive, although each may damage beam-sensitive organic matter to a degree. Perhaps the most important characteristic of these techniques is that they may be performed upon the same sample in a correlated fashion, such that the benefits of each technique are added and form a more complete dataset.

1.1 Structure and Bonding of Kerogen

Kerogen is the term used to describe the residue of carbonaceous matter in biological sedimentary rocks that is insoluble in acids or other solvents (Dow 1977). Unfortunately, this phenomenological definition may be applied to abiological carbonaceous matter in rocks and meteorites (Kerridge 1983; Kerridge et al. 1987; Vandenbroucke and Largeau 2007). A distinction between biogenic and abiotic material is necessary when discussing sedimentary carbonaceous matter predating the emergence of living organisms. Common usage of the term implies a biogenic origin, while descriptive terms such as "kerogen-like" may be applied to abiotic insoluble carbonaceous matter with similar structural and chemical characteristics.

Kerogen does not have a defined crystal structure or fixed chemical composition, but it contains chemical and structural characteristics that may be used for identification and classification. Although it is primarily a collection of cross-linked amorphous hydrocarbon macromolecules, kerogen also contains trace levels of other biotic elements due to its biogenic origin. In general, for every 100 carbon atoms in kerogen, there are between 50 and 200 hydrogen atoms, 3–30 oxygen atoms, and less than six nitrogen and sulfur atoms (Table 1). Phosphorus is relatively rare in

Table 1 Elemental composition of typical kerogens in atomic percent and atomic ratio

Kerogen type	C	H	O	N	S	H/C	O/C	N/C	S/C
Type I	37.6	57.8	3.4	0.8	0.5	1.54	0.090	0.021	0.013
Type I (model)	36.9	60.6	2.2	0.1	0.0	1.64	0.060	0.003	0.001
Type II	39.1	53.3	6.3	0.8	0.6	1.36	0.161	0.020	0.015
Type II (model)	42.2	52.8	3.8	0.5	0.8	1.25	0.089	0.012	0.018
Type III	45.2	44.8	8.8	1.2	0.0	0.99	0.195	0.027	0.000
Type III (model)	46.8	45.8	6.5	0.7	0.2	0.98	0.138	0.016	0.004
Lignite coal	46.0	41.2	10.7	1.5	0.6	0.90	0.233	0.033	0.013

Data compiled from Tissot and Welte (1978) and Behar and Vandenbroucke (1987).

biomolecules, but will be present to some degree in kerogen, estimated at around one phosphorus atom per 1,000 carbon atoms (Nissenbaum 1979). These atomic abundances in kerogen are much lower than that predicted by the composition of extant organisms because biotic elements are progressively removed from recently deposited sedimentary organic matter by a combination of microbial degradation, groundwater leaching, and thermal maturation.

Kerogens are commonly divided into three main groupings (Types I, II, and III) according to atomic H/C and O/C ratios (Van Krevelen 1993) or hydrogen index (Espitalié and Bordenave 1993), although a continuum of compositions likely exists between these end members. General characteristics and sources of kerogen types are described by Tissot and Welte (1978) and Durand (1980). These kerogen types may also be distinguished by observations of their nano-scale molecular structure and the extent of graphitization (Oberlin et al. 1980).

Structurally, kerogens consist of randomly-oriented polycyclic aromatic sheets up to 30 Å in diameter connected by bridging aliphatic hydrocarbon chains (Behar and Vandenbroucke 1987; Tissot and Welte 1978), and may best be thought of as statistical collections of fundamental organic building blocks rather than a crystalline material (Freund et al. 2007). Both the polyaromatic domains and aliphatic bridging hydrocarbons may contain various organic functional groups. Oxygen is commonly present as hydroxyl (–OH), enol (C=C–OH), ketone (C=O), carboxyl (–COOH), and ester (–COOC–) groups on bridging hydrocarbons. Identical organic moieties attached to polyaromatic domains include phenol ($C_{arom.}$–OH), cyclic ketone ($C_{arom.}$=O) and aldehyde ($C_{arom.}$–C=O), and ether bonds in aromatic heterocycles such as pyran and furan ($C_{arom.}$–O–$C_{arom.}$). Nitrogen atoms are most common in aromatic heterocycles such as pyridine ($C_{arom.}$=N–$C_{arom.}$) and pyrrole ($C_{arom.}$–NH–$C_{arom.}$), although amide (O=C–NH_x) and amine (–NH_x) functional groups may be present (Knicker 2004). Sulfur atoms commonly reside in similar aromatic heterocycles, such as thiane and thiophene ($C_{arom.}$–S–$C_{arom.}$), or aliphatic thiol (C–SH) and thioether (C–S–C) functional groups (Calkins 1994; Sinninghe Damsté and de Leeuw 1990).

Thermal maturation of sedimentary organic matter occurs through several phases following initial degradation (Tissot and Welte 1978). During early diagenesis most of the hydrolyzable material is degraded by heterotrophic microbial activity and autolyzing enzymatic activity within cells. However, it is likely that heterotrophic degradation was not a major component in the breakdown of biomolecules in Archean sediments. The majority of nitrogen, sulfur, and presumably phosphorus loss occurs during early diagenesis from hydrolysis reactions (Vandenbroucke and Largeau 2007). These processes create reactive organic species that form cyclic moieties, protecting the organic matter from further degradation. Following burial, organic matter experiences significant volatile loss, primarily as loss of CO_2, H_2O, and low molecular weight hydrocarbons. These diagenetic processes result in carbonization, or condensation of the remaining organic matter into polyaromatic domains cross-linked by aliphatic hydrocarbon chains. With increasing temperature, generally greater than 200°C, the organic matter undergoes catagenesis, the main stage of oil generation

(Tissot and Welte 1978). During catagenesis, bridging hydrocarbons are removed and the average size of polyaromatic domains increases up to 10 Å, generating the canonical kerogen macrostructure (Durand 1980). Metagenesis is the final stage of kerogen maturation, wherein the polyaromatic network is rearranged, releasing significant amounts of methane. Aromatic heteroatoms, which have been mostly protected in polyaromatic domains since early diagenesis, are released as aromatic ring structures are disassembled. At the onset of metagenesis, the H/C and O/C ratios of all kerogen types are indistinguishable. Extensive graphitization may occur during this stage, subject to the initial composition of the organic matter, and can be used to assess the metamorphic grade of the sediments (Buseck and Bo-Jun 1985; Buseck et al. 1988; Wopenka and Pasteris 1993).

1.2 Abiotic Fischer-Tropsch-Type Carbonaceous Matter

Not all sedimentary carbonaceous matter has a biogenic origin. The insoluble organic matter in meteorites (Hayatsu and Anders 1981; Lancet and Anders 1970) and interplanetary dust particles (Llorca 2002), for example, could not have been generated biologically. In the terrestrial regime, carbonaceous films around olivine grains in mantle xenoliths (Mathez 1987; Tingle et al. 1991) and methane seeps within crystalline basement rocks (Sherwood Lollar et al. 2006, 2008) or mid-ocean spreading centers (Holm and Charlou 2001; Kelley and Früh-Green 1999) also appear to have formed in the absence of living organisms. In these cases, an abiotic organic synthesis reaction such as Fischer-Tropsch-type (FTT) synthesis is commonly inferred. FTT synthesis is a naturally-occurring form of industrial Fischer-Tropsch synthesis for generating aliphatic hydrocarbons (Anderson 1984).

Hydrocarbons synthesis via FTT processes requires hydrothermal fluids, a source of H_2 and CO, and an appropriate catalyst. Modern Fischer-Tropsch reactors operate between 200°C and 350°C, above which predominantly graphite crystallites are produced. These temperatures are commonly encountered in seafloor hydrothermal systems or during low grade metamorphism (Lindsay et al. 2005), and the synthesis of aliphatic and aromatic hydrocarbons in these environments may be thermodynamically favorable (Seewald et al. 2006). If high fluxes of dissolved H_2 are present, CO may be generated from CO_2 via the water-gas shift reaction,

$$CO_2 + H_2 \rightarrow CO + H_2O$$

When this occurs, the overall synthesis reaction of hydrocarbons is

$$nCO_2 + (2n + m/2)H_2 \rightarrow C_nH_m + 2nH_2O$$

In seafloor hydrothermal systems and mid-ocean ridges, adequate concentrations of H_2 may be generated by the serpentinization of olivine in ultramafic rocks (Schroeder et al. 2002).

Several geologic materials have been shown to catalyze FTT reactions, including magnetite (Yoshida et al. 1993), hematite (Reymond et al. 1982), chromite (Foustoukos and Seyfried 2004), siderite (McCollom 2003), and pyrite (Cody 2005). Hydrocarbons may be synthesized during the serpentinization of olivine (Berndt et al. 1996), however this result has not been successfully repeated (McCollom and Seewald 2001). Clay minerals may also be important FTT catalysts. Hydrocarbons readily adsorb to clay mineral surfaces (Galwey 1972) and may be protected from degradation within interlayer sites (Williams et al. 2005). FTT experiments using montmorillonite or saponite catalysts have produced many different hydrocarbons, including simple short chain alkanes, long chain alkanes, alkenes, aromatic compounds, alkanoic acids, simple lipids, and other oxygenated compounds (McCollom et al. 1999; Williams et al. 2005). Yields in such experiments are maximized between 150°C and 250°C and contain significant proportions of unresolved branched and cyclic organic compounds (Rushdi and Simoneit 2001). Once a satisfactory amount of FTT hydrocarbons have been generated, condensation reactions between organic molecules may create more complex products (Rushdi and Simoneit 2006; Simoneit et al. 2007). Nitrogen and sulfur atoms can be incorporated into synthesis products if ammonia or CS_2, respectively, are present (Rushdi and Simoneit 2004, 2005).

Prior to the emergence of living organisms on Earth, FTT reactions may have been prevalent in seafloor hydrothermal systems. Without well-defined microfossils or stromatolitic features, it becomes necessary to distinguish biogenic kerogen from abiotic FTT carbonaceous matter. Isotopic fractionation of carbon towards lighter isotopes is commonly used to identify biogenic material, but FTT reactions have been shown to generate similar isotopic depletions (Horita and Berndt 1999; Lancet and Anders 1970; McCollom and Seewald 2006). Since the structure and composition of kerogen is somewhat dependent upon the initial biomolecules, it is possible that abiotic kerogen-like material derived from FTT carbonaceous matter may have a distinct structure and composition. Unfortunately no systematic studies of the thermal maturation of FTT carbonaceous matter have been performed. Because it is initially lacking in complex biomolecules, it is entirely possible that most FTT material could be quickly degraded and fully eliminated during diagenesis and catagenesis. Alternatively, thermal maturation of FTT material may generate chemically and structurally distinct carbonaceous matter. Although the characteristics of abiotic kerogen-like carbonaceous matter are not determined, those of *bona fide* kerogen are well known and may be used as a fundamental condition for assessing biogenicity, rather than a defining test between biogenic and abiotic carbonaceous matter. This "necessary but insufficient" condition must be combined with other analyses in order to make a robust assessment of the biogenicity of an Early Archean carbonaceous sample.

1.3 Putative Archean Microfossils and Stromatolites from Western Australia

Some of the oldest sedimentary rocks on Earth can be found in the Archean Pilbara Craton of Western Australia, containing units dating from 3.655–2.400 Ga (Bagas 2002; Hickman 1983). The southern half of the craton is dominated by metasediments and volcanic rocks of the Neoarchean to Early Proterozoic Hamersly Basin, including the stromatolitic 2.78–2.63 Ga Fortescue Group (Thorne and Trendall 2001; Walter 1983). The northern half contains an older, typical Barberton-cycle granite-greenstone terrane comprised of several ovoid granitoid intrusions and complexes surrounded by a coeval or younger layered greenstone succession of metamorphosed volcanic and sedimentary rocks. The greenstone belts are composed primarily of the 3.515–2.950 Ga Pilbara Supergroup. The depositional facies of these units are quite variable and may be truncated by tectonic structures. The lower portions of the supergroup consist of predominantly volcanic rocks of the Coonterunah, Warrawoona, and Sulphur Springs Groups, while the upper parts of the sequence consist primarily of sedimentary rocks from the Gorge Creek and De Grey Groups (Hickman 1983; Van Kranendonk 2006). The greenstone belts have experienced regional greenschist-facies metamorphism primarily due to deformation and intrusion by Archean granitoid domes, but metamorphic grade varies locally from prehnite-pumpellyite to amphibolites facies (Van Kranendonk 2007).

The volcanic rocks of the 3.515–3.315 Ga Warrawoona Group frequently include thin siliceous units and cross-cutting quartz veins, within which several microfossil, pseudofossil, and stromatolite occurrences have been reported (Schopf 2006; Schopf et al. 2007). In 1987, several taxa of microbe-like features were described from a local chert deposit within the Apex Basalt at Chinaman Creek, near the town of Marble Bar (Schopf and Packer 1987). The features appear to be contained within mm-sized, round, carbon-rich, siliceous clasts in a black sedimentary chert, indicating the putative microfossils were at least as old as the rocks containing them. Since the age of the Apex Basalt is constrained between the underlying 3.465 Ga Duffer Formation and the overlying 3.458 Ga Panorama Formation (Thorpe et al. 1992), the Apex microbe-like features were interpreted to represent the oldest well-preserved fossil evidence of living organisms on Earth.

Stromatolite-like structures are located within the Strelley Pool Chert, a 15–30 m thick, regional, siliceous, carbonate unit about 50 km west of the town of Marble Bar, demarcating the top of the Archean Warrawoona Group (Bagas 2002; Hickman 2008). The unit lies unconformably above the 3.458–3.426 Ga Panorama Formation and below the 3.350 Ga Euro Basalt (Thorpe et al. 1992), making the Strelley Pool Chert slightly younger than the Apex cherts. In several localities, small, conical, laminated stromatolitic structures have been documented within the siliceous carbonates of the lower Strelley Pool Chert (Allwood et al. 2006a; Hofmann et al. 1999; Lowe 1980). These structures are less than 15 cm in height with apical angles averaging 70–80°, and are generally oriented vertically on paleoslopes (Allwood et al. 2006a). They are morphologically similar to Proterozoic *Conophyton* group stromatolites, and therefore would indicate the presence of phototrophic mat-building microbial life at ~3.45 Ga (Schopf 2006; Walter 1983). The structures are commonly found within layers containing evaporite minerals or evidence of evaporite replacement, indicating formation in a shallow water environment (Allwood et al. 2007). Laminated structures have been reported in older Pilbara rocks but are less well-preserved and have fewer biological characteristics (Buick et al. 1981), leaving the Strelley Pool structures as the oldest putative stromatolites on Earth.

Both the Apex microfossil locality and at least one locality of stromatolites in the Strelley Pool Chert are associated with black, carbon-rich, cross-cutting chert dikes (see Sections 4.5 and 4.6 for details). The presence of reduced carbonaceous matter within such chert dikes strongly suggests the possibility of ancient hydrothermal FFT processes, which may have created or influenced the nearby microfossils or stromatolites (Brasier et al. 2002, 2005; Lindsay et al. 2005). An assessment of the biogenicity of this ancient carbonaceous matter will have direct implications on the biogenicity of related microfossil-like features or stromatolitic structures. The two techniques used in this study, scanning transmission X-ray microscopy (STXM) and transmission electron microscopy (TEM), have very high, sub-μm spatial resolution, which is essential for the in situ analysis of ~100 nm carbonaceous grain boundary films surrounding quartz microcrystals in these cherts. The major results of this study have been previously published in De Gregorio and Sharp (2006) and De Gregorio et al. (2009), while this article provides additional details regarding the analytical methods used and a comparison with abiotic organic samples produced by FTT experiments.

2 Methods

2.1 Samples and Standards

Several rock fragments of Precambrian black chert were acquired (Table 2), including a sample from the brecciated Apex dike chert at the Chinaman Creek locality (Brasier et al. 2005) and a sample of Strelley Pool dike chert at the Strelley Pool West locality (Lindsay et al. 2005). These samples were chosen because they represent potentially biogenic carbonaceous matter preserved in ancient hydrothermal veins. A sample of black chert from the ~1.9 Ga Lower Algal Member of the Gunflint Formation at Schrieber Beach, Ontario, known to contain *bona fide* microfossils, was used as a biogenic standard for ancient chert-hosted kerogen (Moreau and Sharp 2004). Two FTT synthesis products were acquired as a standard for natural abiotic synthesis in a hydrothermal system (Table 2). Both were synthesized under hydrothermal conditions using formic acid (CH_2O_2) or oxalic acid ($C_2H_2O_4$) precursors, which decompose to a mixture of CO_2, CO, and H_2 above 150°C (Rushdi and Simoneit 2001). Two additional abiotic carbon standards, amorphous carbon and disordered semi-graphite, were created as needed in the laboratory.

Table 2 Samples used in this study

Sample	Thickness (nm)	TEM	STXM[a]
Apex chert fragments[b]			
I.E10	90	CM200; EM400	11/07/2004 – 11/08/2004
II.A6	90		02/26/2006 – 03/01/2006
II.C9	90	CM200	08/16/2005 – 08/18/2005
III.D6	200	2010F	
Strelley Pool Chert fragments[c]			
III.C1	90		04/22/2005 – 04/23/2005
III.D2	200	2010F	
III.D3	90	CM200; 2010F	09/15/2005 – 09/16/2005
Gunflint chert fragments[d]			
II.B1	200	2010F	
III.A4	90		04/20/2005 – 04/21/2005
III.B1	90	CM200	04/21/2005
175°C FTT liquid extract[e]			
II.D6	n.a.	CM200	08/30/2004
250°C FTT liquid extract[f]			
II.D8	n.a.	CM200	09/10/2004

[a]STXM data from the following dates may be downloaded from http://xray1.physics.sunysb.edu/.
[b]Sample 1057 courtesy of Dr. L. Paul Knauth.
[c]Sample SW401 courtesy of Dr. John Lindsay.
[d]Sample courtesy of Dr. L. Paul Knauth.
[e]Sample FTT-FA-A1(3)-P6.
[f]Sample FTT-OA-p62-8.

Thick amorphous carbon films were deposited on TEM grids by arc-evaporation in a standard carbon-coater apparatus, in which a carbon vapor is generated by passing a high current through the contact between two sharpened graphite rods in vacuum. Low order semi-graphite was created by pulverizing a portion of a graphite rod and dispersing the powder on a TEM grid in deionized water. This process creates randomly-oriented graphite crystallites with disordered lattice spacings and dislocations.

2.2 *Sulfur Embedding and Ultramicrotomy*

Traditional techniques for preparing electron and X-ray transparent samples for high resolution analysis include precision ion sputtering or wedge polishing. However, these techniques are not recommended for geological samples containing carbonaceous matter. Carbonaceous matter may be damaged by incident ion beams (Sviridov 2002), irrevocably altering the structure and bonding in the sample. Wedge polishing using an abrasive, on the other hand, will preferentially remove relatively soft carbonaceous matter. Appropriate sample preparation methods should avoid electron or ion beams, be performed at low temperatures, and be relatively carbon-free to avoid potential laboratory contamination of the carbonaceous matter.

Ultramicrotomy involves removing thin slices of material from a sample with a sharp diamond edge, and is a useful technique for preparing geological samples (Fig. 1; Malis and Steele 1990). Sections with sub-100 nm thickness can be prepared from most geological specimens. Harder materials such as quartz and chert may require thicker sectioning. During sectioning, mechanical "stick-slip" vibrations at the contact between the sample and the diamond edge can create parallel artifacts known as "chattering". In softer, biological samples, chattering takes the form of periodic changes in section thickness perpendicular to the cutting direction (Sitte 1984), but in harder, brittle samples, chatter causes roughly parallel fractures in the section (Malis and Steele 1990). While chattering is generally undesired, it may be beneficial by disaggregating a brittle sample, separating carbonaceous grain boundary films from a mineral matrix. Chattering may also cause the resulting sections to be slightly thicker than intended.

Chert samples were first crushed into a powder with an optimum size of 0.1 mm, although grains as large as 1 mm may be sectioned. Grains smaller than 0.01 mm will be difficult to visualize in most microtome optics. Unfortunately, putative microfossil features cannot be selected from this powdered sample, so the ancient carbonaceous matter in the final prepared sections cannot be directly correlated with any specific microbe-like morphology. However, in brecciated cherts samples, the oldest chert generations were isolated, assuming that, if present, this fraction will contain ancient biogenic material rather than later contaminants.

In traditional microtomy, samples are embedded in epoxy to stabilize them during sectioning. However, embedding epoxies are carbon-based and may contaminate carbonaceous samples. Instead, carbonaceous grains in this study were embedded

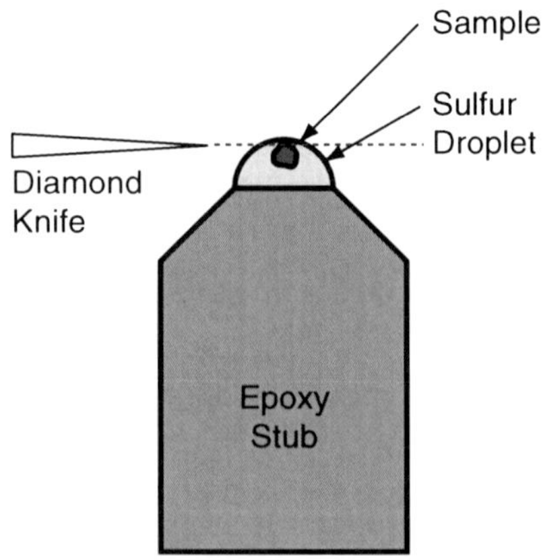

Fig. 1 Schematic diagram of ultramicrotomy of a sulfur-embedded sample

in molten sulfur droplets. High purity sulfur powder will melt at 115°C (Thackray 1970) but can remain liquid in a supercooled state well below 100°C (Meyer 1976). At 159.4°C liquid sulfur undergoes a λ phase transition in which polymerization begins (Meyer 1976). The viscosity increases dramatically and the color darkens. This darker, polymerized liquid can also remain in a metastable supercooled state. If the sample grain is inserted into polymerized liquid sulfur, the increased viscosity helps to keep the grain near the top of the droplet. Because of the metastable nature of supercooled liquid sulfur, the sulfur may crystallize as soon as the sample grain touches the droplet, in which case it may be required to remelt the sulfur in order to completely embed the grain. Manipulation of sample grains during sulfur embedding were done using a clean glass or tungsten micro-needle with a <10 μm tip. Embedding can be done by hand but a micromanipulator system improves the reliability of the operation.

After the embedding sulfur has crystallized, the solid droplet was attached to a standard epoxy stub for sectioning. A small amount of epoxy or quick-setting cyanoacrylate adhesive (i.e. superglue) was used provided that it is viscous enough to avoid infiltration of fine fractures or grain boundaries within the sulfur droplet, which can potentially contaminate the sample grain. Before sectioning, the droplet was trimmed so that there were at least two parallel edges. A trapezoidal shape is preferred, with the cutting stroke beginning at the longest base edge (Fig. 2). The sulfur-embedded chert grains were sectioned with a diamond knife to a thickness of 90 or 200 nm. Ultramicrotomed sections were transferred to 200 mesh, "thin bar", Cu TEM grids with a silicon monoxide (SiO) support film, and the same grids were also used to prepare the abiotic standard materials. Carbon-free SiO

Fig. 2 Photomicrograph of a 100 μm carbon-rich fragment of Strelley Pool Chert embedded in sulfur (*light colored trapezoid*). This sample has been microtomed, and trimming marks are visible on the sides of the epoxy stub

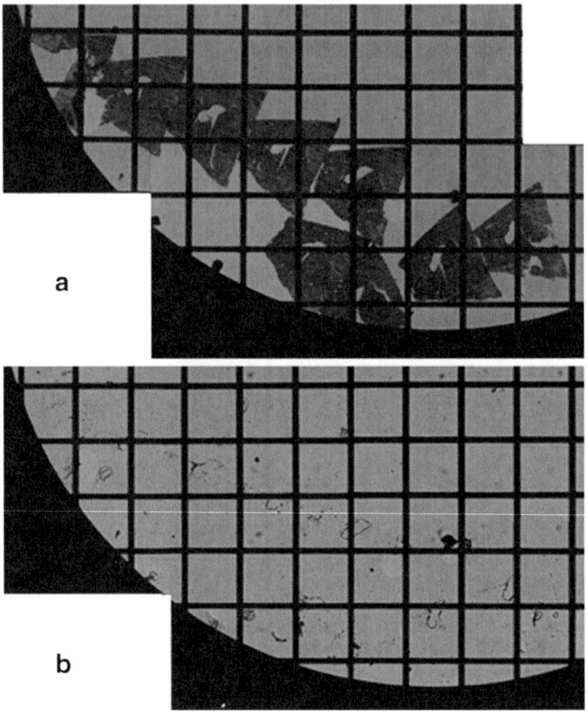

Fig. 3 Photomicrographs of trapezoidal microtome sections of Gunflint Chert embedded in sulfur (**a**) before and (**b**) after heating in a 60°C oven for 24 h. Most of the chert falls out of each section due to chattering, leaving a hole in the section, but microquartz grains held in place by the sulfur are visible after the sulfur has been sublimated. The copper grid bars are spaced approximately 100 μm apart

support films are usually preferable to SiN films since they can be made thinner and do not interfere with nitrogen analyses, although they will interfere with oxygen analyses. Sulfur was removed by sublimation in a 60°C oven for several hours (Fig. 3). The ultramicrotomed sections are now ready for photon-based analytical techniques. Prior to TEM analysis, which is an electron-based technique, a conductive amorphous carbon film was deposited on the grid to mitigate charging effects of the SiO support film or other semi-conducting materials in the sample, such as quartz. This carbon coating was shown to make a negligible contribution for TEM-based ELNES (Section 2.7) spectra. At no point during this process were samples exposed to high energy particle beams or temperatures greater than 120°C.

2.3 *FTT Sample Preparation*

Extracts of FTT reactions were diluted with a small volume of acetone and distilled water. This mixture was transferred to a clean glass vial or a 50 mL centrifuge tube, in which a SiO-supported TEM grid was oriented film-side upwards. The liquid was evaporated over several days in a dessicator, leaving a film of insoluble FTT material on each TEM grid.

2.4 Scanning Transmission X-ray Microscopy (STXM)

Photons with X-ray wavelengths (0.01–10 nm) may have enough energy to activate discrete electronic transitions within individual atoms. Isolated carbon, nitrogen, and oxygen atoms have core shell ground state binding energies ($1s \rightarrow \infty$) of 284.2 eV, 409.9 eV, and 543.1 eV, respectively (Henke et al. 1993). These ionization potentials are shifted slightly to around 291 eV, 405 eV, and 540 eV, respectively, when the electronic orbitals are hybridized for organic bonding (Stöhr 1992). In addition, the unique electronic configurations of σ- and π-bonding for each type of organic bond and functional group further shift the energy levels and allow for anti-bonding σ* and π* states. For example, the $1s \rightarrow \pi^*$ (anti-bonding) transition in aromatic carbon (C=C) can be activated by the absorption of 285.0 eV X-rays, while the same electronic transition in simple carbonyl bonds (C=O) occurs at 286.6 eV. Some characteristic energies for different organic bonds and functional groups are given in Table 3. Photons with energies above these characteristic values ("soft" X-rays) may be absorbed by the electrons in the sample.

In scanning-transmission X-ray microscopy (STXM), soft X-rays generated from a synchrotron source are energy-filtered by a monochromator grating and focused onto the sample by a Fresnel zone plate. In this study, samples were analyzed using the STXM instrument at the X1A1 beam line at the National Synchrotron Light Source at Brookhaven National Laboratory (U.S. Dept. of Energy). Individual photon energies can be selected with an energy resolution of less than 0.1 eV (Winn et al. 2000).

X-ray absorption images are acquired at specific photon energies between 270 and 320 eV by scanning a thin sample across the stationary photon beam and recording the photon flux passing through the sample. Spatial resolution in these images is limited by the width of the X-ray beam. The minimum spatial resolution achievable by zone plate optics is limited only by the width of the outermost zone (Δd_N) by

$$r = 1.22 \Delta d_N$$

and current 25 nm zone plates can focus X-rays down to ~30 nm spot sizes (Jacobsen et al. 1992). Contrast in these images depends on the absorption cross-section (σ), sample density (ρ) in atoms per unit volume, and sample thickness (z) as

$$I = I_0 e^{-\rho \sigma z}$$

Table 3 Common absorption energies for XANES and ELNES

Functional group		Transition	Energy (eV)	Reference
Carbon K				
Alkene	C=C	$1s \rightarrow \pi^*$	284.8	Dhez et al. 2003; Rightor et al. 1997
Aromatic	C=C	$1s \rightarrow \pi^*$	285.0	Stöhr 1992
Imine	C=N	$1s \rightarrow \pi^*$	285.7–285.9	Dhez et al. 2003; Shard et al. 2004
Triazine	$C_3H_3N_3$	$1s \rightarrow \pi^*$	286.2	Apen et al. 1993
Aldehyde	O=CH	$1s \rightarrow \pi^*$	286.3	Hitchcock and Brion 1980
Imidazole	$C_3H_4N_2$	$1s \rightarrow \pi^*$	286.5	Apen et al. 1993
Nitrile	C≡N	$1s \rightarrow \pi^*$	286.6–286.7	Apen et al. 1993; Dhez et al. 2003; Kikuma et al. 1998
Ketone	C=O	$1s \rightarrow \pi^*$	286.6–286.8	Hitchcock and Brion 1980; Urquhart and Ade 2002
Aliphatic	C-C	$1s \rightarrow 3p/s$	287.2–288.3	Hitchcock and Brion 1980; Ishii and Hitchcook 1988
Amide	O=C-NH_x	$1s \rightarrow \pi^*$	288.0–288.1	Bassim et al. 2008; Urquhart and Ade 2002
Carboxyl	O=C-O	$1s \rightarrow \pi^*$	288.5–288.7	Urquhart and Ade 2002
Diamond	C-C	$1s \rightarrow \sigma^*$	289.0	Morar et al. 1985
Alcohol, Ether	C-O	$1s \rightarrow 3p/s$	289.0–289.4	Ishii and Hitchcook 1988
Urea	$(NH_x)_2CO$	$1s \rightarrow \pi^*$	289.4–289.5	Urquhart and Ade 2002, Urquhart et al. 1995
Carbonate	CO_3	$1s \rightarrow \pi^*$	290.3	Zhou et al. 2008
Nitrogen K				
Imine	C=N	$1s \rightarrow \pi^*$	398.8	Apen et al. 1993; Leinweber et al. 2007
Nitrile	C≡N	$1s \rightarrow \pi^*$	399.7–399.9	Leinweber et al. 2007; Shard et al. 2004
Amide	O=C-NH_x	$1s \rightarrow \pi^*$	401.4–401.5	Gordon et al. 2003; Leinweber et al. 2007
Urea	$(NH_x)_2CO$	$1s \rightarrow 3p/s$	401.5–401.7	Urquhart et al. 1995
Pyrrole	C_4H_4NH	$1s \rightarrow \pi^*$	402.2–402.4	Leinweber et al. 2007
Oxygen K				
Carbonyl	C=O	$1s \rightarrow \pi^*$	531.3	Hitchcock and Brion 1980; Urquhart and Ade 2002
Carboxyl	O-C=O	$1s \rightarrow \pi^*$	532.2–532.4	Gordon et al. 2003
Alcohol	OH	$1s \rightarrow \sigma^*$	533.9–534.1	Ishii and Hitchcook 1988
Carbonate	CO_3	$1s \rightarrow \pi^*$	534.0	Madix et al. 1988

where I_0 and I are the incident and transmitted photon flux, respectively (Kirz et al. 1995; Stöhr 1992). The absorption cross-section depends on both the X-ray energy and the electronic structure of the sample and has been tabulated for all major elements, including carbon, nitrogen, and oxygen (Henke et al. 1993). Therefore, STXM image contrast largely represents the composition of the sample. For example, since organic bonds and functional groups do not generally absorb photons with energies below ~284 eV, carbonaceous matter will be invisible in STXM images acquired using photon energies below this value. However, carbonaceous matter will be quite visible in images acquired at 290 eV, where the absorption coefficient for hybridized carbon is largest. Carbonaceous matter can be trivially located in the mineral matrix of geological samples by comparing such pairs of images (Fig. 4). Similar methods may be applied to identify nitrogen-rich and oxygen-rich materials. In addition, the distribution of specific organic bonds and functional groups may also be measured. For example, concentrations of aromatic carbon and graphite in heterogeneous carbonaceous matter like coals may be located by comparing X-ray absorption images acquired at 285 and 290 eV.

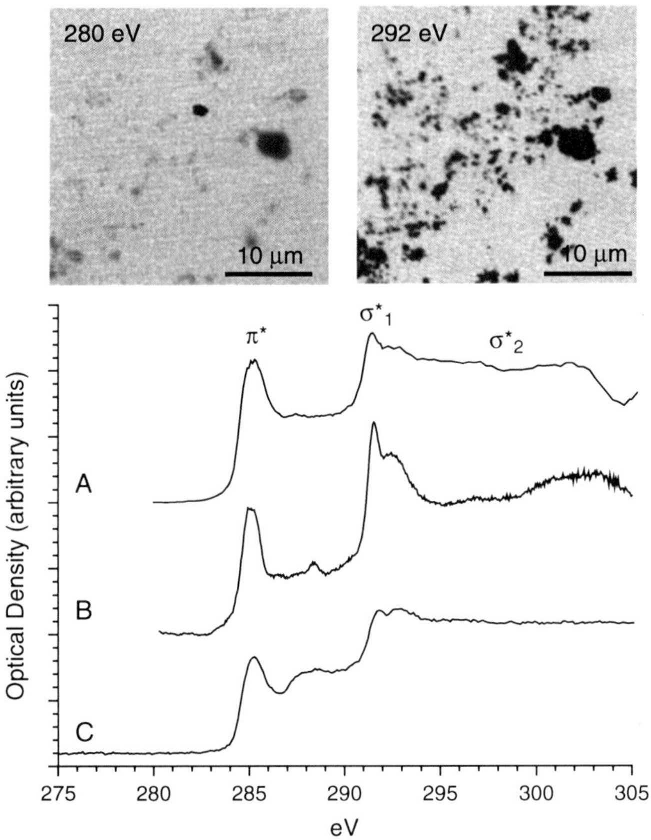

Fig. 4 STXM and XANES of graphite particles. Because carbon atoms do not absorb X-rays below ~284 eV, graphite is largely invisible in STXM images acquired at 280 eV but absorbs strongly in images acquired at 290 eV. (**a**) XANES spectrum of low order graphite particles shown in the above STXM images. (**b**) Reference spectrum of randomly ordered graphite particles. (**c**) Reference spectrum of a single crystal graphite oriented with its c-axis parallel to the X-ray beam, causing a depression of the σ^*_1 peak

2.5 *X-ray Absorption Near-Edge Structure Spectroscopy (XANES)*

A more complete picture of the local structure and bonding of carbonaceous matter may be obtained by recording the amount of X-ray absorption over a range of photon energies for a given position. This was accomplished by acquiring a series of STXM images at multiple energies to create a three-dimensional data "stack" (Jacobsen et al. 2000). The procedure is the same whether the STXM data is a single point, a one-dimensional line, or a full two-dimensional image. The energy step between images may be as small as 0.1 eV, limited by the resolution of the monochromator system (Winn et al. 2000). In a properly aligned stack, X-ray absorption intensities were averaged from an identical region of interest within each image. Contributions from instrument optics and the sample support film can be removed by calculating the optical density (OD) at each energy

$$\mathrm{OD} = -\log\frac{I}{I_0} = \rho\sigma z$$

where I is the average intensity from the region of interest and I_0 is the average intensity from a nearby sample-free region. Regions of interest with similar spectral features were selected using principal component analysis and component cluster analysis (Lerotic et al. 2004, 2005).

A typical C-XANES spectrum of carbonaceous matter contains two main components. The first is a sudden increase in absorption at the ionization potential energy. For core shell excitations, this step is known as the K-edge, and represents electronic transitions to continuum states ($1s \rightarrow \infty$). The second feature of the spectrum is a number of pre-edge peaks at absorption energies below the ionization potential, representing discrete electronic transitions to various anti-bonding π^* and σ^*states (Table 3; Fig. 4).

XANES has several applications for the characterization of kerogen and ancient carbonaceous matter. Because of the intense π^* and σ^*_1 photoabsorptions in graphite, XANES has been utilized to assess the thermal maturity and graphitization of kerogens and coals (Cody et al. 1996, 1998) and the relative proportions of aliphatic and aromatic moieties (Cody et al. 1995; Haberstroh et al. 2006). The strength of C 1*s* photoabsorptions of minerals are affected by the orientation of the sample relative to the incident X-ray beam (Rosenberg et al. 1986), so changes in the intensity of characteristic graphite peaks may be used to determine the crystallinity of sedimentary graphites in order to distinguish different depositional environments (Brandes et al. 2008; Haberstroh et al. 2006). Oxidized functional groups, such as carboxyl, have also been observed in kerogens using XANES (Cody et al. 1996, 1998; Jokic et al. 2003). Relative intensities of these peaks may be used to map the distribution of lignin and polysaccharides in modern and fossil plants (Boyce et al. 2002; Cody et al. 2009). Similar studies have used XANES to map the distribution of microbial organic matter in modern carbonaceous microbialites (Benzerara et al. 2006) and biofilms (Lawrence et al. 2003). Although nitrogen concentrations in kerogen may be low, N-XANES has been used to identify aromatic (Mitra-Kirtley et al. 1993a, b; Mullins et al. 1993) and aliphatic nitrogen compounds (Jokic et al. 2004; Vairavamurthy and Wang 2002). Similar studies using higher energy "hard" X-rays allows the identification of aromatic (Lemelle et al. 2008; Sarret et al. 2002) and aliphatic (Wiltfong et al. 2005) sulfur compounds by S-XANES.

2.6 *Tranmission Electron Microscopy (TEM)*

In transmission electron microscopy (TEM), high energy electrons are focused by magnetic lenses onto a thin sample, after which an image can be formed from scattered and diffracted electrons (Buseck 1992; McLaren 1991). Image resolution in TEM is mainly limited by the wavelength of the electrons in the beam (λ), which is determined by the operating extraction voltage of the instrument, and the spherical aberration of the objective lens (C_s). Using the Scherzer definition of resolution in TEM (O'Keefe 1992), the minimum distance that can be distinguished using an electron beam of wavelength λ can be approximated by

$$d_{\min} = 0.67 C_s^{1/4} \lambda^{3/4}$$

In most modern 200 keV TEMs, $\lambda = 0.00251$ nm and C_s ranges from 0.5–1.0 mm, giving Scherzer resolutions around 2.0 Å. Resolution may be increased by increasing the operating voltage (which decreases λ) or correcting for spherical aberration. Increasing the electron voltage may have negative consequences for beam damage of carbonaceous materials, but current capabilities for spherical aberration correction allow up to 0.28 Å spatial resolution at 200 keV (Krivanek et al. 2008). This capability for sub-nm spatial resolution for both imaging and spectroscopy makes TEM a useful and versatile technique for the characterization of microfossils and ancient organic matter.

In this study, TEM analyses were performed on a Philips CM200-FEG at 200 keV, a JEOL JEM-2010F at 200 keV, or a Philips EM400ST-FEG at 100 keV, all housed at the Center for Solid State Science at Arizona State University. Each of these microscopes utilizes a field emission gun (FEG) electron source for high energy coherence. To avoid beam damage from high energy electrons, TEM imaging was limited to 60 s and any condensed beam spectroscopic analyses such as ELNES (Section 2.7) were limited to 30 s. Samples were carbon coated to mitigate charging effects, but contributions from this carbon coating were negligible.

Coherent electrons can interact with a thin sample in several ways, leading to complex contrast in TEM images (Williams and Carter 1996). Incoherent elastic scattering (i.e. Rutherford scattering) gives rise to mass-thickness contrast. In uniformly thick samples, high atomic mass materials will appear darker than low atomic mass materials, which is useful for the visual identification of organic matter (lower Z) in a mineral matrix such as quartz (higher Z). Mass-thickness will be the dominant form of image contrast in amorphous materials like ungraphitized carbonaceous matter and kerogen (Fig. 5). In crystalline materials, coherent Bragg

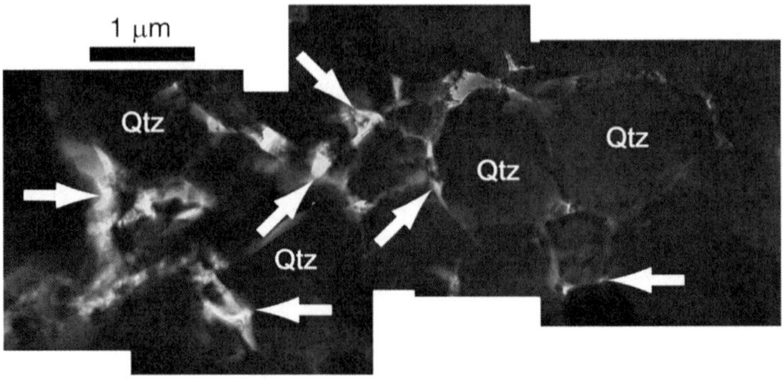

Fig. 5 Bright-field TEM image of ~100 nm carbonaceous films (*arrows*) along quartz (Qtz) grain boundaries in chert. Mass-thickness contrast causes the carbonaceous matter to appear much brighter than quartz (Modified from De Gregorio and Sharp [2006])

diffraction along crystal lattice planes creates diffraction contrast, which usually takes the form of bend contours in bent or strained samples or as thickness fringes in samples of non-uniform thickness and at sample edges. For the analysis of carbonaceous matter associated with microfossils, diffraction contrast is generally undesired and was minimized by taking care to avoid instrument configurations in which only a few diffracted beams contribute to TEM images. Phase contrast arises from Bragg diffraction of individual crystalline materials and gives rise to lattice fringes (Fig. 6). These fringes do not correspond to atom positions in the crystal but rather to the orientations of lattice planes in the crystal, and therefore the lattice fringes will change if the crystal is tilted to a different orientation. Phase contrast, in conjunction with electron diffraction is useful for identifying the alignment of graphite and partially graphitized carbonaceous matter, particularly in relation to adjacent mineral grains which may have served as a substrate for crystal growth. The amount of graphite lattice fringes in TEM images of kerogen samples can be correlated to the thermal maturity of the host rock (Buseck et al. 1988). Fine scale microbial structures can be identified by TEM imaging and compared with microfossil structures or disseminated carbonaceous matter in Precambrian cherts (Glikson et al. 2008).

The above image contrast mechanisms apply to both conventional TEM and scanning TEM (STEM), in which the electron beam is condensed to a small spot near the resolution limit of the instrument and rastered over the sample. Because sample illumination is not parallel in STEM mode, Bragg diffraction is reduced, generating images with primarily mass-thickness contrast. Thus, STEM images of graphitized carbonaceous matter will appear similar to those of amorphous carbonaceous matter. However, carbonaceous matter in a mineral matrix, especially carbonaceous grain boundary films, will appear uniformly brighter than all other materials in the sample. Using an annular detector, only highly scattered electrons are used to contribute to the image, reversing the contrast such that low-Z material like carbonaceous matter appear dark while high-Z materials appear bright.

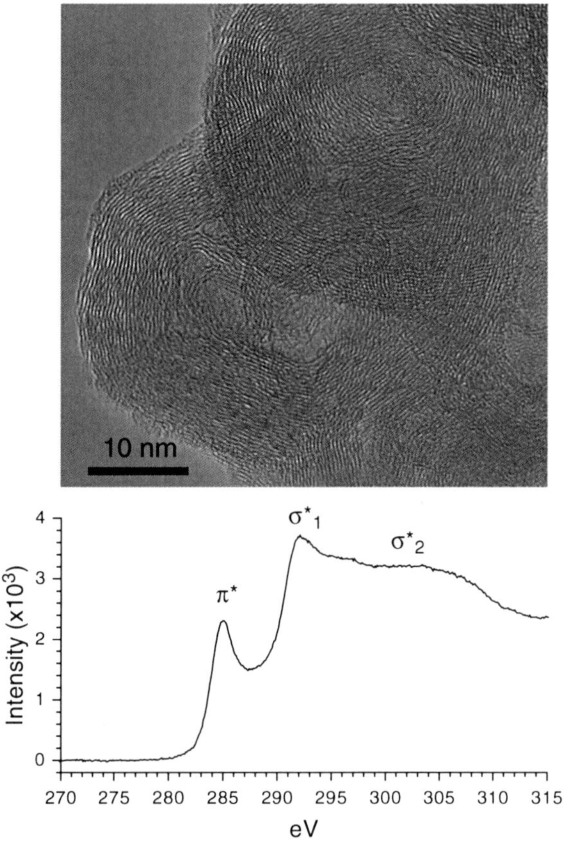

Fig. 6 Bright-field TEM image of low order semi-graphite showing characteristic 3.4 Å lattice fringes, with a corresponding ELNES spectrum

This high-angle annular dark-field (HAADF) imaging in STEM mode may be useful to identify metal atoms or other inorganic materials that have been incorporated into carbonaceous matter. Resolution in STEM mode is greatly increased by using an instrument equipped with a field emission gun (FEG) electron source.

Electron diffraction is the primary method of identifying crystalline materials in TEM. Since each mineral structure is unique, the corresponding electron diffraction patterns (EDPs) are unique to that crystal structure. Aside from the identification of single crystal mineral phases, EDPs may be used to determine the crystallographic orientation of a known mineral or orientation relationships between crystals. The EDP from a crystal in a given orientation is related to the phase contrast lattice fringes in a TEM image by a Fourier transform. Electron diffraction may still give information about poorly-crystalline materials. Randomly-oriented microcrystalline domains in partially graphitized carbonaceous matter may give rise to diffuse rings rather than discrete diffraction spots, with a primary ring radius corresponding to the 3.35 Å spacing between graphene sheets. Structurally amorphous materials will have an EDP dominated by a single central undiffracted spot, surrounded by faint diffuse scattering, usually in the form of broad rings representing the most prevalent interatomic spacings in the material.

TEM instruments can be equipped with an attached energy-dispersive X-ray spectrometer (EDS) allowing for compositional analysis of samples by characteristic X-ray emission. In conventional TEM, EDS analyses are limited to a stationary spot, but STEM allows for compositional mapping. Unfortunately, the fluorescence yield from carbon and other light elements is extremely low (Hink and Paschke 1971), so EDS spectra are dominated by emission peaks from heavier elements like silicon. However, the carbon K_α emission in spectra of carbonaceous matter should be adequate for quantification by the Cliff-Lorimer method, although this is generally limited to calculating the ratio of carbonaceous matter to mineral matrix (C/Si). In studies of microfossils and ancient organic matter, including this one, EDS is most commonly used only to identify and locate carbonaceous material in the sample.

2.7 *Electron Energy-Loss Near-Edge Structure Spectroscopy (ELNES)*

During inelastic scattering of electrons by the sample, a small portion of energy is transferred to crystal lattice vibrational modes (phonons) or to electrons around individual atoms for electronic transitions from ground states to excited states. By dispersing the scattered electrons through a bending magnet, the distribution of electron energy-loss can be measured (Egerton 1996). The majority of electrons are elastically scattered by the sample, and therefore have zero energy loss, but the remainder will show characteristic energy losses depending on the composition and bonding within the sample. The technique, also known as electron energy-loss spectroscopy (EELS) is analogous to XANES except that only a small portion of the electron's energy is absorbed rather than an entire photon. ELNES spectrometers can also generally access a larger energy range than can STXM optics. Energy resolution in ELNES is defined by the full-width half-maximum of the zero-loss peak and depends largely on the initial energy coherence of the electron source. Spectral resolution can also be affected by the choice of spectrometer entrance and exit apertures. A smaller aperture will increase spectral resolution by blocking off-axis electrons but decrease signal-to-noise. Decreasing sample thickness also increases spectral resolution by eliminating multiple scattering interactions. FEG electron sources provide the best energy coherence, and in this study the measured ELNES energy resolution ranged from 0.7 to 1.5 eV at a spectrometer dispersion between 0.05 and 0.3 eV/channel. However, with current aberration correction technology, achievable spectral resolution may be as good as 0.2 eV at 200 keV (Krivanek et al. 2008). Although the practical spectral resolution of ELNES is not as good as that of XANES, the benefit of ELNES is of course the high spatial resolution of the TEM, which is roughly equal to the size of the electron beam on a very thin sample.

Like XANES, ELNES reveals the characteristic energy transitions of electrons in covalent bonds and organic functional groups. In practice, the abundances of non-carbon atoms in carbonaceous matter are commonly too low for reliable ELNES analysis, so most ELNES of microfossil organic matter depends on interpretation of features at the C K-edge. Two dominant features in C-ELNES spectra are the $1s \rightarrow \pi^*$ (anti-bonding) absorption at 285 eV and the broad $1s \rightarrow \sigma^*$ (anti-bonding) absorption above 292 eV, followed by an exponentially-decreasing tail of electronic transitions to continuum states. These two features represent carbon-carbon double bonds (C=C) and single bonds (C–C), respectively. In C-ELNES spectra of graphite and graphitized carbonaceous matter, the π^* absorption is narrow and intense, and the intensity of the σ^* absorption is differentiated into a sharp σ^*_1 excitonic absorption at 292 eV followed by a σ^*_2 plateau between 292 and 305 eV (Fig. 6). Because the electron beam is not polarized in TEM, the crystal orientation of graphite does not affect the π^* and σ^*_1 absorptions as in XANES. C-ELNES spectra of amorphous carbons, such as an evaporated carbon film commonly used to mitigate charging in TEM, contain a well-rounded, almost spherical, σ^* absorption and a broad π^* "shoulder" at 285 eV with about half the intensity as the π^* absorption in graphite. Unfortunately, due to the limited energy resolution of ELNES, additional characteristic energy absorptions from organic functional groups are commonly obscured by the intense π^* and σ^* absorptions. However, if present, energy absorptions will appear at the same resonances as observed by XANES (Table 3).

C-ELNES is useful for assessing the graphitization of carbonaceous matter associated with microfossils. Partially graphitized kerogens that show graphitic 3.4 Å lattice fringes in TEM images also generate graphitic σ^*_1 excitons in C-ELNES spectra, while carbonaceous material lacking graphitic lattice fringes also lack a σ^*_1 absorption (De Gregorio and Sharp 2006; Moreau and Sharp 2004). Kerogen and carbonaceous material containing abundant randomly-oriented polyaromatic domains with no ordered stacking of domains will be represented in C-ELNES spectra by an intense π^* absorption and the absence of a graphitic σ^*_1 absorption. In addition, σ^* intensity will likely be distributed towards lower energies without forming a distinct graphitic σ^*_2 plateau.

3 Results

3.1 Apex Carbonaceous Matter

Correlated STXM and TEM characterization of disseminated carbonaceous matter within Apex chert revealed that the carbonaceous matter is mainly constrained to ~100 nm films along microquartz grain boundaries and at triple junctions (De Gregorio and Sharp 2006; De Gregorio et al. 2009), similar to other TEM studies of carbonaceous matter in Precambrian cherts (Moreau and Sharp 2004). Because quartz has a high hardness, ultramicrotomy during sample preparation commonly mechanically disaggregates the chert grains from the ancient carbonaceous matter, which can be located on the TEM grids by comparing STXM images acquired at 290 eV, where the interaction cross-section for carbon atoms is large, with those acquired at 280 eV, where the interaction cross-section is small (Fig. 7). In TEM images, much of the carbonaceous

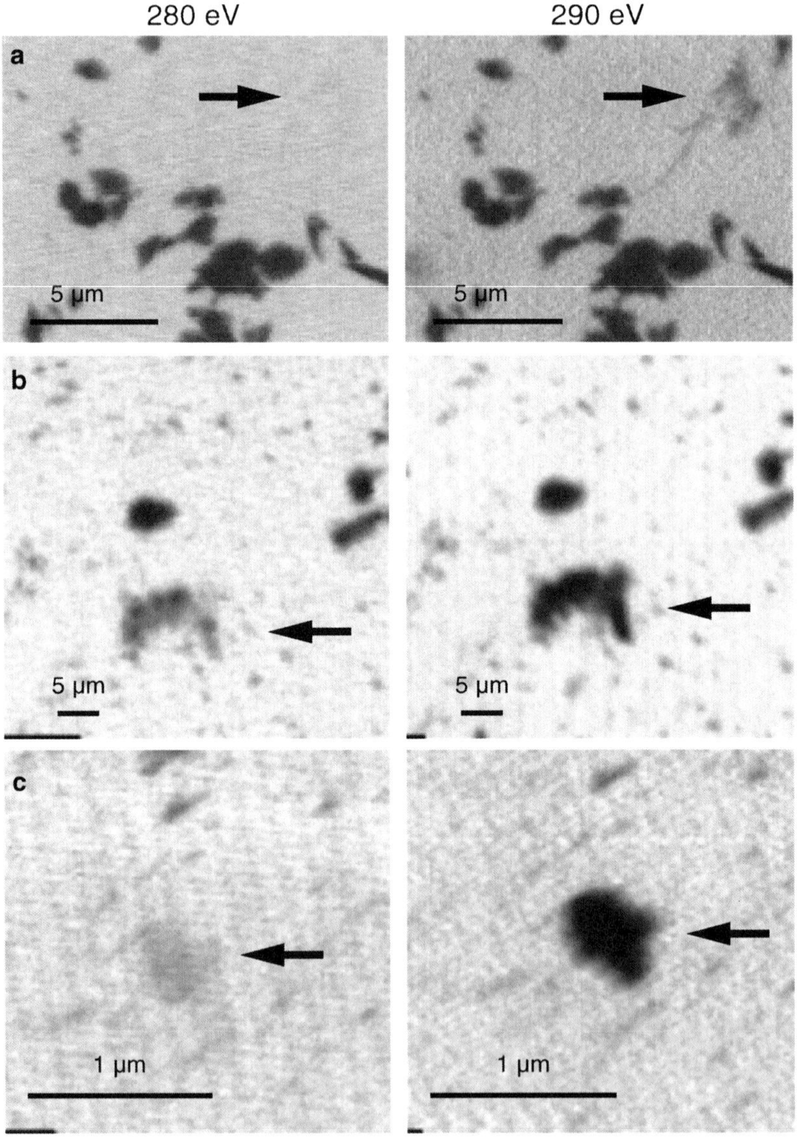

Fig. 7 STXM images of Apex carbonaceous matter. The X-ray absorption properties of carbon atoms changes between 280 and 290 eV, but those of quartz remain relatively constant over this energy range. Sample designations are (**a**) I.E10, (**b**) II.A6, and (**c**) II.C9 (Table 2)

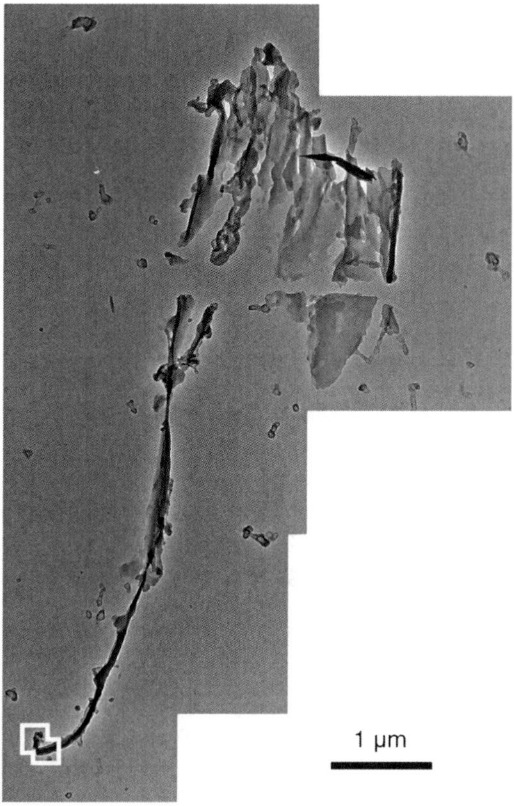

Fig. 8 TEM image mosaic of string-like Apex carbonaceous matter in sample I.E10. The corresponding STXM image is shown in Fig. 7a. The *white box* in the *lower left* is the location of the high-resolution TEM image shown in Fig. 10a

matter exhibits a strip- or string-like morphology (Fig. 8), due to the mechanical separation of carbonaceous grain boundary films from microquartz grains during ultramicrotomy. The widths of these strips are roughly equivalent to the sectioning thickness, while the thicknesses of the strips are equal to the width of the unperturbed grain boundary film. The widths of carbonaceous strips are increased in ultramicrotomed samples sectioned at a thickness of 200 nm (Fig. 9). In these thicker sections the effects of chattering are not as prevalent as in the 90 nm sections, and the carbonaceous films generally remain attached to the microquartz grains. High-resolution TEM images indicate that Apex carbonaceous matter is structurally amorphous, lacking the characteristic, closely-spaced, 3.4 Å lattice fringes visible in low order semi-graphite and graphitized kerogen (Fig. 10).

The local carbon bonding environment and distribution of organic functional groups in Apex carbonaceous matter are indicated by ELNES and XANES spectroscopy. C-ELNES spectra contain a sharp peak at 285 eV, corresponding to a

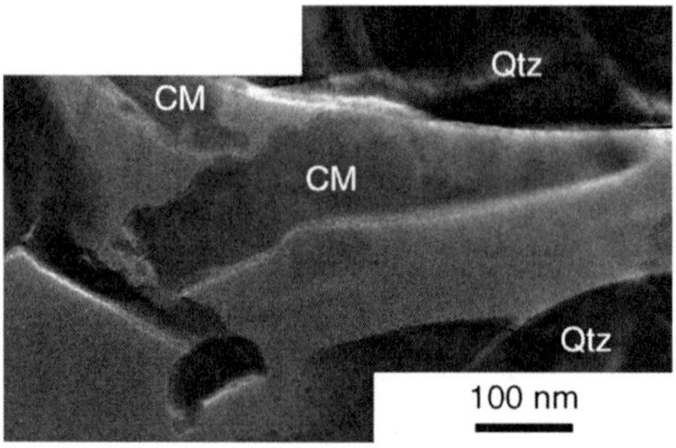

Fig. 9 TEM image of strip-like Apex carbonaceous matter (CM) attached to microquartz grains (Qtz) in sample III.D6. This sample was ultramicrotomed at a thickness of 200 nm

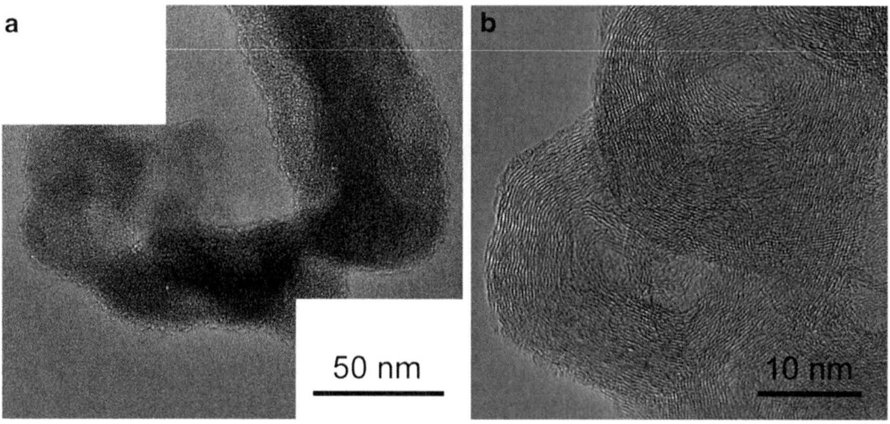

Fig. 10 TEM image mosaic of (**a**) string-like Apex carbonaceous matter from sample I.E10 compared with (**b**) low order graphite containing characteristic 3.4 Å lattice fringes (Modified from De Gregorio et al. [2009])

$1s \rightarrow \pi^*$ electronic transition in aromatic carbon bonds (Fig. 11). The intensity of this peak nearly matches that in semi-graphite, indicating that Apex carbonaceous matter contains a significant amount of aromatic bonding, most likely in polyaromatic domains. However, the characteristic $1s \rightarrow \sigma^*_{\mathrm{l}}$ exciton at 292 eV, present in graphite and low order graphitic carbon, is not observed in Apex carbonaceous matter. The presence of abundant aromatic carbon without long range graphitic ordering signified by a σ^*_{l} peak indicates that the polyaromatic domains in Apex carbonaceous material are randomly oriented with respect to each other, consistent with a kerogen-like macrostructure. At higher energies above the ionization potential, the spectra are dominated by a continuum of $1s \rightarrow \sigma^*$ electronic transitions from covalent σ bonds. In general, the C-ELNES spectra appear similar to that of amorphous carbon except that the shape of the σ^* peak shows a subtle redistribution

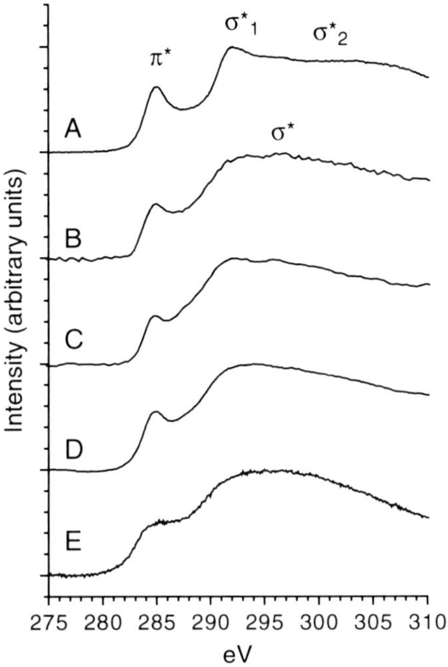

Fig. 11 C-ELNES spectra of Apex carbonaceous matter compared with inorganic standards: (**a**) Low order graphite; (**b, c**) Apex carbonaceous matter in sample I.E10 from De Gregorio et al. (2009); (**d**) Apex carbonaceous grain boundary films from De Gregorio and Sharp (2006); (**e**) Thick amorphous carbon film

of intensity towards lower energies, rather than a nearly hemispherical σ* peak as seen in amorphous carbon. It is likely that this shift in intensity corresponds to singly-bonded functional groups and aliphatic chains attached to polyaromatic domains.

A kerogen-like macrostructure is supported by the corresponding C-XANES spectra (Fig. 12). These spectra contain aromatic π* peaks at 285 eV and lack a graphitic σ^*_1 peak at 291 eV, consistent with ungraphitized polyaromatic carbon. In addition, two intermediate $1s \rightarrow \pi^*$ peaks are present, due to oxygen atoms bonded to carbon (Table 3). The intense peak at 288.5 eV arises from carboxylic acid functional groups (–COOH), while the less intense peak at 286.8 eV is most likely due to the presence of phenol and enol functional groups ($C_{arom.}$–OH) or aromatic ketones ($C_{arom.}$–C=O). Although nitrile functional groups (C≡N) also absorb X-rays at this energy, there is no other indication of high N abundance in the sample, nor do the XANES spectra indicate the presence of other N-containing functional groups. Some of the carbonaceous samples were too thick and electron dense for the efficient transmission of X-rays, causing the resulting XANES spectra to saturate (Fig. 12d). In these spectra, any local bonding information is hidden by the increased photoabsorption of the sample.

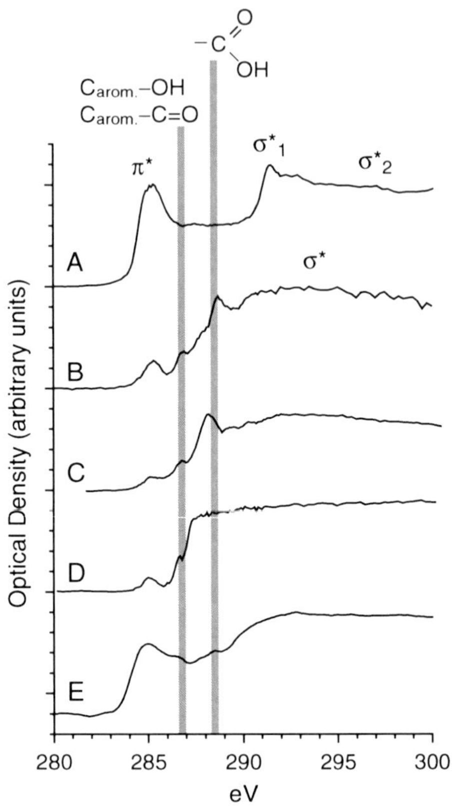

Fig. 12 C-XANES spectra of Apex carbonaceous matter compared with inorganic standards: (**a**) Low order graphite; Apex carbonaceous matter from samples (**b**) I.E10, (**c**) II.A6, and (**d**) II.C9; (**e**) Thick amorphous carbon film

3.2 Strelley Pool Carbonaceous Matter

Strelley Pool Chert samples analyzed in this study contain more carbonaceous matter than the Apex samples. Consequently, this is reflected in ultramicrotomed samples of Strelley Pool Chert by thicker string-like and strip-like sections. STXM images of Strelley Pool sections commonly contain regions of diffuse carbonaceous matter in addition to dense carbonaceous sections (Fig. 13). Correlated high-resolution TEM images reveal that the diffuse carbonaceous matter in STXM images are regions full of string-like carbonaceous matter, commonly still connected to microquartz grains (Fig. 14). Sheet-like carbonaceous matter also occurs when carbonaceous grain boundary films are detached intact from the quartz and lie flat upon the TEM grid support film. In thicker, 200 nm ultramicrotomed sections,

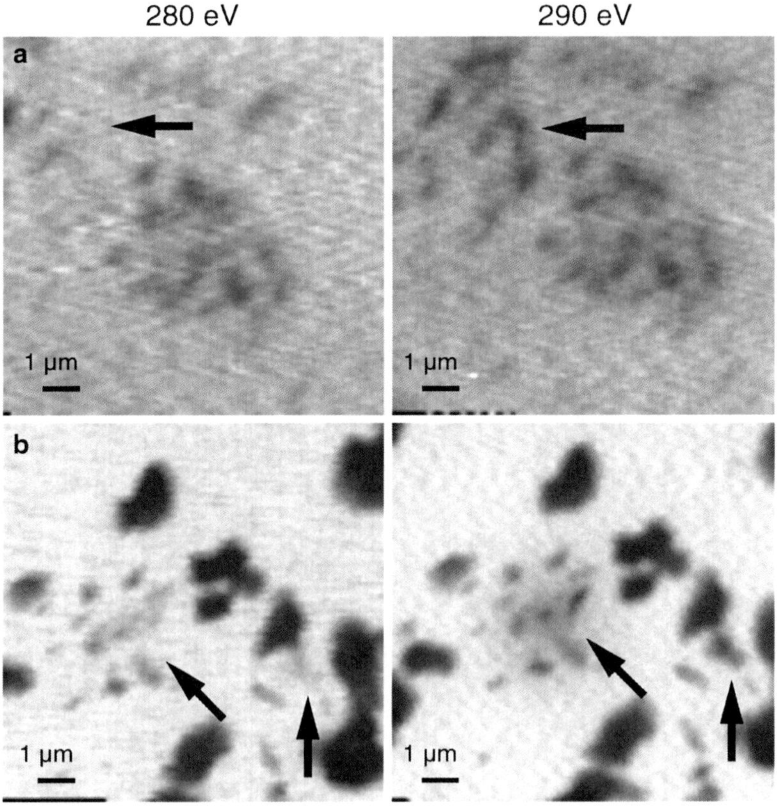

Fig. 13 STXM images of Strelley Pool carbonaceous matter. The X-ray absorption properties of carbon atoms changes between 280 and 290 eV, but those of quartz remain relatively constant over this energy range. Sample designations are (**a**) III.C1 and (**b**) III.D3 (Table 2)

string-like carbonaceous matter is commonly connected in large networks spanning several μms (Fig. 15). Open areas in these networks represent the locations of microquartz grains that were removed during sectioning, leaving the carbonaceous grain boundary films intact. Three carbonaceous films will commonly intersect at a single node as a triple junction between recrystallized quartz grains.

Despite the wider variety of morphology for Strelley Pool carbonaceous matter, it is structurally and chemically similar to Apex carbonaceous matter. C-ELNES spectra of Strelley Pool carbonaceous matter contain strong π^* peaks, indicating a generally amorphous macrostructure with abundant aromatic carbon (Fig. 16a). C-XANES spectra of both string-like and sheet-like carbonaceous matter contain similar peaks characteristic of aromatic carbon (285 eV), aromatic enols and ketones (286.8 eV), and carboxyl functional groups (288.5 eV), similar to those

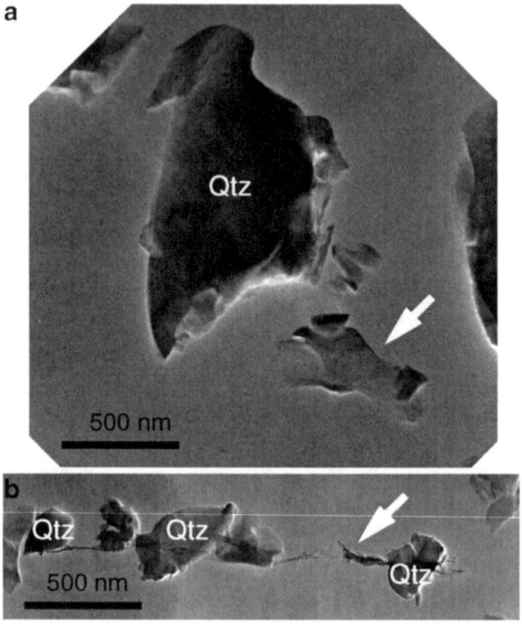

Fig. 14 Correlated TEM images of (**a**) sheet-like and (**b**) string-like carbonaceous matter in Strelley Pool Chert, sample III.D3. The *white arrows* correspond to the *black arrows* in Fig. 13b

observed in Apex carbonaceous matter (Fig. 17). These spectral results are consistent with a kerogen-like macrostructure containing oxygenated polyaromatic domains connected by aliphatic hydrocarbon chains. Since no graphitic 3.4 Å lattice fringes were observed in high-resolution TEM images, the aromatic moieties in Strelley Pool carbonaceous matter are randomly oriented, similar to low maturity kerogen.

3.3 Gunflint Kerogen

Gunflint kerogen in ultramicrotomed chert sections shows similar morphological characteristics to Apex and Strelley Pool carbonaceous matter. String-like and sheet-like carbonaceous matter is present, derived from thin carbonaceous films surrounding microquartz grains. In thicker ultramicrotomed sections, thick strips of kerogen connect quartz crystals (Fig. 18). The macrostructure and local carbon bonding in Gunflint kerogen is similar to that of Apex and Strelley Pool carbonaceous matter. The C-ELNES spectrum (Fig. 16b) contains a strong π^* peak indicative of polyaromatic domains but lacks a graphitic σ^*_1 peak, while the C-XANES spectra (Fig. 17) contain peaks due to aromatic π bonding (285 eV), aromatic enols

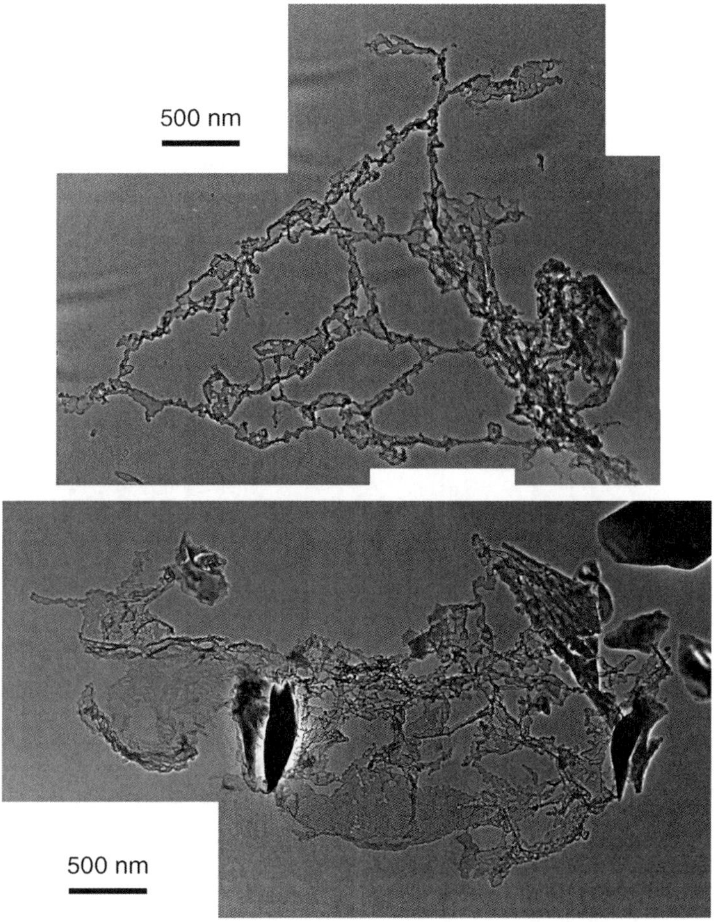

Fig. 15 TEM image mosaics of carbonaceous networks in 200 nm thick ultramicrotomed sections of Strelley Pool Chert, sample III.D2. Networks are created when microquartz grains fall out of the section, leaving the carbonaceous grain boundary films between them intact

and ketones (286.8 eV), and carboxyl functional groups (288.5 eV). These ELNES and XANES spectral features are similar between all three Precambrian samples and consistent with the structure and bonding within a low maturity kerogen.

3.4 Fischer-Tropsch-Type Carbonaceous Matter

The abundances of different organic components in FTT carbonaceous matter were measured by gas chromatography-mass spectrometry (Table 4). As reaction temperature increases, the organic products generally become dominated by aliphatic molecules

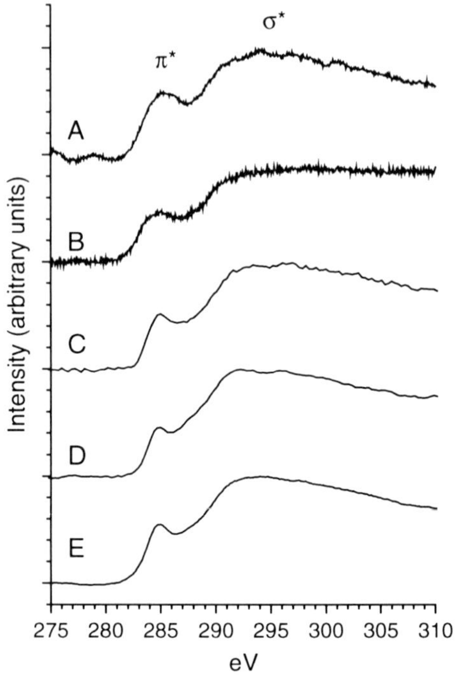

Fig. 16 C-ELNES spectra of ancient carbonaceous matter in (**a**) Strelley Pool (sample III.D2) and (**b**) Gunflint Chert (sample II.B1) compared with (**c-e**) Apex carbonaceous matter. Spectra (**c**) and (**d**) are reprinted from De Gregorio et al. (2009) and spectrum (**e**) is reprinted from De Gregorio and Sharp (2006)

typical of industrial Fischer-Tropsch synthesis, and the weight percent yield of FTT reaction product is maximized near 200°C (Rushdi and Simoneit 2001). The relative concentration of an unresolved complex mixture (UCM) of branched and cyclic hydrocarbons in the FTT products is strongly dependent on the reaction temperature, decreasing markedly as temperature increases up to 350°C, where hydrocarbon "cracking" reactions decrease the abundance of aliphatic compounds. In this study, the FTT sample synthesized at 175°C contains almost 50% (by weight) UCM, while the sample synthesized at 250°C is dominated by aliphatic compounds.

The desiccated FTT liquid extracts form heterogeneous films on TEM grids (Fig. 19). FTT organics synthesized at 175°C generate a nebulous "clumpy" film with few condensed particulate matter. Carbon concentration in this film varies spatially on the order of ~50 nm. FTT organics synthesized at 250°C, on the other hand, separate into two distinct phases. The first phase is deposited as a relatively homogenous film across the entire TEM grid, while the second phase condenses into ~200 nm aggregates of 5–10 nm carbonaceous nanoparticles. These aggregates likely form during desiccation in sample preparation. As the relative concentration of

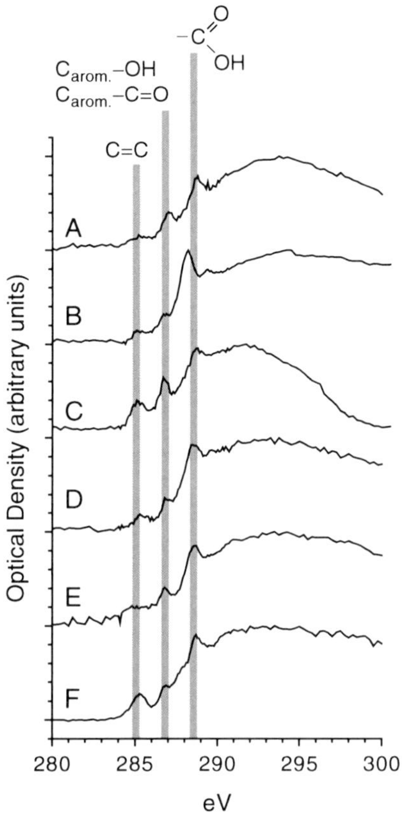

Fig. 17 C-XANES spectra of carbonaceous matter in Strelley Pool Chert samples (**a**) III.C1 (Fig. 13a) and (**b**, **c**) III.D3 (Fig. 14a and b, respectively) and in Gunflint Chert samples (**d**) III.A4 and (**e**) III.B1, compared with (**f**) Apex carbonaceous matter

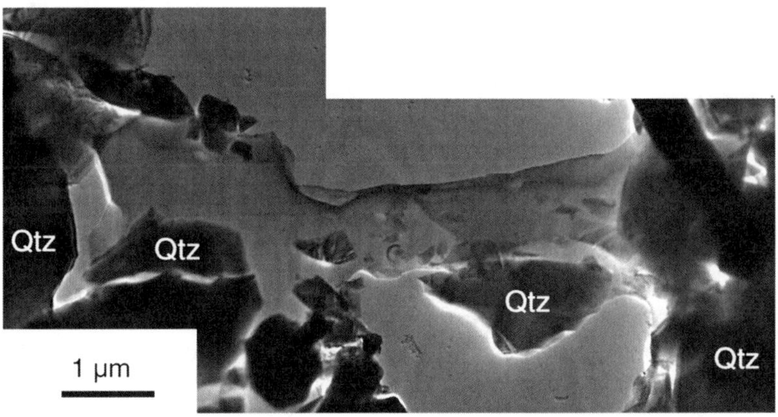

Fig. 18 TEM image mosaic of strip-like carbonaceous matter spanning microquartz grains (Qtz) in 200 nm thick ultramicrotomed sections of Gunflint Chert (sample II.B1)

Table 4 Relative concentrations of principal FTT products (in wt.% per 100 μg)

	Temperature (°C)[a]				This study (°C)	
Compound	150	200	250	300	175	250
n-alkanes	5	17	58	72	10	59
n-alkenes	4	11	1		7	1
n-alkanols	9	24	20		15	17
n-alkanoic acids	7	7	8	20	6	8
UCM[b]	73	37	5	0	49	15

[a]Data compiled from Rushdi and Simoneit (2001).
[b]Unresolved complex mixture of branched and cyclic compounds.

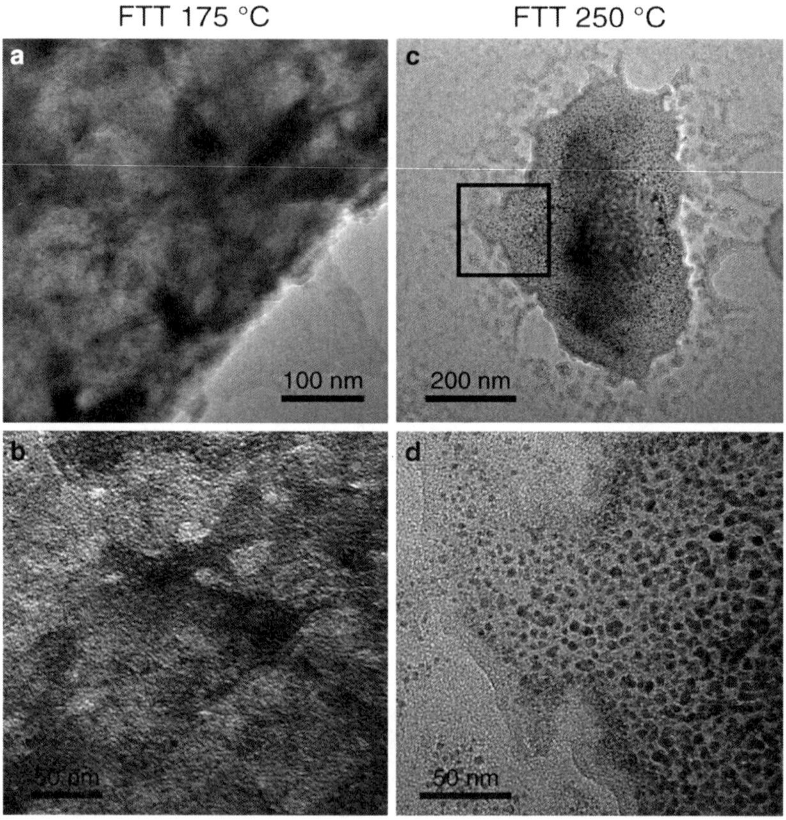

Fig. 19 TEM images of FTT carbonaceous matter synthesized at (**a**, **b**) 175°C and at (**c**, **d**) 250°C. The location of (**d**) is shown by the box in (**c**)

hydrocarbons increases, particulate matter within the 250°C extract begins to aggregate and precipitate out of solution. In addition, hydrocarbons begin to segregate into hydrophobic and hydrophilic phases. The hydrophobic phase is composed of abundant non-polar alkane compounds, while alkanols, alkanoic acids, and UCM dominate the hydrophobic phase.

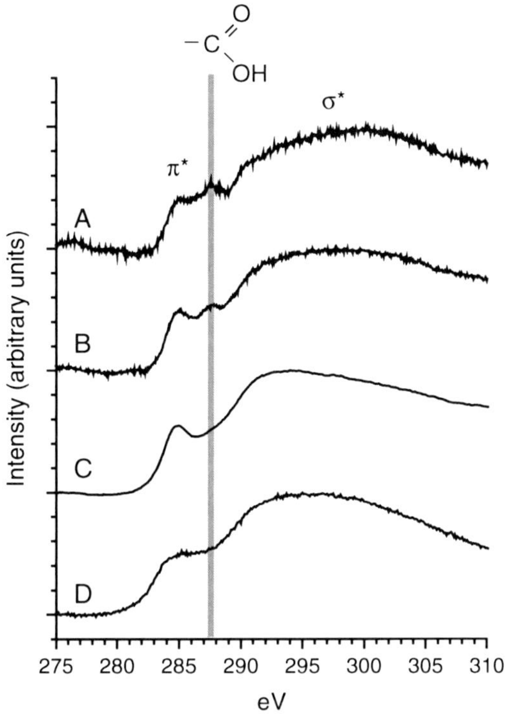

Fig. 20 C-ELNES spectra of FTT carbonaceous matter synthesized at (**a**) 175°C and (**b**) 250°C, compared with (**c**) Apex carbonaceous matter and (**d**) an amorphous carbon film. The two FTT samples contain abundant carboxyl functional groups not present in the other two samples

Both ELNES and XANES spectra of FTT carbonaceous matter indicate the presence of abundant carboxyl functional groups. Energy loss due to carboxyl functional groups are not often observed in C-ELNES spectra because of lower instrumental resolution and overlapping from the surrounding π* and σ* peaks. However, small carboxyl peaks are observed in the spectra of both FTT samples (Fig. 20). In addition the shape of the σ* peak is more hemispherical and reminiscent of amorphous carbon than that of the Precambrian carbonaceous matter, in which much of the σ* intensity is concentrated at lower energies. C-ELNES spectra of the 250°C FTT sample indicates that particulate aggregates contain enhanced carboxyl functionality over the relatively homogenous film between aggregates, suggesting that the majority of the oxidized hydrocarbons and UCM in the sample are contained within the particulate material. The overabundance of carbonyl functional groups in both samples is also apparent in C-XANES spectra (Fig. 21). Although the two FTT extracts have different compositions, their respective C-ELNES and C-XANES spectra are similar. These spectroscopic techniques highlight the similar abundances of alkanols (containing –OH groups) and alkanoic acids (containing –COOH groups) in

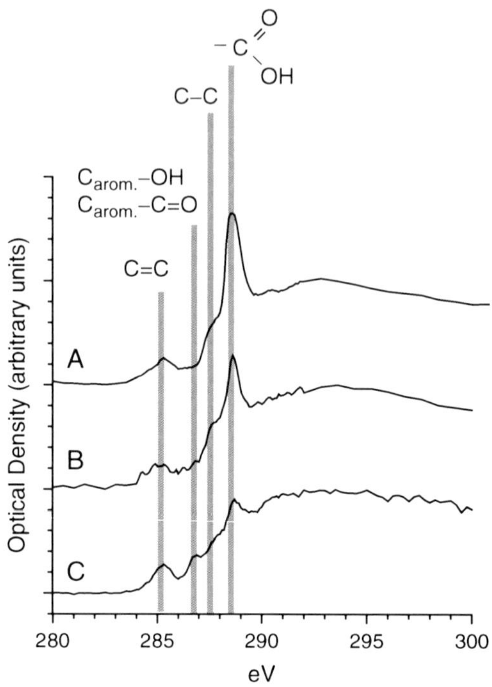

Fig. 21 C-XANES spectra of FTT carbonaceous matter synthesized at (**a**) 175°C and (**b**) 250°C, compared with (**c**) Apex carbonaceous matter. Both FTT samples are distinct from Apex carbonaceous matter by their overabundance of carboxyl functional groups and lack of aromatic enol and ketone functional groups

both samples. A small difference may be present around 287.8 eV, where photoabsorption for the 1s → 3*s*/*p* transition in C–C single bonds occur. The C-XANES spectrum from the 250°C extract appears to have slightly more intensity at this energy than in the spectrum from the 175°C extract. Since the 175°C extract contains more aromatic material in the form of UCM, the 250°C extract will have a somewhat higher proportion of aliphatic material, even though the overall aliphatic abundance in both samples is similar.

4 Discussion

4.1 Beam Damage of Carbonaceous Matter by STXM and TEM

Reliable interpretation of XANES and ELNES data from carbonaceous matter requires an assessment of potential alteration of the sample during the spectroscopic analyses. STXM, which utilizes 270–580 eV X-rays, deposits a much lower energy dose on the sample than 100–300 keV electrons in TEM, and therefore generally represents a much less damaging environment for beam-sensitive samples. During XANES, photoabsorption of low energy X-rays promotes core electrons to anti-bonding or continuum states, making it possible to alter the structure and bonding in carbonaceous matter. Rather than simply breaking bonds, the local electronic configuration may change such that new covalent bonds or functional groups with lower energy states form. In predominantly aromatic carbonaceous matter, such as kerogen and humic acids, which are generally more stable under the beam, C-XANES spectra do not visibly change during analyses, but small amounts of amine ($–NH_x$) and amide ($O=C–NH_x$) functional groups may be altered into nitrile ($C≡N$) functional groups, while heterocyclic nitrogen abundance appears unchanged (Leinweber et al. 2007). Similar effects are observed for X-ray damage of amino acids (Zubavichus et al. 2004). Fortunately, these structural changes are minor for the typical XANES acquisition times and photon doses used in this study. In aliphatic carbonaceous matter and many polymers, however, aromatic bonding and functional groups are observed after significant exposure to the X-ray beam (Rightor et al. 1997; Zhang et al. 1995). Cellulose and polysaccharide molecules in wood and plant matter contain abundant hydroxyl (–OH) and ester (C–O–C) bonds that are quickly altered into aromatic enol ($C_{arom.}$–OH) and vinyl ketone ($C_{arom.}$–C=O) functional groups (Cody et al. 2009).

Because TEM utilizes high energy electrons, it has a higher potential than STXM for damage and alteration of carbonaceous material, and therefore all TEM analyses should be performed after STXM. Unlike photons from synchrotron or laser sources, electrons have mass and can cause atomic displacements, or "knock-on" damage, in addition to radiolysis (Egerton et al. 2004). Local heating by a condensed electron beam can also damage sensitive organic matter (Egerton and Rauf 1999). These types of inelastic scattering can cause mass loss, break bonds, and change the overall chemical composition of carbonaceous matter (Egerton et al. 1987). Amorphization of poorly graphitized carbonaceous matter can occur in seconds and is visible in high resolution TEM images. Chemical and bonding changes are visible in C-ELNES spectra. For dominantly aromatic carbonaceous matter, the intensity of the aromatic π^* absorption increases slightly while the shape of the σ^* peak becomes more rounded, similar to pure amorphous carbon (Fig. 22). However, for dominantly aliphatic carbonaceous matter and polymers, beam damage can cause dramatic changes, creating abundant aromatic bonding and other aromatic functional groups by the cross-linking of adjacent hydrocarbon chains (Fig. 23). This difference in radiolysis sensitivity between aromatic and aliphatic molecules arises from the delocalized π resonances in polyaromatic domains, which can more effectively distribute a given dose of deposited energy than aliphatic σ bonds (Isaacson et al. 1973).

Damage in TEM may be minimized by several methods (Isaacson 1979). The overall electron dose incident on the sample may be reduced by operating the microscope at the lowest magnification possible using an uncondensed beam. Selecting a smaller spot size, which adjusts the strength of the first condenser lens,

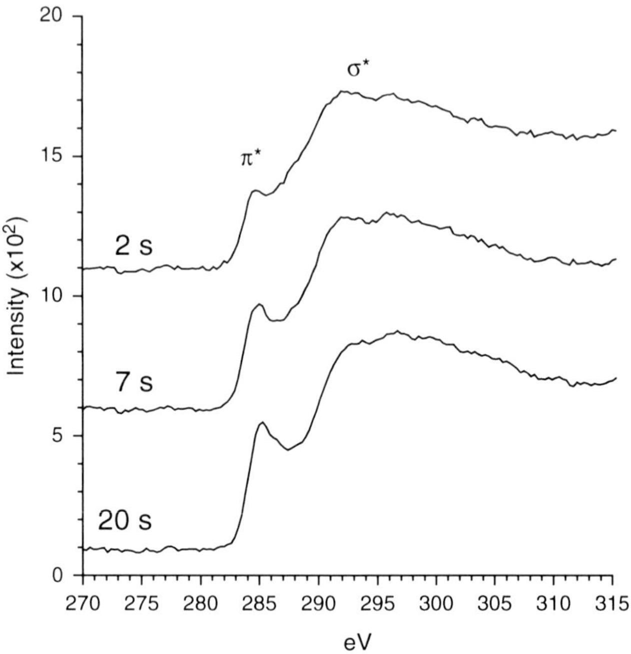

Fig. 22 C-ELNES spectra of aromatic carbonaceous matter during exposure to a condensed electron beam in TEM. The intensity of the π* absorption due to C=C bonds significantly increases in only 20 s, while intensity of the σ* absorption shifts towards higher energies

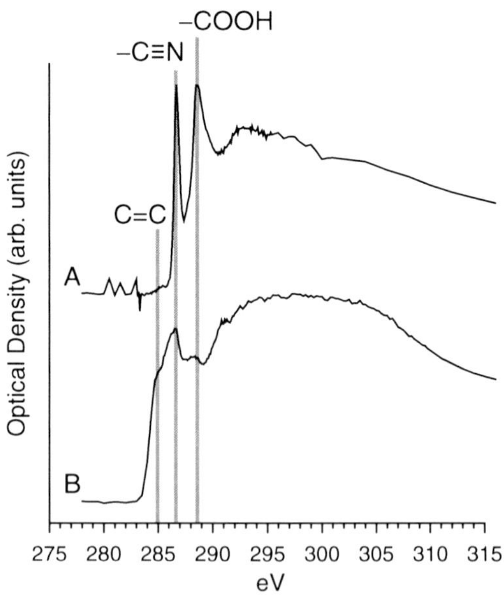

Fig. 23 C-XANES spectra of nitrogen-rich aliphatic organic matter (**a**) before and (**b**) after TEM imaging. Nitrile (286.7 eV) and carboxyl (288.5 eV) functional groups are destroyed by electronic beam damage and aromatic carbon (285.0 eV) is produced

also decreases illumination on the sample in addition to increasing the angular coherence of the beam. Dwell time for analyses requiring a condensed beam, such as ELNES and EDS, should be strictly minimized. Operating the instrument in STEM mode with a continuously moving beam generally delivers a smaller electron dose than conventional TEM imaging of the same area. Radiolysis damage may be reduced by using a higher operating voltage, although this will increase the amount of knock-on damage. Specimen heating and, to a lesser extent, radiolysis can be reduced by using a liquid-N_2-cooled sample holder, although this will introduce sample drift from thermal expansion or contraction of the holder until it reaches thermal equilibrium. In practice, beam damage of carbonaceous samples cannot be avoided in the TEM, but can only be mitigated as much as possible.

4.2 *Spectral Differences Between XANES and ELNES*

There are significant differences between the XANES and ELNES spectra acquired from the same carbonaceous samples. The presence of carbonyl (288.5 eV) and phenol/ketone (286.8 eV) peaks in XANES spectra but not in ELNES spectra is partly due to the enhanced energy resolution afforded by the synchrotron radiation source. However, it is also likely that exposure to the electron beam in TEM may remove oxygen atoms, which would decrease the intensity of any carboxyl and phenol peaks in ELNES spectra such that they are not resolvable. Similar spectral differences between ELNES and XANES have been observed in correlated studies of diesel soot and have been attributed to beam damage of the sample during TEM analysis (Braun et al. 2005, 2009). In addition, the aromatic π^* peak (285 eV) appears to be less intense in the XANES spectra relative to the carbon K-edge. Although the intensity of the same π^* peak in ELNES spectra may be enhanced by beam damage in TEM, it is likely that, in this case, orientation effects are causing a depression of the π^* peak in XANES spectra. Since this ancient carbonaceous matter is located along grain boundaries in the host rock, it is possible that the majority of the polyaromatic domains are likely to be oriented relatively parallel to those grain boundaries. When separated from the grain boundaries, the carbonaceous matter will lie flat on the support film of the TEM grid, and the polyaromatic domains will then be roughly parallel to the surface of the grid. In this orientation, the interaction cross-section between the incident X-rays and the aromatic π bonds is small because the polarization of the X-rays is roughly perpendicular to the hybridized π orbitals (Belavin et al. 2006; Rosenberg et al. 1986; Skytt et al. 1994). A similar depression of $1s \rightarrow \pi^*$ photoabsorptions are observed in XANES spectra of natural graphite deposited in hydrothermal veins (Brandes et al. 2008).

4.3 Structure and Bonding of Precambrian Carbonaceous Matter

The similarity between C-ELNES and C-XANES spectra of Apex carbonaceous matter and Gunflint kerogen, led De Gregorio et al. (2009) to suggest that Apex carbonaceous matter may have a kerogen-like macrostructure of randomly oriented polyaromatic domains connected by aliphatic hydrocarbon chains. A similar argument may be made for Strelley Pool carbonaceous matter. All three Precambrian carbonaceous materials are ungraphitized in high-resolution TEM images, but contain significant aromatic carbon bonding. This observation agrees with the aromatic π^* peaks in both C-ELNES and C-XANES spectra without a characteristic graphitic σ^*_1 exciton, which arises once graphitization begins. Although the Precambrian carbonaceous samples are ungraphitized, the abundant polyaromatic domains can give rise to graphitic resonances in laser Raman spectroscopy, which have been observed in both Apex (Brasier et al. 2002; Schopf et al. 2002, 2007) and Strelley Pool samples (Allwood et al. 2006b, 2007). In addition, C-XANES spectra from each sample indicate the presence of aromatic enol (–OH) and ketone (–C=O) functional groups, likely attached to the edges of polyaromatic moieties. These C-XANES spectra also include photoabsorptions due to carboxyl (–COOH) functional groups. The ELNES and XANES spectra indicate that Apex and Strelley Pool carbonaceous matter and Gunflint kerogen are more than simple hydrocarbon macromolecules. These Precambrian materials are spectrally distinct from both graphite and amorphous carbon, and their C-XANES spectra indicated the presence of oxygenated functional groups within their complex macrostructure.

However, biogenicity cannot be determined from structure and bonding alone. For example, the C-XANES of Apex and Strelley Pool carbonaceous matter also share spectral similarities with clearly abiotic organic residue from carbonaceous chondrite meteorites (Cody et al. 2008) and interplanetary dust particles (Flynn et al. 2003; Keller et al. 2004). De Gregorio et al. (2009) paired their TEM and STXM data with secondary ion mass spectrometry measurements of nitrogen, sulfur, and phosphorus abundances, showing a striking similarity between atomic N/C, S/C, and P/C in Apex carbonaceous matter and Gunflint kerogen, despite their formation at different locations and time periods. This structural and chemical similarity strongly suggests a single origin for both samples. Since Gunflint kerogen is biogenic, Apex and Strelley Pool carbonaceous matter likely represent the degraded remains of Archean microorganisms.

4.4 Comparison with FTT Carbonaceous Matter

Precambrian kerogen and kerogen-like carbonaceous matter can be clearly distinguished from the FTT carbonaceous products analyzed in this study by ELNES and XANES. In particular, both C-ELNES and C-XANES spectra of FTT material contain much larger absorptions due to carbonyl functional groups (–COOH) than in Precambrian carbonaceous matter. In addition, C-XANES spectra of FTT material show no evidence of aromatic enol (–OH) and ketone (–C=O) present in the Precambrian samples. The spectroscopic characterization of FTT carbonaceous matter is more consistent with aliphatic hydrocarbons modified by hydroxyl and carboxyl functional groups rather than an aromatic kerogen-like macrostructure.

Although FTT carbonaceous matter is structurally and chemically distinct from Precambrian chert-hosted carbonaceous matter, this does not necessarily mean that FTT synthesis within an ancient hydrothermal system could not create an abiotic kerogen-like material. The FTT samples analyzed in this study were synthesized without a natural catalyst, presumably reacting with metal carbonyl complexes on the surface of the stainless steel reaction vessel, but mineral catalysts may preferentially generate different types of organic molecules. Abiotic carbon reduction in a natural setting like a hydrothermal vent may utilize several different mineral catalysts, increasing the diversity of organic bonding in the FTT products. Some FTT reactions tend to synthesize more alkenes containing carbon double bonds (Leif and Simoneit 2000), generating more aromatic organic matter. Other light elements common in biomolecules, such as nitrogen, sulfur, and phosphorus, may also be incorporated into FTT carbonaceous matter if appropriate organic precursors are present in hydrothermal fluids (Rushdi and Simoneit 2004, 2005). Simple inorganic aqueous species, such as HS^- or NH_4^+, are less likely to be incorporated into FTT carbonaceous matter, but the efficiency of this process is unknown. More research into the reaction pathways and ubiquity of FTT processes on the early Earth are necessary, yet it appears unlikely that FTT synthesis as we currently understand it could produce carbonaceous matter with the structural and compositional complexity of kerogen.

The fate of FTT carbonaceous matter during diagenesis and thermal maturation is also largely unknown. During early diagenesis the organic remains of living organisms undergo significant volatile loss. Insoluble FTT carbonaceous matter would presumably also lose volatile elements during diagenesis, while the more soluble molecules would likely be quickly transported away. The residual FTT material would lose most of the abundant carboxyl functionality and begin to form aromatic structures. Polyaromatic domains may form more easily from FTT material dominated by alkenes. Although these processes would decrease the intensity of carboxyl peaks in ELNES and XANES spectra of matured FTT carbonaceous matter, it is not known if this process would also generate significant amounts of aromatic enol and ketone functional groups, as observed in the Precambrian kerogen and kerogen-like samples. However, these hypothesized processes have yet to be tested experimentally. Since the maturation of biological material into kerogen is relatively well known, the most parsimonious explanation of the sub-μm TEM and STXM characterization of Apex and Strelley Pool carbonaceous matter is that they represent the degraded remains of Archean microorganisms.

4.5 Controversy Surrounding the Apex Microfossils

Until 2002, microbe-like features discovered within the 3.458–3.465 Ga Apex Basalt cherts were hailed as the most ancient microfossils on Earth. Eleven distinct filamentous microfossil taxa have been described based upon morphological characteristics (Schopf 1992, 1993). The microbe-like features resemble prokaryote trichromes, averaging 3 μm in diameter and 30–80 μm, with limited size ranges for each taxon. Individual filaments contain regularly sized, septate medial cells and apparent terminal cells (Schopf 1992), although fine cellular features are not preserved. Their morphological similarity to modern cyanobacteria led to the conclusion that oxygen producing photoautotrophic organisms existed as far back as 3.5 Ga. In addition, bulk rock analyses of carbon within the chert yielded a light $\delta^{13}C$ isotopic signature ranging from −22‰ to −30‰ (Brasier et al. 2002; Strauss and Moore 1992), which is typical for photoautotrophic microorganisms (Schidlowski 1988) and consistent with well-accepted Proterozoic microfossils, such as those from the 850 Ma Bitter Springs Formation in Central Australia (House et al. 2000).

However, concerns were eventually raised regarding the biogenicity of the Apex microfossils. Reexamination of several of the holotype thin sections using automated imaging software revealed that many microbe-like features contain additional branched and twisted portions that appears out of focus or inconspicuous in the original, manually-acquired images (Brasier et al. 2002). Furthermore, only around 40% of all the putative microfossils are contained within primary chert fabrics (Brasier et al. 2002, 2005). Brasier et al. (2005) suggested that much of the sinuous and arcuate nature of the features may be generated by adsorption to silica spherules during chert formation, and that the apparent septation and seemingly cellular morphology may be mimicked by hydrocarbon adsorption around individual calcite rhombs, later replaced by silica. Alternatively, a microbe-like morphology could be abiotically created by sequestration of carbonaceous matter within aggregates of silica-witherite crystallites (García-Ruiz et al. 2003). Proponents of the biogenic interpretation for the Apex microfossils suggest that since the region has experienced some metamorphism, any *bona fide* microfossils could have been severely altered, leaving only a generic cellular morphology, similar to degraded filamentous microfossils in hydrothermally altered Silurian cherts from the Bardo Mountains in Poland (Kaźmierczak and Kremer 2002). By admission, less than 1% of the Apex microbe-like features are well preserved and morphologically complete for taxonomic description (Schopf 1993).

Subsequent field mapping of the chert outcrop at Chinaman Creek has also led to implications against the biogenicity of the putative Apex microfossils. Although briefly mentioned in previous reports, the Apex chert includes two intersecting chert units, only one of which is conformable with the Apex Basalt flows (Brasier et al. 2002, 2005; Van Kranendonk 2006). The parallel chert unit is a gray, cross-bedded stratiform chert that most likely formed in a shallow sea environment. The second chert unit is a brecciated siliceous dike cross-cutting the Apex pillow basalts for approximately 1 km along a local fault zone before merging with the stratiform chert. It is within the dike chert and not the stratiform chert that the putative microfossils were recovered. Although the stratiform chert is discontinuous due to faulting, its stratigraphic position throughout the area is relatively constant.

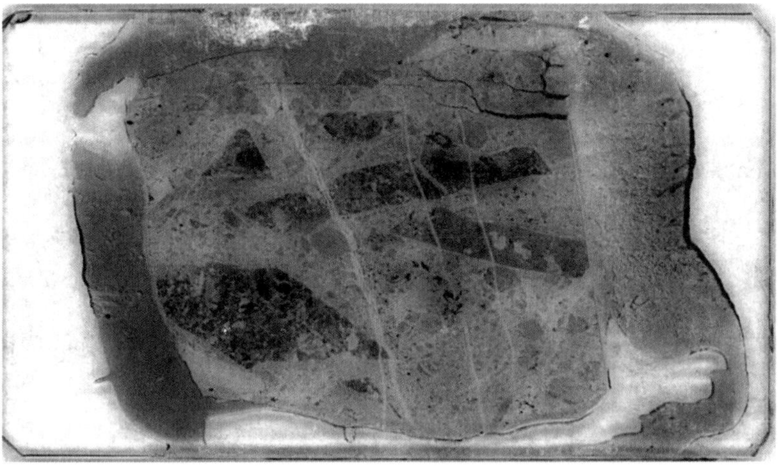

Fig. 24 Thin section of brecciated Apex chert from the putative microfossil locality at Chinaman Creek. At least five major generations of chert precipitation, veining, and fissure fill are visible. The glass slide is approximately 1″ × 2″

The black chert dike most likely represents a relic hydrothermal vent, driven by a coeval granitoid pluton, that deposited the stratiform chert on the ancient Archean seafloor (Brasier et al. 2002, 2005). Swarms of chert dikes associated with bedded chert units are present elsewhere in the Pilbara Supergroup (Ueno et al. 2004; Van Kranendonk 2006) and can be correlated with the activity of granitoid complexes (Lindsay et al. 2005). The multiple brecciation events apparent in hand samples and thin sections of Apex chert (Fig. 24) could have been caused by multiple injections of hydrothermal fluids (Jébrak 1997) or regional deformation while the hydrothermal system was still active (Van Kranendonk 2006). Recrystallization and coarsening of the dike chert must have occurred during regional metamorphism because the $\delta^{18}O$ isotopic composition of +13.7‰ to +14.7‰ is consistent with a paleotemperature of 250–350°C for prehnite-pumpellyite metamorphic fluids (Brasier et al. 2002; Smith et al. 1982).

The implication of a hydrothermal origin for the carbon-rich dike chert is that the microbe-like features within it may not be of biogenic origin. Although phylogenetic studies using ribosomal RNA indicate that hyperthermophiles lie at the root of the genetic tree of life (Stetter 2006; Woese et al. 1990), modern hyperthermophiles have always been observed residing at the periphery of hydrothermal springs, never inside them where the fluid temperature is too high. Rather, an abiotic organic synthesis such as Fischer-Tropsch-type (FTT) reactions could explain the presence of carbonaceous matter preserved within an ancient hydrothermal vent (Brasier et al. 2002). In an attempt to characterize the macromolecular structure of the Apex carbonaceous matter, both Schopf et al. (2002, 2007) and Brasier et al. (2002) performed in situ laser Raman spectroscopy. They observed both G (ordered) and D (disordered) peaks consistent with low maturity kerogen and many other disordered carbonaceous materials (Pasteris and Wopenka 2003), but the original researchers interpreted their results as evidence of "graphitized kerogen" (Schopf et al. 2002) and "amorphous graphite" (Brasier et al. 2002). Further bulk organic analyses (Derenne et al. 2008) and high resolution transmission electron microscopy (De Gregorio and Sharp 2006) of Apex carbonaceous matter confirm that it is most consistent with an amorphous macrostructure composed of aromatic and aliphatic domains, similar to low maturity kerogen.

4.6 Controversy Surrounding the Strelley Pool Stromatolites

Like the putative Apex microfossils, the stromatolite-like structures within the 3.350–3.458 Ga Strelley Pool Chert have also undergone intense debate regarding their biogenicity. The *Conophyton*-like structures are limited to the cherty lower member of the Strelley Pool Chert, consisting of finely laminated alternating gray chert, black carbonaceous chert, and dark gray carbonate-rich chert, in many cases grading laterally into one another (Lindsay et al. 2005; Lowe 1980). The lower member also includes silicified evaporitic facies, indicating deposition in a low energy, subtidal or intertidal environment, such as a restricted coastal lagoon (Lowe 1980, 1994). This paleoenvironment would have been favorable for mat-building phototrophic organisms.

The biogenicity of the Strelley Pool stomatolites has been contested on several accounts since they were first reported (Lowe 1980). They lack explicit microfossil remains or a well-defined axial zone common to *Conophyton* group stromatolites (Lowe 1994). Some researchers have argued that the laminae are smooth, isopachous and continuous between the structures (Lindsay et al. 2005; Lowe 1994), implying low sedimentation and chemical precipitation rates, where the abiotic formation of conical structures may be possible (Grotzinger and Rothman 1996). Other researchers dispute this claim, identifying stromatolitic structures with crenulated and corrugated laminae, thinning non-isopachous laminae, and morphometric diversity correlating with depositional paleoenvironment (Allwood et al. 2006a, 2007). Finely laminated corrugations and wrinkle structures may be produced abiotically during the hardening of a colloidal siliceous ooze (McLoughlin et al. 2008), but many of the other biogenic features common to Strelley Pool stromatolites have not yet been reproduced experimentally (Grotzinger and Knoll 1999).

In at least one locality, the Strelley Pool Chert overlies a system of black, carbon-rich dike cherts, similar to those intersecting the Apex Basalt at Chinaman Creek, suggesting the possibility of hydrothermal controls on chert petrogenesis and stromatolite formation (Lindsay et al. 2005). The thermal source of the ancient hydrothermal vents was the nearby 3.47 Ga Carlindi granitoid complex. However, in general it appears that stromatolite abundance and diversity decreases in proximity to hydrothermal activity (Allwood et al. 2007). The carbonaceous matter preserved within the black chert dikes has a $\delta^{13}C$ isotopic composition ranging from −26.9‰ to −34.9‰, similar to that of the Apex dikes (Lindsay et al. 2005). Laser

Raman spectroscopy, Fourier transform infrared spectroscopy, solid state ^{13}C nuclear magnetic resonance spectroscopy, and pyrolysis gas chromatography-mass spectrometry of carbonaceous matter within the Strelley Pool dike chert indicate an amorphous structure composed of large, disordered polyaromatic domains, more similar to low maturity kerogen than fluid deposited graphite (Allwood et al. 2006b; Marshall et al. 2007). Considering all the available evidence, a stronger case can be made for the biogenicity of the Strelley Pool stromatolites than for the Apex microfossils.

4.7 Archean Hydrothermal Microbial Communities

If the ancient carbonaceous matter preserved within the Apex and Strelley Pool dike cherts represents *bona fide* kerogen, then it is unclear why it should be preserved inside a relic hydrothermal vent, up to 1 km below the paleosurface, rather than within a bedded, sedimentary chert. Hyperthermophilic microorganisms could reside in microbial communities along the periphery of a hydrothermal spring, but the fluid temperatures within the vent would be too high for such microorganisms withstand. This necessarily requires transport of biogenic material into the vent. If primary productivity were high, then copious biogenic matter and deceased organisms would be generated and may be deposited along the walls inside the ancient vent, driven against the net outward flux of fluid by gravitational settling. The brecciated nature of the Apex dike chert indicates either motion along a local fault or cycles of fluid heating. As the temperature of hydrothermal fluids rises and falls, thermal expansion and contraction of siliceous sinter deposits may cause significant spallation and brecciation (Jébrak 1997). As the fluid temperature decreases, microorganisms may migrate inwards into the vent to access greater concentrations of nutrients.

It is also possible, although less likely, that biogenic material from a deeper sedimentary unit was transported upwards through hydrothermal feeder dikes of the Apex Basalt and Strelley Pool Chert. However, no evidence of hydrocarbon source rocks has been observed stratigraphically beneath the Apex and Strelley Pool Chert units. Biogenic organic matter may be concentrated in hydrothermal veins by the circulation of hydrothermal fluids in the surrounding surface rocks. The organic matter may then be protected from further transport by silica deposition along the walls of the vent.

Much is still unknown about the habitat and metabolism of the earliest living species. This case study of Apex and Strelley Pool carbonaceous matter indicates that biogenic material from Archean microorganisms is preserved in ancient hydrothermal dikes, suggesting that the microorganisms associated with this kerogen were hyperthermophilic chemoautotrophs. Similar conclusions have been reached from studies of carbonaceous matter in black hydrothermal dike cherts penetrating the 3.458 Ga Dresser Formation (Glikson et al. 2008; Ueno et al. 2004).

5 Summary

Generally, carbonaceous matter comprises only a small portion of Precambrian rocks but has the largest utility for assessing the biogenicity of putative microfossils and/or stromatolites in these units. Since biomolecules undergo a relatively specific pathway of thermal maturation into kerogen, the macrostructure, local bonding environment, and organic functional group diversity allow for comparisons with ancient carbonaceous matter of indeterminate origin. Common structural and chemical characteristics of kerogen–randomly oriented polyaromatic domains connected by aliphatic hydrocarbon chains, enol (–OH) and ketone (–C=O) functional groups attached to aromatic moieties, prevalent carboxyl (–COOH) functional groups, and other biotic elements (e.g. N, S, and P) present in aromatic heterocycles–should be observed in all biogenic carbonaceous matter. Scanning-transmission X-ray microscopy, and transmission electron microscopy are useful analytical tools for characterizing the structure and bonding of ancient carbonaceous matter at μm to sub-μm spatial scales relevant for in situ analyses. For example, in the case study of carbonaceous matter preserved in Archean Apex and Strelley Pool Cherts, correlated TEM and STXM were used to show the structural and chemical similarity between ~100 nm thick carbonaceous grain boundary films in these samples and kerogen from the microfossiliferous Gunflint Formation, implying that the Archean kerogen-like carbonaceous matter derived from early hyperthermophilic microorganisms.

Although characterization of the structure and bonding of ancient carbonaceous matter is useful for assessing the biogenicity of potential microfossils and stromatolites, it is a necessary but insufficient criterion (Pasteris and Wopenka 2003). The ubiquity of abiotic carbon reduction and FTT synthesis in Archean hydrothermal systems is undetermined, and the thermal maturation history of FTT carbonaceous products has not yet been established. Therefore, the macrostructure and composition of ancient FTT material potentially present in Archean sediments is not known. More experimental studies of FTT processes, catalysts, and organic products, as well as their alteration during burial, diagenesis, and metamorphism, is necessary in order to adequately assess the "abiogenicity" of Archean carbonaceous matter to the same degree we can attest to its biogenicity through comparison with kerogen.

References

Allwood AC, Walter MR, Kamber BS et al (2006a) Stromatolite reef from the Early Archaean era of Australia. Nature 441:714–718

Allwood AC, Walter MR, Marshall CP (2006b) Raman spectroscopy reveals thermal palaeoenvironments of c.3.5 billion-year-old organic matter. Vib Spectrosc 41:190–197

Allwood AC, Walter MR, Burch IW et al (2007) 3.43 billion-year-old stromatolite reef from the Pilbara Craton of Western Australia: ecosystem-scale insights to early life on Earth. Precambr Res 158:198–227

Anderson RB (1984) The Fischer-Tropsch synthesis. Academic Press, Orlando

Apen E, Hitchcock AP, Gland JL (1993) Experimental studies of the core excitation of imidazole, 4, 5-dicyanoimidazole, and *s*-triazine. J Phys Chem 97:6859–6866
Bagas L (2002) Stratigraphic revision of the Warrawoona and Gorge Creek Groups in the Kelly greenstone belt, Pilbara Craton, Western Australia. In: 2001–02 Annual Review. Geological Survey of Western Australia, Perth, Western Australia
Bassim ND, De Gregorio BT, Stroud RM et al (2008) Study of FIB damage in carbonaceous materials using XANES. Microsc Microanal 14:1008–1009
Behar F, Vandenbroucke M (1987) Chemical modelling of kerogens. Org Geochem 11:15–24
Belavin VV, Okotrub AV, Bulusheva LG et al (2006) Determining misorientation of graphite grains from the angular dependence of X-ray emission spectra. J Exp Theor Phys 103:604–610
Benzerara K, Menguy N, López-García P et al (2006) Nanoscale detection of organic signatures in carbonate microbialites. Proc Natl Acad Sci USA 103:9440–9445
Berndt ME, Allen DE, Seyfried WE (1996) Reduction of CO_2 during serpentinization of olivine at 300°C and 500 bar. Geology 24:351–354
Boyce CK, Cody GD, Feser M et al (2002) Organic chemical differentiation within fossil plant cell walls detected with X-ray spectromicroscopy. Geology 30:1039–1042
Brandes JA, Cody GD, Rumble D et al (2008) Carbon K-edge XANES spectromicroscopy of natural graphite. Carbon 46:1424–1434
Brasier MD, Green OR, Jephcoat AP et al (2002) Questioning the evidence for Earth's oldest fossils. Nature 416:76–81
Brasier MD, Green OR, Lindsay JF et al (2005) Critical testing of Earth's oldest putative fossil assemblage from the 3.5 Ga Apex chert, Chinaman Creek, Western Australia. Precambr Res 140:55–102
Braun A, Huggins FE, Shah N et al (2005) Advantages of soft X-ray absorption over TEM-EELS for solid carbon studies–a comparative study on diesel soot with EELS and NEXAFS. Carbon 43:117–124
Braun A, Kubatova A, Wirick S et al (2009) Radiation damage from EELS and NEXAFS in diesel soot and diesel soot extracts. J Electron Spectrosc Relat Phenom 170:42–48
Buick R, Dunlop JSR, Groves DI (1981) Stromatolite recognition in ancient rocks: an appraisal of irregularly laminated structures in an Early Archean chert-barite unit from North Pole, Western Australia. Alcheringa 5:161–181
Buseck PR (ed) (1992) Minerals and reactions at the atomic scale. Mineralogical Society of America, Washington, DC
Buseck PR, Bo-Jun H (1985) Conversion of carbonaceous material to graphite during metamorphism. Geochim Cosmochim Acta 49:2003–2016
Buseck PR, Bo-Jun H, Miner B (1988) Structural order and disorder in Precambrian kerogens. Org Geochem 12:221–234
Calkins WH (1994) The chemical forms of sulfur in coal: a review. Fuel 73:475–484
Cody GD (2005) Geochemical connections to primitive metabolism. Elements 1:139–143
Cody GD, Botto RE, Ade H et al (1995) C-NEXAFS microanalysis and scanning X-ray microscopy of microheterogeneities in a high-volatile A bituminous coal. Energy Fuels 9:75–83
Cody GD, Botto RE, Ade H et al (1996) The application of soft X-ray microscopy to the in-situ analysis of sporinite in coal. Int J Coal Geol 32:69–86
Cody GD, Ade H, Wirick S et al (1998) Determination of chemical-structural changes in vitrinite accompanying luminescence alteration using C-NEXAFS analysis. Org Geochem 28:441–455
Cody GD, Alexander CMO'D, Yabuta H et al (2008) Organic thermometry for chondritic parent bodies. Earth Planet Sci Lett 272:446–455
Cody GD, Brandes JA, Jacobsen C et al (2009) Soft X-ray induced chemical modification of polysaccharides in vascular plant cell walls. J Electron Spectrosc Relat Phenom 170:57–64
De Gregorio BT, Sharp TG (2006) The structure and distribution of carbon in 3.5 Ga Apex chert: implications for the biogenicity of Earth's oldest putative microfossils. Am Mineral 91:784–789

De Gregorio BT, Sharp TG, Flynn GJ et al (2009) A biogenic origin for Earth's oldest putative microfossils. Geology 37:631–634
Derenne S, Robert F, Skrzypczak-Bonduelle A et al (2008) Molecular evidence for life in the 3.5 billion year old Warrawoona chert. Earth Planet Sci Lett 272:476–480
Dhez O, Ade H, Urquhart SG (2003) Calibrated NEXAFS spectra of some common polymers. J Electron Spectrosc Relat Phenom 128:85–96
Dow WG (1977) Kerogen studies and geological interpretations. J Geochem Explor 7:79–99
Durand B (ed) (1980) Kerogen: insoluble organic matter from sedimentary rocks. Editions TECHNIP, Paris
Egerton RF (1996) Electron energy-loss spectroscopy in the electron microscope, 2nd edn. Plenum Press, New York
Egerton RF, Rauf I (1999) Dose-rate dependence of electron-induced mass loss from organic specimens. Ultramicroscopy 80:247–254
Egerton RF, Crozier PA, Rice P (1987) Electron energy-loss spectroscopy and chemical change. Ultramicroscopy 23:305–312
Egerton RF, Li P, Malac M (2004) Radiation damage in the TEM and SEM. Micron 35:399–409
Espitalié J, Bordenave ML (1993) Rock-Eval pyrolysis. In: Bordenave ML (ed) Applied petroleum geochemistry. Editions TECHNIP, Paris
Flynn GJ, Keller LP, Feser M et al (2003) The origin of organic matter in the solar system: evidence from the interplanetary dust particles. Geochim Cosmochim Acta 67:4791–4806
Foustoukos DI, Seyfried WE (2004) Hydrocarbons in hydrothermal vent fluids: the role of chromium-bearing catalysts. Science 304:1002–1005
Freund H, Walters CC, Kelemen SR et al (2007) Predicting oil and gas compositional yields via chemical structure-chemical yield modeling (CS-CYM): Part 1 – concepts and implementation. Org Geochem 38:288–305
Galwey AK (1972) The rate of hydrocarbon desorption from mineral surfaces and the contribution of heterogeneous catalytic-type processes to petroleum genesis. Geochim Cosmochim Acta 36:1115–1130
García-Ruiz JM, Hyde ST, Carnerup AM et al (2003) Self-assembled silica-carbonate structures and detection of ancient microfossils. Science 302:1194–1197
Glikson M, Duck LJ, Golding SD et al (2008) Microbial remains in some earliest Earth rocks: comparison with a potential modern analogue. Precambr Res 164:187–200
Gordon ML, Cooper G, Morin C et al (2003) Inner-shell excitation spectroscopy of the peptide bond: Comparison of the C 1s, N 1s, and O 1s spectra of glycine, glycyl-glycine, and glycyl-glycyl-glycine. J Phys Chem A 107:6144–6159
Grotzinger JP, Knoll AH (1999) Stromatolites in Precambrian carbonates: evolutionary mileposts or environmental dipsticks? Annu Rev Earth Planet Sci 27:313–358
Grotzinger JP, Rothman DH (1996) An abiotic model for stromatolite morphogenesis. Nature 383:423–425
Haberstroh PR, Brandes JA, Gélinas Y et al (2006) Chemical composition of the graphitic black carbon fraction in riverine and marine sediments at sub-micron scales using carbon X-ray spectromicroscopy. Geochim Cosmochim Acta 70:1483–1494
Hayatsu R, Anders E (1981) Organic compounds in meteorites and their origins. Top Curr Chem 99:1–37
Henke BL, Gullikson EM, Davis JC (1993) X-Ray interactions: photoabsorption, scattering, transmission, and reflection at E = 50–30,000 eV, Z = 1–92. At Data Nucl Data Tables 54: 181–342
Hickman AH (1983) Geology of the Pilbara block and its environs. Geological Survey of Western Australia, Perth, Western Australia
Hickman AH (2008) Regional review of the 3426–3350 Ma Strelley Pool Formation, Pilbara Craton, Western Australia. Geological Survey of Western Australia, Perth, Western Australia
Hink W, Paschke H (1971) *K*-shell-fluorescence yield for carbon and other light elements. Phys Rev A 4:507

Hitchcock AP, Brion CE (1980) Inner-shell excitation of formaldehyde, acetaldehyde and acetone studied by electron impact. J Electron Spectrosc Relat Phenom 19:231–250
Hofmann HJ, Grey K, Hickman AH et al (1999) Origin of 3.45 Ga coniform stromatolites in Warrawoona Group, Western Australia. Geol Soc Am Bull 111:1256–1262
Holm NG, Charlou JL (2001) Initial indications of abiotic formation of hydrocarbons in the Rainbow ultramafic hydrothermal system, Mid-Atlantic Ridge. Earth Planet Sci Lett 191:1–8
Horita J, Berndt ME (1999) Abiogenic methane formation and isotopic fractionation under hydrothermal conditions. Science 285:1055–1057
House CH, Schopf JW, McKeegan KD et al (2000) Carbon isotopic composition of individual Precambrian microfossils. Geology 28:707–710
Isaacson M (1979) Electron beam induced damage of organic solids: implications for analytical electron microscopy. Ultramicroscopy 4:193–199
Isaacson M, Johnson D, Crewe AV (1973) Electron beam excitation and damage of biological molecules; its implications for specimen damage in electron microscopy. Radiat Res 55:205–224
Ishii I, Hitchcook AP (1988) The oscillator strengths for C1*s* and O1*s* excitation of some saturated and unsaturated organic alcohols, acids and esters. J Electron Spectrosc Relat Phenom 46:55–84
Jacobsen C, Kirz J, Williams S (1992) Resolution in soft X-ray microscopes. Ultramicroscopy 47:55–79
Jacobsen C, Wirick S, Flynn GJ et al (2000) Soft X-ray spectroscopy from image sequences with sub-100 nm spatial resolution. J Microsc 197:173–184
Jébrak M (1997) Hydrothermal breccias in vein-type ore deposits: a review of mechanisms, morphology and size distribution. Ore Geol Rev 12:111–134
Jokic A, Cutler JN, Ponomarenko E et al (2003) Organic carbon and sulphur compounds in wetland soils: insights on structure and transformation processes using K-edge XANES and NMR spectroscopy. Geochim Cosmochim Acta 67:2585–2597
Jokic A, Schulten H, Cutler JN et al (2004) A significant abiotic pathway for the formation of unknown nitrogen in nature. Geophys Res Lett 31:L05502
Kaźmierczak J, Kremer B (2002) Thermal alteration of the Earth's oldest fossils. Nature 420:477–478
Keller LP, Messenger S, Flynn GJ et al (2004) The nature of molecular cloud material in interplanetary dust. Geochim Cosmochim Acta 68:2577–2589
Kelley DS, Früh-Green GL (1999) Abiogenic methane in deep-seated mid-ocean ridge environments: insights from stable isotope analyses. J Geophys Res 104:10439–10460
Kerridge JF (1983) Isotopic composition of carbonaceous-chondrite kerogen: evidence for an interstellar origin of organic matter in meteorites. Earth Planet Sci Lett 64:186–200
Kerridge JF, Chang S, Shipp R (1987) Isotopic characterisation of kerogen-like material in the Murchison carbonaceous chondrite. Geochim Cosmochim Acta 51:2527–2540
Kikuma J, Warwick T, Shin H-J et al (1998) Chemical state analysis of heat-treated polyacrylonitrile fiber using soft X-ray spectromicroscopy. J Electron Spectrosc Relat Phenom 94:271–278
Kirz J, Jacobsen C, Howells M (1995) Soft X-ray microscopes and their biological applications. Q Rev Biophys 28:33–130
Knicker H (2004) Stabilization of N-compounds in soil and organic-matter-rich sediments–What is the difference? Mar Chem 92:167–195
Krivanek OL, Corbin GJ, Dellby N et al (2008) An electron microscope for the aberration-corrected era. Ultramicroscopy 108:179–195
Lancet MS, Anders E (1970) Carbon isotope fractionation in the Fischer-Tropsch synthesis and in meteorites. Science 170:980–982
Lawrence JR, Swerhone GDW, Leppard GG et al (2003) Scanning transmission X-ray, laser scanning, and transmission electron microscopy mapping of the exopolymeric matrix of microbial biofilms. Appl Environ Microbiol 69:5543–5554
Leif RN, Simoneit BRT (2000) The role of alkenes produced during hydrous pyrolysis of a shale. Org Geochem 31:1189–1208

Leinweber P, Kruse J, Walley FL et al (2007) Nitrogen K-edge XANES–an overview of reference compounds used to identify 'unknown' organic nitrogen in environmental samples. J Synchrotron Radiat 14:500–511

Lemelle L, Labrot P, Salomé M et al (2008) In situ imaging of organic sulfur in 700–800 My-old Neoproterozoic microfossils using X-ray spectromicroscopy at the S K-edge. Org Geochem 39:188–202

Lerotic M, Jacobsen C, Schäfer T et al (2004) Cluster analysis of soft X-ray spectromicroscopy data. Ultramicroscopy 100:35–57

Lerotic M, Jacobsen C, Gillow JB et al (2005) Cluster analysis in soft X-ray spectromicroscopy: finding the patterns in complex specimens. J Electron Spectrosc Relat Phenom 144–147:1137–1143

Lindsay JF, Brasier MD, McLoughlin N et al (2005) The problem of deep carbon–an Archean paradox. Precambr Res 143:1–22

Llorca J (2002) Are organic molecules produced by nebular Fischer-Tropsch processes preserved in comets? Adv Space Res 30:1469–1472

Lowe DR (1980) Stromatolites 3,400-Myr old from the Archean of Western Australia. Nature 284:441–443

Lowe DR (1994) Abiological origin of described stromatolites older than 3.2 Ga. Geology 22:387–390

Madix RJ, Solomon JL, Stöhr J (1988) The orientation of the carbonate anion on Ag(110). Surf Sci 197:L253–L259

Malis TF, Steele D (1990) Ultramicrotomy for materials science. In: Specimen preparation for transmission electron microscopy II, MRS symposium proceedings, vol 199. Materials Research Society, Pittsburgh, pp 3–42

Marshall CP, Love GD, Snape CE et al (2007) Structural characterization of kerogen in 3.4 Ga Archean cherts from the Pilbara Craton, Western Australia. Precambr Res 155:1–23

Mathez EA (1987) Carbonaceous matter in mantle xenoliths: composition and relevance to the isotopes. Geochim Cosmochim Acta 51:2339–2347

McCollom TM (2003) Formation of meteorite hydrocarbons from thermal decomposition of siderite ($FeCO_3$). Geochim Cosmochim Acta 67:311–317

McCollom TM, Seewald JS (2001) A reassessment of the potential for reduction of dissolved CO_2 to hydrocarbons during serpentinization of olivine. Geochim Cosmochim Acta 65:3769–3778

McCollom TM, Seewald JS (2006) Carbon isotope composition of organic compounds produced by abiotic synthesis under hydrothermal conditions. Earth Planet Sci Lett 243:74–84

McCollom TM, Ritter G, Simoneit BRT (1999) Lipid synthesis under hydrothermal conditions by Fischer-Tropsch-type reactions. Orig Life Evol Biosph 29:153–166

McLaren AC (1991) Transmission electron microscopy of minerals and rocks. Cambridge University Press, Cambridge, UK

McLoughlin N, Wilson LA, Brasier MD (2008) Growth of synthetic stromatolites and wrinkle structures in the absence of microbes–implications for the early fossil record. Geobiology 6:95–105

Meyer B (1976) Elemental sulfur. Chem Rev 76:367–388

Mitra-Kirtley S, Mullins OC, Branthaver JF et al (1993a) Nitrogen chemistry of kerogens and bitumens from X-ray absorption near-edge structure spectroscopy. Energy Fuels 7:1128–1134

Mitra-Kirtley S, Mullins OC, Van Elp J et al (1993b) Determination of the nitrogen chemical structures in petroleum asphaltenes using XANES spectroscopy. J Am Chem Soc 115:252–258

Morar JF, Himpsel FJ, Hollinger G et al (1985) Observation of a C-1*s* core exciton in diamond. Phys Rev Lett 54:1960–1963

Moreau JW, Sharp TG (2004) A transmission electron microscopy study of silica and kerogen biosignatures in 1.9 Ga Gunflint microfossils. Astrobiology 4:196–210

Mullins OC, Mitra-Kirtley S, Van Elp J et al (1993) Molecular structure of nitrogen in coal from XANES spectroscopy. Appl Spectrosc 47:1268–1275

Nissenbaum A (1979) Phosphorus in marine and non-marine humic substances. Geochim Cosmochim Acta 43:1973–1978

O'Keefe MA (1992) "Resolution" in high-resolution electron microscopy. Ultramicroscopy 47:282–297

Oberlin A, Boulmier JL, Villey M (1980) Electron microscopic study of kerogen microtexture: selected criteria for determining the evolution path and evolution stage of kerogen. In: Durand B (ed) Kerogen. Editions TECHNIP, Paris

Pasteris JD, Wopenka B (2003) Necessary, but not sufficient: Raman identification of disordered carbon as a signature of ancient life. Astrobiology 3:727–738

Reymond JP, Mérieudeau P, Teichner SJ (1982) Changes in the surface structure and composition of an iron catalyst of reduced or unreduced Fe_2O_3 during the reaction of carbon monoxide and hydrogen. J Catal 75:39–48

Rightor EG, Hitchcock AP, Ade H et al (1997) Spectromicroscopy of poly(ethylene terephthalate): comparison of spectra and radiation damage rates in X-ray absorption and electron energy loss. J Phys Chem B 101:1950–1960

Rosenberg RA, Love PJ, Rehn V (1986) Polarization-dependent C(*K*) near-edge x-ray-absorption fine structure of graphite. Phys Rev B 33:4034–4037

Rushdi AI, Simoneit BRT (2001) Lipid formation by aqueous Fischer-Tropsch-type synthesis over a temperature range of 100 to 400°C. Orig Life Evol Biosph 31:103–118

Rushdi AI, Simoneit BRT (2004) Condensation reactions and formation of amides, esters, and nitriles under hydrothermal conditions. Astrobiology 4:211–224

Rushdi AI, Simoneit BRT (2005) Abiotic synthesis of organic compounds from carbon disulfide under hydrothermal conditions. Astrobiology 5:749–768

Rushdi AI, Simoneit BRT (2006) Abiotic condensation synthesis of glyceride lipids and wax esters under simulated hydrothermal conditions. Orig Life Evol Biosph 36:93–108

Sarret G, Mongenot T, Connan J et al (2002) Sulfur speciation in kerogens of the Orbagnoux deposit (Upper Kimmeridgian, Jura) by XANES spectroscopy and pyrolysis. Org Geochem 33:877–895

Schidlowski M (1988) A 3,800-million-year isotopic record of life from carbon in sedimentary rocks. Nature 333:313–318

Schopf JW (1992) Paleobiology of the Archean. In: Schopf JW, Klein C (eds) The Proterozoic biosphere: a multidisciplinary study. Cambridge University Press, Cambridge

Schopf JW (1993) Microfossils of the early Archean Apex chert: new evidence of the antiquity of life. Science 260:640–646

Schopf JW (2006) Fossil evidence of Archaean life. Phil Trans R Soc B 361:869–885

Schopf JW, Packer B (1987) Early Archean (3.3-billion to 3.5-billion-year-old) microfossils from Warrawoona Group, Australia. Science 237:70–73

Schopf JW, Kudryavtsev AB, Agresti DG et al (2002) Laser-Raman imagery of Earth's earliest fossils. Nature 416:73–76

Schopf JW, Kudryavtsev AB, Czaja AD et al (2007) Evidence of Archean life: stromatolites and microfossils. Precambr Res 158:141–155

Schroeder T, John B, Frost BR (2002) Geologic implications of seawater circulation through peridotite exposed at slow-spreading mid-ocean ridges. Geology 30:367–370

Seewald JS, Zolotov MY, McCollom TM (2006) Experimental investigation of single carbon compounds under hydrothermal conditions. Geochim Cosmochim Acta 70:446–460

Shard AG, Whittle JD, Beck AJ et al (2004) A NEXAFS examination of unsaturation in plasma polymers of allylamine and propylamine. J Phys Chem B 108:12472–12480

Sherwood Lollar B, Lacrampe-Couloume G, Slater G et al (2006) Unravelling abiogenic and biogenic sources of methane in the Earth's deep subsurface. Chem Geol 226:328–339

Sherwood Lollar B, Lacrampe-Couloume G, Voglesonger K et al (2008) Isotopic signatures of CH_4 and higher hydrocarbon gases from Precambrian Shield sites: a model for abiogenic polymerization of hydrocarbons. Geochim Cosmochim Acta 72:4778–4795

Simoneit BRT, Rushdi AI, Deamer DW (2007) Abiotic formation of acylglycerols under simulated hydrothermal conditions and self-assembly properties of such lipid products. Adv Space Res 40:1649–1656

Sinninghe Damsté JS, de Leeuw JW (1990) Analysis, structure and geochemical significance of organically-bound sulphur in the geosphere: state of the art and future research. Org Geochem 16:1077–1101

Sitte H (1984) Process of ultrathin sectioning. In: Revel JP, Barnard T, Haggis GH et al (eds) The science of biological specimen preparation for microscopy and microanalysis. Scanning Electron Microscopy, Chicago, IL

Skytt P, Glans P, Mancini DC et al (1994) Angle-resolved soft-x-ray fluorescence and absorption study of graphite. Phys Rev B 50:10457–10461

Smith RE, Perdrix JL, Parks TC (1982) Burial metamorphism in the Hamersley Basin, Western Australia. J Petrol 23:75–102

Stetter KO (2006) Hyperthermophiles in the history of life. Philos Trans R Soc B 361: 1837–1843

Stöhr J (1992) NEXAFS spectroscopy. Springer, Berlin

Strauss H, Moore TB (1992) Abundances and isotopic compositions of carbon and sulfur species in whole rock and kerogen samples. In: Schopf JW, Klein C (eds) The Proterozoic biosphere: a multidisciplinary study. Cambridge University Press, Cambridge

Sviridov DV (2002) Chemical aspects of implantation of high-energy ions into polymeric materials. Russ Chem Rev 71:315–327

Thackray M (1970) Melting point intervals of sulfur allotropes. J Chem Eng Data 15:495–497

Thorne AM, Trendall AF (2001) Geology of the Fortescue Group, Pilbara Craton, Western Australia. Geological Survey of Western Australia, Perth, Western Australia

Thorpe R, Hickman AH, Davis D et al (1992) U-Pb zircon geochronology of Archaean felsic units in the Marble Bar region, Pilbara Craton, Western Australia. Precambr Res 56:169–189

Tingle TN, Mathez EA, Michael F Jr (1991) Carbonaceous matter in peridotites and basalts studied by XPS, SALI, and LEED. Geochim Cosmochim Acta 55:1345–1352

Tissot BP, Welte DH (1978) Petroleum formation and occurrence, 2nd edn. Springer, Berlin

Ueno Y, Yoshioka H, Maruyama S et al (2004) Carbon isotope and petrography of kerogens in 3.5-Ga hydrothermal silica dikes in the North Pole area, Western Australia. Geochim Cosmochim Acta 68:573–589

Urquhart SG, Ade H (2002) Trends in the carbonyl core (C 1S, O 1S) $\rightarrow \pi^*_{C=O}$ transition in the near-edge X-ray absorption fine structure spectra of organic molecules. J Phys Chem B 106:8531–8538

Urquhart SG, Hitchcock AP, Priester RD et al (1995) Analysis of polyurethanes using core excitation spectroscopy. Part II: inner shell spectra of ether, urea and carbamate model compounds. J Polym Sci B 33:1603–1620

Vairavamurthy A, Wang S (2002) Organic nitrogen in geomacromolecules: insights on speciation and transformation with K-edge XANES spectroscopy. Environ Sci Technol 36:3050–3056

Van Kranendonk MJ (2006) Volcanic degassing, hydrothermal circulation and the flourishing of early life on Earth: a review of the evidence from c. 3490–3240 Ma rocks of the Pilbara Supergroup, Pilbara Craton, Western Australia. Earth Sci Rev 74:197–240

Van Kranendonk MJ (2007) A review of the evidence for putative Paleoarchean life in the Pilbara Craton, Western Australia. In: Van Kranendonk MJ, Smithies RH, Bennett VC (eds) Earth's oldest rocks, developments in Precambrian geology, vol 15. Elsevier, Amsterdam, pp 855–877

Van Krevelen DW (1993) Coal: typology – physics – chemistry – constitution, 3rd edn. Elsevier, Amsterdam

Vandenbroucke M, Largeau C (2007) Kerogen origin, evolution and structure. Org Geochem 38:719–833

Walter MR (1983) Archean stromatolites: evidence of Earth's earliest benthos. In: Schopf JW (ed) Earth's earliest biosphere: its origin and evolution. Princeton University Press, Princeton, NJ

Williams DS, Carter CB (1996) Transmission electron microscopy. Plenum Press, New York

Williams LB, Canfield B, Voglesonger KM et al (2005) Organic molecules formed in a "primordial womb". Geology 33:913–916

Wiltfong R, Mitra-Kirtley S, Mullins OC et al (2005) Sulfur speciation in different kerogens by XANES spectroscopy. Energy Fuels 19:1971–1976

Winn B, Ade H, Buckley C et al (2000) Illumination for coherent soft X-ray applications: the new X1A beamline at the NSLS. J Synchrotron Radiat 7:395–404

Woese CR, Kandler O, Wheelis ML (1990) Towards a natural system of organisms: proposal for the domains Archaea, Bacteria, and Eucarya. Proc Natl Acad Sci USA 87:4576–4579

Wopenka B, Pasteris JD (1993) Structural characterization of kerogens to granulite-facies graphite: applicability of Raman microprobe spectroscopy. Am Mineral 78:533–557

Yoshida T, Nishizawa K, Tabata M et al (1993) Methanation of CO_2 with H_2-reduced magnetite. J Mater Sci 28:1220–1226

Zhang X, Jacobsen C, Lindaas S et al (1995) Exposure strategies for polymethyl methacrylate from *in situ* x-ray absorption near edge structure spectroscopy. J Vac Sci Technol B 13:1477–1483

Zhou D, Metzler RA, Tyliszczak T et al (2008) Assignment of polarization-dependent peaks in carbon K-edge spectra from biogenic and geologic aragonite. J Phys Chem B 112:13128–13135

Zubavichus Y, Fuchs O, Weinhardt L et al (2004) Soft X-ray-induced decomposition of amino acids: an XPS, mass spectrometry, and NEXAFS study. Radiat Res 161:346–358

What Can Carbon Isotopes Tell Us About Sources of Reduced Carbon in Rocks from the Early Earth?

Thomas M. McCollom

Abstract Carbon isotopes have been widely used as a tool to search for evidence of life in rocks from the early Earth. Many of the oldest rocks on Earth have been found to contain isotopically light reduced carbon, which many researchers have interpreted to have a biological origin. While light carbon isotopic compositions are consistent with biologically produced organic matter, alternative non-biological sources that might also explain the data have not been given thorough consideration. A review of the isotopic composition of potential non-biological sources of reduced carbon to the early Earth indicates that many have light isotopic signatures that encompass the values found in ancient rocks, indicating that the reduced carbon in rocks from the early Archean could plausibly derive from non-biological sources. These observations indicate that the evidence for life provided by carbon isotopes remains far from conclusive, and additional criteria need to be applied before it can be convincingly determined whether the reduced carbon found in the oldest rocks on Earth has a biological or non-biological origin.

Keywords Biosignatures • Carbon isotopes • Origin of life • Archean

1 Introduction

Determination of the time when life first arose on our planet has been a major objective of earth scientists for more than a century, and has significant implications for understanding how and where life originated. In recent decades, research on the earliest evidence of life has focused on the oldest known rocks on Earth dating from the early Archean era, and particularly on ancient rocks from southern West

T.M. McCollom (✉)
CU Center for Astrobiology & Laboratory for Atmospheric and Space Physics, University of Colorado, Campus Box 392, Boulder, CO 80309-0392, USA
e-mail: mccollom@lasp.colorado.edu

S.D. Golding and M. Glikson (eds.), *Earliest Life on Earth: Habitats, Environments and Methods of Detection*, DOI 10.1007/978-90-481-8794-2_11,

Greenland and Western Australia. In both cases, it has been argued that carbonaceous materials in the rocks provide persuasive evidence for a biological origin, indicating that advanced life had developed prior to 3.85 billion years ago (Ga) (e.g., Schopf and Packer 1987; Schidlowski 1988; Schopf 1993; Mojzsis et al. 1996).

Identification of evidence for life in rocks from the early Earth depends on the ability to recognize remnant vestiges of organisms that existed long ago or their metabolic products, often referred to collectively as 'biosignatures.' The concept of a biosignature is predicated on the assumption that biological processes generate products with a limited and definable set of characteristics (Fig. 1). At the same time, purely abiotic processes produce their own defined set of characteristics. The challenge to scientists seeking to differentiate between biological and non-biological products in geologic samples is that there is often significant overlap in the characteristics of products generated by biotic and abiotic processes (Fig. 1). A further complicating factor in the search for evidence of past life in ancient rocks is that physical and chemical processes over time can transform both biological and non-biological products in ways that obscure differences that were originally present when they formed.

The origin of products that can be generated by either biological or non-biological processes is inherently ambiguous because the ultimate source of the product cannot be uniquely identified. As a consequence, only those products that can be demonstrated to form exclusively through biological processes and that could not plausibly form by any non-biological process truly constitute biosignatures (Fig. 1). Conversely, products that can be uniquely attributed to non-biological processes to the exclusion of biological products can be identified as "abiosignatures." Compounds that can form by either biological or non-biological processes can be consistent with either origin, but cannot be used to differentiate between these sources.

For some time, carbon isotopes have been employed as a primary tool to examine rocks from the early Earth for evidence of ancient life. Initial surveys of the carbon isotopic composition of natural samples established that biologically generated organic compounds are typically depleted in ^{13}C relative to precursor carbon sources (e.g., Nier and Gulbransen 1939; Craig 1953a). Scientists soon seized on this observation to assert that isotopically light graphitic carbon found in ancient rocks could be attributed to a biogenic source (Wickman 1941; Wickman and von Ubisch 1951;

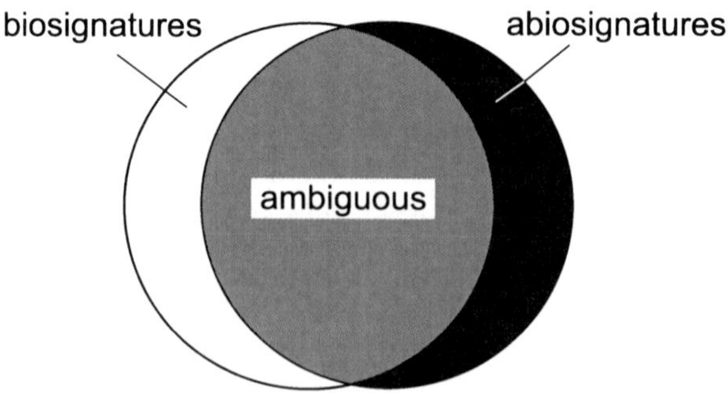

Fig. 1 Schematic Venn diagram illustrating the concept of a biosignature (see text)

Rankama 1948, 1954). One of the pioneers of the field, Harmon Craig, quickly responded to such claims with a caution against this use of carbon isotopes (Craig 1953a, b, 1954). The crux of Craig's argument was that one could not confidently assign a biologic origin to light carbon isotope values without eliminating the possibility that non-biological processes could produce the same isotopic signature. In other words, a biological source could only be established by demonstrating that the isotopic values fall outside the region of overlap between biological and non-biological sources (i.e., the "ambiguous" area depicted in Fig. 1). With regard to graphite in geologic deposits, Craig argued that there was insufficient data available at the time to eliminate the possibility that chemical precipitation of graphite from carbon-bearing fluids could produce depletions in ^{13}C comparable to those of biological processes, so it could not be confidently concluded that light isotopic values originated from biological sources. Similar arguments could be extended to consideration of other potential non-biological sources.

Despite the cautionary note struck by Craig, many scientists since that time have continued to claim that light carbon isotopic values in ancient rocks indicate a biological origin without persuasively demonstrating the implausibility of non-biological inputs. As Brasier et al. (2006) observed in reference to the search for the earliest evidence of life in the geologic record: "we have tended to rely too much upon evidence that is 'consistent with' microbial processes, without falsifying or rejecting (*sensu* Popper 1959) other possible non-biological scenarios that may likewise be consistent." Nevertheless, rejecting alternative non-biological explanations is required to build a convincing case for evidence of early life.

In the intervening decades since Craig published his cautionary arguments, considerable progress has been made in understanding the isotopic fractionation produced by many biological and non-biological processes, and the isotopic compositions of organic matter for a variety of potential contributing sources to the early Earth have been reported. While this knowledge remains far from complete, the data that have been acquired to date allow for more detailed evaluation of potential sources for reduced carbon in ancient rocks than was possible even a few years ago. This chapter provides a brief overview of current constraints on the isotopic compositions of the most likely sources of organic matter on the surface of the early Earth, and considers the implications of these constraints for differentiating between biotic and abiotic contributions to the reduced carbon in ancient geologic deposits, focusing in particular on the oldest known rocks from southern West Greenland and Western Australia.

2 Sources of Reduced Carbon and Their Isotopic Compositions

This section provides a brief overview of current constraints on the carbon isotopic composition for a number of potential major contributors of organic matter to the early Earth. For brevity, only the most salient points are summarized here, and the references cited below should be consulted for more detailed information. A compilation of the inferred ranges of isotopic compositions for these sources is shown in Fig. 2.

2.1 Biological Sources

Autotrophic organisms fix inorganic carbon into biomass that is composed predominantly of proteins, nucleic acids, saccharides, and lipids. Except for the lipids, these components tend to be highly labile and are generally decomposed prior to lithification of geologic sediments. As a consequence, biogenic organic matter preserved in geologic deposits is primarily derived from the lipid component of biomass. Through a series of thermal and chemical reactions, the original biolipids are transformed over time into kerogen, an insoluble, amorphous, high-molecular-weight material composed of cross-linked aromatic and aliphatic components (see Hayes et al. 1983). If the deposits are exposed to especially prolonged heating or high temperatures, the kerogen becomes increasingly aromatic in nature, and may be gradually converted into the mineral graphite. In cases where these reactions have progressed to only a limited extent, structural information from the original biolipids may still remain, and can provide insight into the origin of the organic matter (Eigenbrode 2008). However, the oldest geologic deposits on Earth have experienced sufficient heating that any biomolecular structural information that was once present has been destroyed.

Fixation of inorganic carbon by most biological processes preferentially incorporates ^{12}C relative to ^{13}C, resulting in biomass that in most cases is significantly depleted in ^{13}C compared to the original carbon source. For plants and other photosynthetic microorganisms, the magnitude of this fractionation relative to atmospheric CO_2 is −11‰ to −24‰ (Fig. 2). As a result, organic matter in recent sediments consistently has an average isotopic composition near −25‰, reflecting a predominant input from photosynthetic organic matter (Hayes et al. 1983; Schidlowski 1988). Heterotrophic organisms and autotrophic organisms employing alternative carbon fixation pathways can generate biomass with a much wider range of isotopic compositions (e.g., Schidlowski 1988; House et al. 2003) (Fig. 2), but the quantitative contribution from these organisms to sedimentary organic matter appears to minor, at least in recent eras.

Early diagenesis of bioorganic matter apparently produces a minor depletion of about 2‰ in the ^{13}C content of organic matter, but further diagenesis and mild metamorphism does not appear to have an appreciable impact on the isotopic composition of organic matter in geologic deposits (Hayes et al. 1983). Consequently, organic matter in most geologic samples dominated by biological inputs has been found to have an isotopic composition around −27‰. High degrees of metamorphism at temperatures exceeding a few hundred degrees C and conversion of kerogen to graphite can result in ^{13}C depletions of several permil or more in the

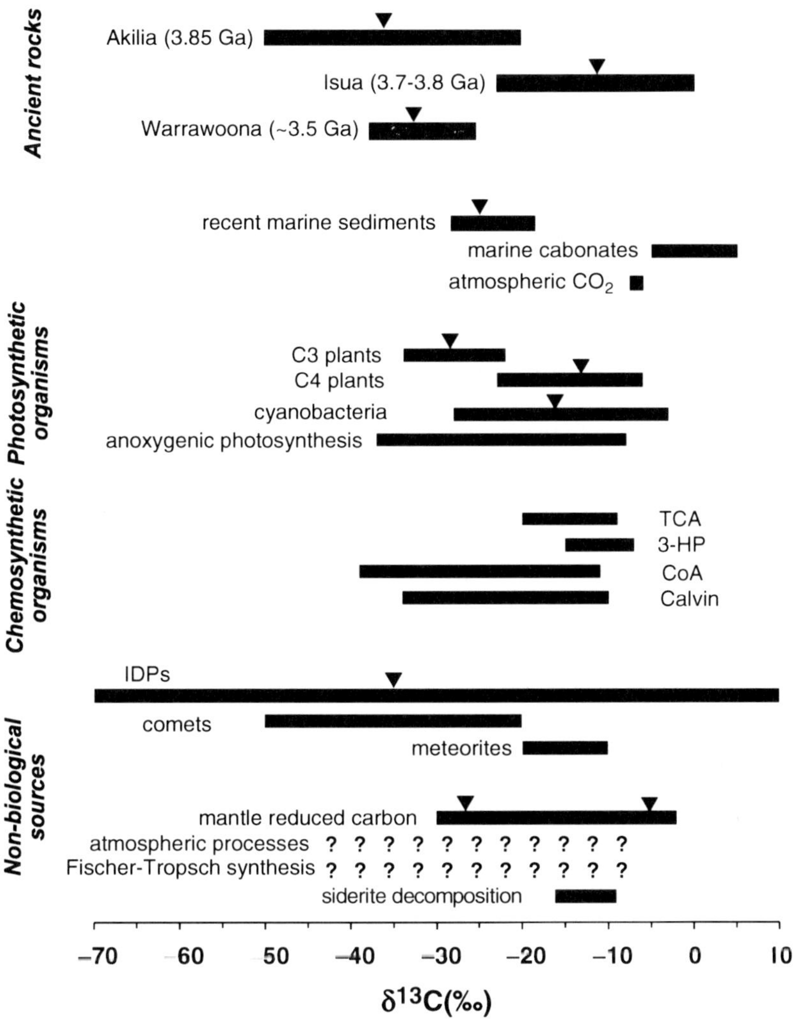

Fig. 2 Comparison of the isotopic compositions of reduced carbon in rocks from the early Archean with those of representative biological and non-biological sources of organic matter. Data sources: Akilia (Mojzsis et al. 1996); Isua Supracrustal Belt (Schidlowski et al. 1979; Rosing 1999; Ueno et al. 2002; van Zuilen et al. 2002); Warrawoona Group (Strauss and Moore 1992; Ueno et al. 2004; Brasier et al. 2002); photosynthetic organisms (Schidlowski 1988); chemosynthetic organisms (calculated from fractionation factors reported by House et al. 2003, assuming atmospheric CO_2 as a carbon source); interplanetary dust particles (IDPs) (Messenger 2000; Messenger et al. 2003); comets (McKeegan et al. 2006); meteorites (Cronin and Chang 1993; Alexander et al. 1998; Sephton et al. 2003); mantle carbon (Deines 2002), and siderite decomposition (van Zuilen et al. 2002, 2003). Carbon fixation pathways for chemoautotrophs include the reductive tricarboxylic acid cycle (TCA), Calvin cycle (Calvin), acetyl-CoA pathway (CoA), and 3-hydroxypropionate cycle (3-HP). Limited data preclude accurate estimation of the isotopic compositions of organic matter produced by atmospheric processes and Fischer-Tropsch synthesis at the present time. *Inverted triangles* indicate approximate mean values

remaining reduced carbon (Hayes et al. 1983; Schidlowski 1988; Ueno et al. 2002) produced by net loss of heavy carbon in the form of CO_2 and isotope exchange with heavy inorganic reservoirs (carbonates) within the rocks.

2.2 *Abiotic Sources*

There are a number of different abiotic sources that could have potentially contributed significant amounts of organic matter to the early Earth. These include carbonaceous materials with both exogenous (meteorites, comets, interplanetary dust particles) and endogenous (atmosphere, fluid-rock reactions, etc.) origins. These sources are considered in separate sections below.

2.2.1 Exogenous Inputs

Many types of meteorites are relatively rich in organic compounds, and impacts of these meteorites can deliver this organic matter to the surface of the Earth. The chemical and isotopic composition of organic matter in meteorites is fairly well constrained by laboratory analysis of meteorite samples. Meteorites can contain up to several weight percent carbon, most of which (~70–99%) is present in the form of insoluble, amorphous, macromolecular organic material with a high aromatic content. The bulk isotopic composition of the insoluble material generally falls within the range of −10‰ to −20‰ (Cronin and Chang 1993; Alexander et al. 1998; Sephton et al. 2003). In addition, some meteorites, particularly the carbonaceous chondrites, may contain up to ~30 wt% of soluble organic compounds, including aliphatic hydrocarbons, free aromatic compounds, and amino acids. In most cases, these compounds have heavier isotopic compositions than the bulk macromolecular material, with values as high as +40‰ reported for amino acids in the Murchison meteorite (Cronin and Chang 1993; Pizzarello et al. 1991; Krishnamurthy et al. 1992). These heavy soluble compounds, however, seem less likely to be incorporated into geologic deposits than the macromolecular carbon, both because they may be transported out of sedimentary deposits and because they are more likely to be degraded by biological or chemical processes.

Impacts of comets would also have contributed organic compounds to the early Earth, (Chyba et al., 1990) but the structural and isotopic composition of the organic matter in comets is generally less well defined than that of meteorites owing to the scarcity of primary materials available for laboratory analysis. While terrestrial spectroscopic studies and spacecraft fly-bys have contributed to the characterization of cometary organic matter, the most direct evidence for its composition comes from recent analyses of cometary particles returned from comet 81P/Wild 2 by the Stardust mission (Sandford et al. 2006; Sandford 2008). Analyses of these particles indicates that the carbonaceous fraction contains abundant polycyclic aromatic hydrocarbons (PAH), but also includes aliphatic moieties and functional groups containing O and N, such as amines and carboxylic acids (Sandford 2008; Cody et al. 2008). Overall, the carbonaceous matter includes a smaller fraction of aromatic compounds and higher O/C and N/C ratios than meteoritic organic matter. Reported isotopic compositions of bulk carbon in the comet particles fall in the range of −20‰ to −50‰ (McKeegan et al. 2006), but isotopic compositions of the different carbon reservoirs represented by this bulk value are not yet available.

In addition to larger impacts of comets and meteorites, the surface of the Earth has been subject to a continuous infall of small interplanetary dust particles (IDPs) throughout its history. The IDPs contain a substantial fraction of organic matter; for example, one study found an average carbon content of ~12% in a suite of IDPs (Thomas et al. 1993). The carbonaceous matter is apparently dominated by an assortment of polycyclic aromatic hydrocarbons (PAH), but aliphatic compounds are also present (Sandford 2008, and references therein). Bulk carbon isotopic compositions of individual IDPs exhibit a wide range of values from −70‰ to +10‰, with an average composition around −35‰ to −40‰ (Messenger 2000; Messenger et al. 2003). Presumably, these values represent mostly organic carbon, but the isotopic compositions of individual carbon phases within IDPs have not yet been reported.

Estimates of the fluxes of organic matter to the surface of the early Earth from exogenous sources require a number of assumptions and extrapolations that impart a great deal of uncertainty to the results. However, estimates indicate that IDPs are likely to be the largest exogenous source of organic carbon, with estimated fluxes to the early Earth on the order of 10^4–10^7 kg C per year (Chyba and Sagan 1992; Pasek and Lauretta 2008). Input of organic matter from comets and meteorites would apparently have been 2–3 orders of magnitude lower. However, fluxes from all of these sources would have increased by several orders of magnitude during the Late Heavy Bombardment (LHB) between 3.8 and 4 Ga. During the LHB, fluxes of organic carbon have been estimated to be about 10^9 kg per year from IDPs and 10^5 from carbonaceous chondrites, which is equivalent to ~10 kg per km^2 per year of organic C if distributed evenly around the surface of the Earth (Pasek and Lauretta 2008). For comparison, the flux of organic matter that is buried in recent sediments is on the order of 10^{11} kg per year (Hayes et al. 1983; Berner 2003), suggesting that exogenous fluxes of organic matter in the Archean would have represented only a small fraction of the material being incorporated into modern sediments. While these relatively low fluxes might preclude the formation of organic-rich sediments comparable to those forming in recent eras, they would not prevent the possibility that exogenous organic matter may have accumulated in surface sediments over time in sufficient quantities to account for the trace amounts of reduced carbon present in ancient rocks.

Of course, organic matter from IDPs would have been distributed globally while that from cometary and meteoritic impacts may have been concentrated in regions adjacent to the impact, which may have affected the potential for accumulation and preservation in sediments. Another consideration with regard to the incorporation of exogenous organic matter into geologic deposits on the early Earth is preservation. On the modern Earth, organic matter exposed at the surface is degraded on relatively short time scales, predominantly by biological processes. On the early Earth, however, the lack of oxidants would have allowed organic matter delivered from exogenous sources to persist for much longer periods. As a consequence, there may have been a greater potential for transport and accumulation of organic matter in surficial sediments.

2.2.2 Endogenous Inputs

Atmospheric processes have long been considered as a potential source of organic matter on the early Earth, particularly with regard to the formation of prebiotic compounds that may have contributed to the origin of life (Chang et al. 1983). Laboratory experiments designed to examine the synthesis of organic compounds from lightning and other energetic atmospheric processes have produced an assortment of organic compounds, as well as high-molecular-weight insoluble organic residues (e.g., Miller 1953; Chang et al. 1983). Very few studies have reported the isotopic composition of these compounds (see Horita 2005). In one spark discharge experiment using pure CH_4 gas as a carbon source, however, light hydrocarbons formed during the experiment were depleted in ^{13}C by 10–25‰ relative to CH_4, with the extent of fractionation increasing with increasing molecular size (Chang et al. 1983). Similarly, Des Marais et al. (1981) found light hydrocarbons produced by energetic discharges in CH_4 gas to be depleted by 11–14‰ relative to the initial methane. In a spark discharge experiment with a CH_4–NH_3 gas mixture, insoluble organic matter was found to be depleted by ~10‰ relative to CH_4, but amino acids were depleted by as much as 20‰ (Chang et al. 1983). While all of these results indicate depletions relative to CH_4 for organic matter produced by lightning and high energy atmospheric processes, uncertainties in the abundance and isotopic composition of atmospheric CH_4 on the early Earth make it difficult to use the results to constrain the isotopic composition of organic matter reaching the surface from this source.

In addition to these high energy inputs, if the atmosphere of the early Earth had CH_4/CO_2 ratios $\geq$ ~1, photochemical reactions involving CH_4 may have produced organic hazes that contributed to accumulations of organic matter on the planet's surface (Pavlov et al. 2001). Methane has been proposed as a greenhouse gas on the early Earth, and potential sources include both geologic processes such as serpentinization and biologic inputs from methanogenic organisms, once these organisms evolved. Chang et al. (1983) measured the isotopic compositions of hydrocarbons up to C_6 formed during UV irradiation of CH_4-bearing gases, and reported that the products were depleted in ^{13}C by up to 11‰ relative to the starting CH_4. The amount of the fractionation increased with number of carbon in the products, and presumably the magnitude of the fractionation might continue to increase as the size of the molecules increases to form condensable molecules (Pavlov et al. 2001). Again, it is difficult to use these measurements to infer constraints on the overall isotopic composition of organic matter delivered to the surface from photochemical hazes on the early Earth owing to uncertainties in the isotopic composition of CH_4 in the Archean atmosphere. However, it appears likely that such organic matter would be strongly depleted in ^{13}C relative to inorganic carbon present in the atmosphere and ocean at the time.

Reduced carbon compounds would also have been produced within the solid part of the planet on the early Earth. Owing to the oxidation state of the mantle, carbon in volcanic gases exsolving from mantle melts is predominantly in the form of CO_2, with only minor amounts of reduced species (e.g., Mathez 1984; Kelley 1996). Nevertheless, mantle rocks that have been displaced to the surface by tectonic or volcanic processes contain small amounts of graphite and, in some cases, trace concentrations of organic compounds (see review by Deines 2002). Bulk carbon isotope measurements of mantle rocks exhibit a bimodal distribution with peaks at −5‰ and −25‰, although the isotopic compositions of different carbon forms represented by these bulk values are not known (Deines 2002). Erosion of mantle rocks could potentially have contributed reduced carbon to surface sediments on the early Earth, although it is unclear whether mantle rocks would have been sufficiently widespread to make a large contribution to global reservoirs of reduced carbon.

Fluid-rock reactions within the crust can result in conditions that are more strongly reducing than those prevailing in the mantle, and lower crustal temperatures also favor the thermodynamic stability of organic matter relative to CO_2 (see McCollom 2008). As a result, CO_2 from mantle fluids or in circulating hydrothermal fluids can potentially be reduced to organic matter within the Earth's crust. Within terrestrial geologic systems, the most widely invoked reaction for the formation of abiotic organic matter from CO_2 is Fischer-Tropsch-type synthesis (FTS) (McCollom and Seewald 2007). Fischer-Tropsch synthesis is a surface-catalyzed process for the reduction of simple inorganic carbon sources (CO_2 and CO) to organic compounds through reaction with H_2, and products range from CH_4 and other volatile light hydrocarbons to high-molecular-weight aliphatic precipitates ("wax"). Fischer-Tropsch-type synthesis has been invoked as a source of abiotic compounds in variety of geologic settings including volcanic fumaroles, hydrothermal vent fluids, serpentinites, and crystalline rocks (e.g., Abrajano et al. 1990; Sherwood Lollar et al. 1993, 2002, 2006, 2008; Salvi and Williams-Jones 1997, 2006; Kelley and Früh-Green 1999; Charlou et al. 2002; Kelley et al. 2005; Fiebig et al. 2007; Proskurowksi et al. 2008).

While FTS has been extensively studied owing to its use in industrial applications, very few of these studies have reported the isotopic compositions of reaction products (see recent reviews by Horita 2005; McCollom and Seewald 2007). Experimental studies of isotopic fractionation during FTS have examined only a small fraction of the range of conditions under which the reaction may proceed in geologic systems, but several significant trends have become apparent. In particular, all studies to date have generated non-volatile hydrocarbons that are substantially depleted in ^{13}C relative to their inorganic precursors (which is CO or HCOOH in most laboratory experiments). The magnitude of the fractionation is variable depending on experimental conditions, but reported depletions predominantly fall in the range of 16–33‰ (e.g., Lancet and Anders 1970; McCollom and Seewald 2006; Taran et al. 2007). Because these experimental values represent fractionation relative to CO and the isotopic composition of CO has rarely been reported for the relevant geologic systems, it is difficult to use the observed fractionations to predict the absolute isotopic compositions of organic matter that might be generated from reduction of CO_2 in natural systems. However, it is apparent that FTS can generate organic matter that is strongly depleted in ^{13}C relative to inorganic carbon, and can potentially produce reduced carbon with isotopic compositions that overlap the ^{13}C-depleted isotopic compositions of biologically produced organic compounds (McCollom and Seewald 2006).

In a related process, Horita and Berndt (1999) studied the reduction of CO_2 to CH_4 catalyzed by NiFe-alloy under hydrothermal conditions. NiFe-alloys are fairly common products of the hydrothermal alteration of ultramafic rocks, which may have been relatively more widespread in near-surface environments on the early Earth than they are today. The CH_4 generated was depleted in ^{13}C by up to 49‰ at 200°C and 29‰ at 300°C, relative to the initial CO_2. Of course, further processing would be required to convert the CH_4 produced in these experiments to non-volatile organic matter that might accumulate in geologic deposits, but the results provide further evidence that abiotic reactions can produce strongly ^{13}C-depleted reduced carbon.

Methane and other volatile hydrocarbons that likely formed by FTS or other abiotic reactions within the Earth's crust have recently been documented in a number of geologic samples, including fluids within crystalline rocks and marine hydrothermal systems (e.g., Charlou et al. 2002; Fiebig et al. 2007; Proskurowksi et al. 2008; Sherwood Lollar et al. 2008; McCollom 2008, and references therein). Measured carbon isotopic compositions for these hydrocarbon gases range from −5‰ to −45‰. Presumably, the processes that formed these volatile compounds might also produce less volatile products that could accumulate as condensed organic matter in geologic deposits, but non-volatile compounds with a demonstrably abiotic origin have not yet been identified, and it remains uncertain whether the isotopic compositions of the volatile gases is representative of less volatile compounds that might be created in the same environment. However, recent experimental studies suggest that FTS produces non-volatile products with an isotopic composition close to that of the light hydrocarbons (McCollom and Seewald 2006; Taran et al. 2007).

Another proposed source of reduced carbon in ancient rocks is the thermal decomposition of carbonate minerals during metamorphism. For example, the co-occurrence of graphite particles together with the minerals siderite and magnetite in some ~3.8 Ga rocks from the Isua Supracrsutal Belt (ISB) in Greenland has been attributed to thermal decomposition of siderite, which can be represented by the reaction:

$$\underset{\text{siderite}}{6FeCO_3} \rightarrow \underset{\text{magnetite}}{2Fe_3O_4} + 5CO_2 + \underset{\text{graphite}}{C}$$

(Perry and Ahmad 1977; van Zuilen et al. 2002, 2003). Laboratory experiments have confirmed that thermal decomposition of siderite generates highly aromatic organic matter (McCollom 2003), but the isotopic fractionation of this process has not yet been investigated experimentally. Graphite particles associated with carbonates at Isua have isotopic compositions in the range of −9‰ to −16‰ and are depleted in ^{13}C by 6–10‰ relative to the co-existing carbonate minerals, which is similar to the inferred isotopic equilibrium between these phases at 500°C (Perry and Ahmad 1977; van Zuilen et al. 2002, 2003). Whether the isotopic compositions actually reflect equilibrium and whether such values are representative of other reaction conditions requires further research.

3 Carbon Isotopes and the Early Geologic Record

Two requirements must be met in order to confidently assign a biological origin to reduced carbon in rocks from the early Earth (Craig 1953a, b, 1954; Brasier et al. 2006). First, it must be shown that the characteristics of the carbonaceous deposits are consistent with a biological source. Second, it must also be demonstrated that the deposits are highly unlikely to form by any plausible non-biological mechanism. In the case of carbon isotopes, the observations summarized in the previous section indicate that clear differentiation between biological and non-biological sources of carbonaceous matter on the basis of isotopic composition may be problematic. In particular, it is evident that there is considerable overlap between the isotopic composition of organic matter produced by biological processes and that of a number of potential non-biological sources of reduced carbon (Fig. 2). In the remainder of this section, the implications of these considerations in interpretation of the earliest proposed evidence for life in the geologic record are discussed.

3.1 Isotopic Composition of Organic Matter Through Time

The isotopic composition of carbon in geologic samples exhibits a remarkable consistency throughout the history of the Earth, extending back to the most ancient rocks (Fig. 3). The isotopic composition of inorganic carbon, in the form of carbon ate minerals, varies by only a few per mil around 0‰, suggesting that the isotopic composition of seawater has remained essentially constant for at least the last 3.5 Ga (Hayes et al. 1983; Schidlowski 1988). Over the same period, reduced carbon is consistently depleted by about 25–30‰ in ^{13}C relative to the carbonates, with few exceptions. Graphitic carbon in rocks older than ~3.7 Ga from southern Greenland deviate from this trend, but still have isotopic compositions that are significantly depleted in ^{13}C relative to coexisting carbonates. Another notable departure in the historic record occurs at around 2.7 Ga where organic matter has anomalously light isotopic values. The cause of this excursion is uncertain, but appears to be attributable to recycling of CH_4 by methanotrophic organisms (Hayes et al. 1983).

The magnitude of the fractionation between inorganic and organic carbon present in most of the historic record is similar to that observed in modern sediments, where the isotopic composition is dominated by input from photosynthetic organic matter. One logical hypothesis to explain the historic record, therefore, is that the same processes that dominate sedimentary organic inputs today have prevailed throughout

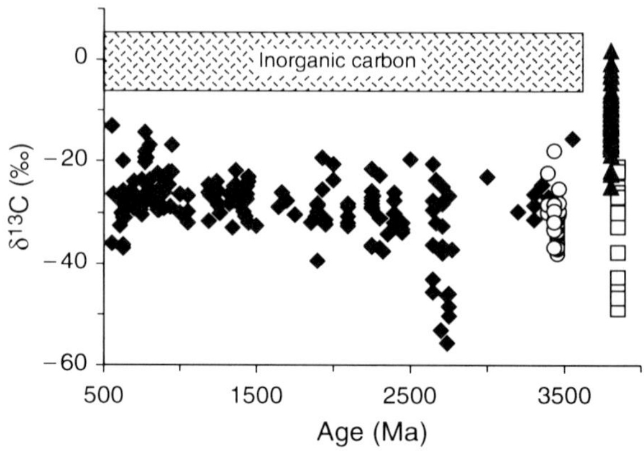

Fig. 3 Isotopic composition of reduced carbon in geologic deposits through Earth's early history. Data for relatively recent deposits (*black diamonds*) are from a compilation by Hayes et al. (1983), and are representative of a much more extensive data set. *Open circles* represent data from the Warrawoona Group, Australia (Strauss and Moore 1992; Ueno et al. 2004; Brasier et al. 2002), *filled triangles* are data from the Isua Supracrustal Belt, Greenland (Schidlowski et al. 1979; Rosing 1999; Ueno et al. 2002; van Zuilen et al. 2002), and empty squares are data for Akilia Island, Greenland (Mojzsis et al. 1996)

the geologic record to at least 3.5 Ga (e.g., Hayes et al. 1983; Schidlowski 1988). Since organic carbon in modern sediments is largely attributable to input of organic matter produced through carbon fixation by photosynthetic organisms, this would imply that biological sources, and possibly photosynthetic organisms, have dominated organic matter contributions to geologic deposits for the last 3.5 billion years, and maybe longer. Of course, this hypothesis would not preclude the possibility that some local deposits might contain reduced carbon from non-biological sources, but, overall, a predominance of biological inputs is inferred in this hypothesis.

The isotopic composition of organic matter through most of the historic record is certainly consistent with the composition expected for biological organic matter based on measurements of modern sediments with an unquestionable biological origin. For the most ancient rocks, however, it remains uncertain whether this is a unique interpretation for the isotopic trends, or whether the isotopic composition could also be explained by contributions from non-biological sources that produce a similar isotopic signature. In many relatively recent deposits, isotopically light reduced carbon is accompanied by ancillary indicators of biogenicity such as cellular structures and organic molecular biosignatures that support the isotopic interpretation of a biological origin for the carbon (Summons et al. 2008). Such supporting evidence is commonly missing or equivocal in the most ancient deposits (older than 3 Ga or so), either because it was never there or because it has been removed by subsequent geologic processes.

On the modern Earth, the contribution of organic matter from non-biological sources represents only a very small fraction of the flux of organic compounds from biological sources, although there may be instances where non-biological compounds predominate locally (e.g., Proskurowksi et al. 2008). It is likely, however, that the relative contribution of non-biological sources was much more significant in Earth's early history, both because the fluxes from non-biological sources were likely much larger in the past (e.g., Chyba and Sagan 1992; Pasek and Lauretta 2008) and because the flux of organic compounds from biological carbon fixation was presumably lower. Indeed, prior to the origin of life, non-biological sources would have provided the only inputs to organic matter.

Inevitably, there was some point in Earth's history when a transition occurred between non-biologically dominated organic fluxes and biologically dominated fluxes. However, the time of this transition and whether it is preserved in the geologic record are not yet clear. Nevertheless, it appears possible that this transition could have occurred within the time period spanned by the oldest rocks in the geologic record, and that these rocks may document a time when non-biological sources represented a significant fraction, or even dominated, fluxes of organic matter to the Earth's surface. This would require that non-biological sources of organic matter to have isotopic compositions similar to those observed in ancient rocks, but the constraints outlined in the previous section seem to allow this possibility.

3.2 *Reduced Carbon in Rocks from Southwest Greenland*

The oldest sediments deposited at the surface of the Earth that have so far been identified in the geologic record are found in southern West Greenland, including outcrops in the Isua Supracrustal Belt (ISB) dated at 3.7–3.8 Ga and slightly older rocks (~3.85 Ga) on Akilia Island. Graphite particles within these rocks are observed to have ^{13}C-depleted isotopic compositions, which has led a number of authors to confidently assert that the carbon is a remnant of biological organic matter deposited contemporaneously with the enclosing rocks (e.g., Schidlowski et al. 1979; Schidlowski 1988, 2001; Mojzsis et al. 1996; Rosing 1999). If these claims are accurate, the graphite deposits would represent the oldest known traces of life, and establish that life on Earth had evolved prior to 3.85 Ga.

Because the rocks at Isua and Akilia have been subjected to extensive metamorphism, structural features or other evidence that might buttress a biological interpretation would have been obliterated long ago. As a consequence, the isotopic composition of reduced carbon is essentially the only evidence that has been provided to support a biological interpretation. Isotopic compositions of reduced carbon from Isua range from about 0‰ to −27‰, with an average around −12‰ (Fig. 3). While these values are generally heavier than those found in younger deposits that have been attributed to biological sources, it has been inferred that the current isotopic compositions of reduced carbon in rocks from this locality have been reset from substantially lighter values during metamorphism (Schidlowski et al. 1979; Schidlowski 1988; Rosing 1999; Ueno et al. 2002; but see Naraoka et al. 1996, for a dissenting opinion). Isotopic compositions of graphite in rocks from Akilia range from −20‰ to −50‰ with an average near −37‰ (Mojzsis et al. 1996).

The rocks at Akilia have also been subjected to metamorphism, but closed system conditions during metamorphism may have precluded the type of isotopic alteration proposed to have occurred in rocks from Isua.

The isotopic compositions of graphite at Akilia as well as the presumed pre-metamorphic compositions of reduced carbon at Isua are both consistent with the values observed for biologically generated organic carbon (Figs. 2 and 3). However, the attribution of carbonaceous deposits from these sites to a biological origin presumes that non-biological sources could not plausibly account for the observed isotopic values. But the publications claiming a biological origin for the deposits give only very limited consideration to alternative non-biological interpretations and their arguments are unconvincing, particularly in light of additional constraints that have become available since the original claims were made.

For example, Schidlowski (1988) concluded that alternatives to a biological interpretation for graphite at Isua could be "safely dismissed", but considered only potential non-biological inputs from Fischer-Tropsch synthesis and atmospheric discharges. Fischer-Tropsch synthesis was rejected as a possible source based on a presumed fractionation of −50‰ to −100‰ for the process. Although the work of Lancet and Anders (1970) is cited as the source of these values, the values quoted by Schidlowski (1988) are actually much larger than the fractionations between the carbon source and organic products reported by those authors (see above). Furthermore, the more recent studies summarized above indicate much lower fractionations are likely to prevail during abiotic organic synthesis in natural geologic systems. Energetic discharges in the atmosphere were also rejected as a possible source even though the experimental data available at the time (Chang et al. 1983) encompass most of the reported isotopic measurements for graphite at Isua. Instead, atmospheric inputs were excluded as a source of the relatively heavy isotope compositions measured at Isua because the absence of similar heavy isotopic signatures in younger deposits implied that atmospheric inputs were insignificant. Such reasoning, however, is inherently flawed, since it presumes carbon inputs at Isua can be judged by deposits that formed hundreds of millions of years later at other localities. Schidlowski (1988) did not consider the possibility of contributions from exogenous sources.

Similarly, Mojzsis et al. (1996) concluded that "no known abiotic process can explain" the isotopic compositions of graphite particles at Akilia, but they considered only a very limited scope of possible non-biological contributors. Mojzsis and colleagues rejected possible inputs of organic matter from energetic discharges in the atmosphere and from meteorites because they claim that reduced carbon from these sources would be too heavy to account for the extent of ^{13}C depletions observed in the rocks at Akilia. This argument, however, is insufficient to dismiss the possibility of a non-biological origin for the graphitic carbon, for two reasons. First, Mojzsis et al. quote a value of −10‰ to represent organic matter produced by atmospheric discharges, citing the experimental results of Chang et al. (1983). But, the isotopic compositions of organic products in those experiments are dependent on the isotopic composition of the CH_4 gas used, and the value quoted refers to experimental results obtained using CH_4 with a composition of 0‰. In other experiments using lighter initial CH_4 compositions, organic compounds with significantly lighter isotopic compositions (to ≤50‰) were produced. Since the isotopic composition of CH_4 in the early atmosphere is unknown, a value of −10‰ cannot be assumed to be an accurate representation of organic matter from atmospheric sources on the early Earth. Second, while organic carbon from meteorites may indeed be too heavy to account for the graphite observed at Akilia, other exogenous sources (comets, IDPs) have organic matter with isotopic compositions that clearly overlap the range of values observed at Akilia, and endogenous sources (e.g., photochemical reactions, Fischer-Tropsch synthesis) also appear capable of producing organic matter with ^{13}C-depleted isotopic compositions in this same range.

Mojzsis et al. (1996) also presented model calculations examining whether open system degassing from an original reduced carbon source with an isotopic composition ≥ −10‰ (their presumed lower limit for non-biological sources) could account for the isotopically light graphite observed, and concluded that an unrealistically large fraction of the original carbon would have to be removed to account for the observed isotopic compositions. However, their model calculations contained errors, and more appropriate models indicate that the observed compositions could be achieved with plausible fractions of degassing (Eiler et al. 1997). Furthermore, if such models are expanded to allow for initial carbon values reflective of lighter carbon sources such as meteorites, the observed range of isotopic values can be accounted for with relatively small fractions of degassing. Consequently, such processes cannot be dismissed as a potential contributor to the isotopic compositions of graphite at Akilia.

In contrast to the conclusions of these and other studies, there appear to be several non-biological sources that could plausibly account for the isotopic compositions of graphite from Akilia and from Isua, whether one considers the current isotopic values or the presumed lighter pre-metamorphic compositions (Fig. 2). This suggests that the confidence with which the graphite in the early Archean rocks from Greenland has been attributed to a biological source is unjustified, and that more serious consideration needs to be given to the possibility that the graphite in these rocks might include substantial contributions from non-biological sources. At the very least, a convincing attribution of the graphite to biological sources requires more convincing demonstration that non-biological sources cannot account for the isotopic compositions.

3.3 Carbon Isotopes and Putative Microfossils in the Warrawoona Group

Rocks from the ~3.45 Ga Apex Chert and adjacent formations within the Warrawoona Group of Western Australia have been proposed to contain the oldest fossil traces of life on Earth (Schopf and Packer 1987; Schopf 1993). The proposed fossil evidence consists of microscopic carbonaceous particles whose structures resemble remnants of cyanobacteria preserved in more recent sediments. The isotopic compositions of these carbonaceous materials cluster within a few permil of −33‰ (Fig. 3).

In contrast to the rocks from Greenland where carbon isotopic composition is the primary evidence cited to support a biological origin, the interpretation of a biological origin for the carbonaceous deposits in the Warrawoona Group is primarily based on their morphology and inferred geologic setting, with the isotopic measurements employed to support this interpretation (Schopf and Packer 1987; Schopf 1993). Nevertheless, the consistency of the isotopic composition of the carbonaceous materials at Warrawoona with more recent deposits has been used to infer that the putative microfossils represent oxygenic photosynthetic cyanobacteria, which would require evolution of this complex and critical biological process prior to 3.45 Ga (Schopf 1993).

Recent studies, however, have contested both the inferred geological setting of the chert deposits and the biological origin of the structural features observed in the carbonaceous matter (Brasier et al. 2002, 2006; Garcia-Ruiz et al. 2003; Lindsay et al. 2005). Instead of the original interpretation that the carbon-bearing cherts were deposited as chemical precipitates from seawater that accumulated as sediments on the seafloor, an alternative interpretation infers a hydrothermal origin for the chert deposits, with the morphologic structures of the carbonaceous particles arising during recrystallization. The carbonaceous matter in this interpretation is inferred to be synthesized by Fischer-Tropsch-type reactions within the hydrothermal system. These reactions probably would have occurred deep within the system where circulating fluids reacted with ultramafic rocks to generate strongly reducing conditions, and were then transported and deposited in a near-surface environment.

The measured isotopic compositions of the carbonaceous matter in the geologic deposits within the Warrawoona Group could apparently be consistent with either of these proposed interpretations. Although the average isotopic composition of the carbon deposits (~−33‰) is depleted in ^{13}C by a few per mil relative to the average biological matter in more recent deposits, it is within the range of the historic record and of known carbon fixation pathways (Figs. 2 and 3). The isotopic composition of organic matter generated by Fischer-Tropsch-type reactions within hydrothermal systems remains poorly constrained, but the available data allow for the production of ^{13}C-depletions comparable to those observed in the Warrawoona rocks. Experimental FTS studies have generated ^{13}C depletions of 16–33‰ relative to the original carbon source, although it must be kept in mind that these values reflect depletions relative to the CO or HCOOH used as the carbon source rather than CO_2 (Lancet and Anders 1970; McCollom and Seewald 2006; Taran et al. 2007). Measured isotopic compositions of light hydrocarbons with an apparent abiotic origin in natural systems range from −5‰ to −45‰, encompassing the values observed for the Warrawoona deposits (e.g., Sherwood Lollar et al. 2008; Proskurowksi et al. 2008). Since it is apparently possible to explain the isotopic composition of the carbonaceous material by either a biological or hydrothermal origin, the isotopic compositions do not provide a means of differentiating between the proposed origins and cannot be confidently used to support either interpretation.

It is worth noting that if the carbonaceous matter in the Apex Chert and other rocks from the Warrawoona Group does have a hydrothermal origin and this could be convincingly demonstrated by other means, this result would have an enormous impact on interpretations of the origin of organic matter in the early isotopic record. Such a finding would demonstrate that non-biological processes are capable of producing an isotopic signature in geologic deposits that is essentially identical to that generated by biological sources. If the ~3.5 Ga Warrawoona deposits with isotopic compositions around −33‰ have a non-biological origin, some of the less ancient geologic deposits might also be non-biological, which would undermine one of the principal arguments used to advocate for the widespread distribution of life predating the early Archean.

4 Concluding Remarks

Most of the scientific community has come to accept that life on Earth originated prior to the onset of the geologic record at ~3.85 Ga, which is largely based on the proposed evidence for life in rocks from Isua/Akilia in Greenland and the Apex Chert and adjacent formations in Australia. The confidence placed in the veracity of these interpretations has led to categorical statements such as: "With the currently available geological record at hand, there is no doubt that microbial (prokaryotic) ecosystems have been prolific on the Archaean Earth since 3.5, if not 3.8 Gyr ago" (Schidlowski 1993). Considerations like those discussed here indicate that the actual geologic record is more nuanced and less definitive than such statements would allow. From the standpoint of carbon isotopes, the evidence for a biological origin for reduced carbon in the earliest rock record is less persuasive than has been portrayed by many authors. Recent studies suggest that other proposed biosignatures in ancient rocks may also be open to debate (Brasier et al. 2002, 2006; Garcia-Ruiz et al. 2003; Lindsay et al. 2005). While biological processes remain a viable explanation for the features found in these ancient rocks (e.g., Glikson et al. 2008), it seems apparent that alternative explanations that do not require biology may also be plausible. Unless and until these alternatives can be confidently rejected, a biological explanation should not be accepted as definitive by scientists, and due consideration should be given to alternative interpretations for both the source of organic matter in ancient rocks and the timing of the origin of life.

With regard to the isotopic data, several lines of investigation should be pursued to gain greater clarity on these issues. While several studies have documented the occurrence of probable abiotic organic compounds in active geologic systems (e.g., Sherwood Lollar et al. 2008; Proskurowksi et al. 2008), further field investigations are needed to understand the extent of such processes, the full complement of compounds produced, and the conditions under which they are generated. Also, although recent studies of the chemical and isotopic compositions of organic matter in IDPs and comets have greatly improved constraints on their contributions, more work is needed to characterize the organic component of these sources. Fluxes of organic matter from extraterrestrial sources and their capacity to survive and accumulate in geologic deposits remain uncertain and would benefit from greater study. Additional laboratory experiments are needed to more precisely constrain the conditions under which organic compounds can be synthesized in geologic systems and in the early atmosphere, and to characterize the isotopic composition of the products. With such data in hand, we may be able to more clearly evaluate whether rocks from the early Earth preserve evidence of life.

Acknowledgments This author's research on this subject and preparation of this manuscript have been supported by the US National Science Foundation Earth Sciences Directorate through grants OCE-0550800 and EAR-0636056.

References

Abrajano TA, Sturchio NC, Kennedy BM et al (1990) Geochemistry of reduced gas related to serpentinization of the Zambales ophiolite, Philippines. Appl Geochem 5:625–630

Alexander CMO'd, Russell SS, Arden JW et al (1998) The origin of chondritic macromolecular organic matter: a carbon and nitrogen isotope study. Earth Planet Sci 33:603–622

Berner RA (2003) The long-term carbon cycle, fossil fuels and atmospheric composition. Nature 426:323–326

Brasier MD, Green OR, Jephcoat AP et al (2002) Questioning the evidence for Earth's oldest fossils. Nature 416:76–81

Brasier M, McLaughlin N, Green O, Wacey D (2006) A fresh look at the fossil evidence for early Archaean cellular life. Phil Trans R Soc B 361:887–902

Chang S, Des Marais D, Mack R et al (1983) Prebiotic organic synthesis and the origin of life. In: Schopf JW (ed) Earth's earliest biosphere: its origin and evolution. Princeton University Press, Princeton, NJ

Charlou JL, Donval JP, Fouquet Y et al (2002) Geochemistry of high H_2 and CH_4 vent fluids issuing from ultramafic rocks at the Rainbow hydrothermal field (36°14′N, MAR). Chem Geol 191:345–359

Chyba C, Sagan C (1992) Endogenous production, exogenous delivery and impact-shock synthesis of organic molecules: an inventory for the origins of life. Nature 355:125–132

Chyba CF, Thomas PJ, Brookshaw L, Sagan C (1990) Cometary delivery of organic molecules to the early Earth. Science 249:366–373

Cody GD, Ade H, Alexander CMO'd et al (2008) Quantitative organic and light-element analysis of comet 81P/Wild 2 particle using C-, N-, and O-μ-XANES. Meteor Planet Sci 43:353–365

Craig H (1953a) The geochemistry of the stable carbon isotopes. Geochim Cosmochim Acta 3:53–92

Craig H (1953b) Corycium defunctum: the non-indicative properties of isotopes and review articles. Econ Geol 48:600–603

Craig H (1954) Geochemical implications of the isotopic composition of carbon in ancient rocks. Geochim Cosmochim Acta 6:186–196

Cronin JR, Chang S (1993) Organic matter in meteorites: molecular and isotopic analyses of the Murchison meteorite. In: Greenburg JM, Pirronello V (eds) Chemistry of life's origins. Kluwer, Dordrecht, The Netherlands

Deines P (2002) The carbon isotope geochemistry of mantle xenoliths. Earth Sci Rev 58:247–278

Des Marais DJ, Donchin JH, Nehring NL, Truesdell AH (1981) Molecular isotopic evidence for the origin of geothermal hydrocarbons. Nature 292:826–828

Eigenbrode J (2008) Fossil lipids for life-detection: a case study from the early Earth record. Space Sci Rev 135:161–185

Eiler JM, Mojzsis SJ, Arrhenius G (1997) Carbon isotope evidence for early life. Nature 386:665

Fiebig J, Woodland AB, Spangenberg J, Oschmann W (2007) Natural evidence for rapid abiogenic hydrothermal generation of CH_4. Geochim Cosmochim Acta 71:3028–3039

Garcia-Ruiz JM, Hyde ST, Carnerup AM et al (2003) Self-assembled silica-carbonate structures and detection of ancient microfossils. Science 302:1194–1197
Glikson M, Duck LJ, Golding SD et al (2008) Microbial remains in some earliest Earth rocks: comparison with a potential modern analog. Precambr Res 164:187–200
Hayes JM, Kaplan JR, Wedeking KW (1983) Precambrian organic geochemistry, preservation of the record. In: Schopf JW (ed) Earth's earliest biosphere: its origin and evolution. Princeton University Press, Princeton, NJ
Horita J (2005) Some perspectives on isotope biosignatures for early life. Chem Geol 218:171–186
Horita J, Berndt ME (1999) Abiogenic methane formation and isotopic fractionation under hydrothermal conditions. Science 285:1055–1057
House CH, Schopf JW, Stetter KO (2003) Carbon isotope fractionation by Archaeans and other thermophilic prokaryotes. Org Geochem 34:345–356
Kelley DS (1996) Methane-rich fluids in the oceanic crust. J Geophys Res 101:2943–2962
Kelley DS, Früh-Green GL (1999) Abiogenic methane in deep-seated mid-ocean ridge environments: insights from stable isotope analyses. J Geophys Res 104:10439–10460
Kelley DS, Karston JA, Früh-Green GL et al (2005) A serpentinite-hosted ecosystem: The Lost City hydrothermal field. Science 307:1428–1434
Krishnamurthy RV, Epsteint S, Cronin JR et al (1992) Isotopic and molecular analyses of hydrocarbons and moncarboxylic acids of the Murchison meteorite. Geochim Cosmochim Acta 56:4045–4058
Lancet MS, Anders EA (1970) Carbon isotope fractionation in the Fischer-Tropsch synthesis and in meteorites. Science 170:980–982
Lindsay JF, Brasier MD, McLoughlin N et al (2005) The problem of deep carbon – an Archean paradox. Precambr Res 143:1–22
Mathez EA (1984) Influence of degassing on oxidation states of basaltic magma. Nature 310:371–375
McCollom TM (2003) Formation of meteorite hydrocarbons by thermal decomposition of siderite ($FeCO_3$). Geochim Cosmochim Acta 67:311–317
McCollom TM (2008) Observational, experimental, and theoretical constraints on carbon cycling in mid-ocean ridge hydrothermal systems. In: Lowell RP, Seewald J, Perfit MR, Metaxas A (eds) Modeling hydrothermal processes at oceanic spreading centers: magma to microbe. American Geophysical Union, Washington, DC
McCollom TM, Seewald JS (2006) Carbon isotope composition of organic compounds produced by abiotic synthesis under hydrothermal conditions. Earth Planet Sci Lett 243:74–84
McCollom TM, Seewald JS (2007) Abiotic synthesis of organic compounds in deep-sea hydrothermal environments. Chem Rev 107:382–401
McKeegan KD, Aléon J, Bradley J et al (2006) Isotopic compositions of cometary matter returned by Stardust. Science 314:1724–1728
Messenger S (2000) Identification of molecular-cloud material in interplanetary dust particles. Nature 404:968–971
Messenger S, Staderman FJ, Floss C et al (2003) Isotopic signatures of presolar materials in interplanetary dust. Space Sci Rev 106:155–172
Miller SL (1953) A production of amino acids under possible primitive Earth conditions. Science 117:528–529
Mojzsis SJ, Arrhenius G, McKeegan KD et al (1996) Evidence for life on Earth before 3,800 million years ago. Nature 384:55–59
Naraoka H, Ohtake M, Maruyama S, Ohmoto H (1996) Non-biogenic graphite in 3.8-Ga metamorphic rocks from the Isua district, Greenland. Chem Geol 133:251–260
Nier AO, Gulbransen EA (1939) Variations in the relative abundance of the carbon isotopes. J Am Chem Soc 61:697–698
Pasek M, Lauretta D (2008) Extraterrestrial flux of potentially prebiotic C, N, and P to the early Earth. Orig Life Evol Biosph 38:5–21
Pavlov AA, Kasting JF, Eigenbrode JL, Freeman KH (2001) Organic haze in Earth's early atmosphere: source of low-13C Late Archean kerogens. Geology 29:1003–1006

Perry EC Jr, Ahmad SN (1977) Carbon isotope composition of graphite and carbonate minerals from 3.8-AE metamorphosed sediments, Isukasia, Greenland. Earth Planet Sci Lett 36:280–284

Pizzarello S, Krishnamurthy RV, Epstein S, Cronin JR (1991) Isotopic analyses of amino acids from the Murchison meteorite. Geochim Cosmochim Acta 55:905–910

Popper K (1959) The logic of scientific discovery. Hutchinson, London

Proskurowksi G, Lilley MD, Seewald JS et al (2008) Abiogenic hydrocarbon production at Lost City Hydrothermal Field. Science 319:604–607

Rankama K (1948) New evidence of the origin of Pre-Cambrian carbon. Bull Geol Soc Am 59:389–416

Rankama K (1954) The isotopic constitution of carbon in ancient rocks as an indicator of its biogenic or nonbiogenic origin. Geochim Cosmochim Acta 5:142–152

Rosing MT (1999) ^{13}C-Depleted carbon microparticles in >3700-Ma sea-floor sedimentary rocks from West Greenland. Science 283:674–676

Salvi S, Williams-Jones AE (1997) Fischer-Tropsch synthesis of hydrocarbons during sub-solidus alteration of the Strange Lake peralkaline granite, Quebec/Laborador, Canada. Geochim Cosmochim Acta 61:83–99

Salvi S, Williams-Jones AE (2006) Alteration, HFSE mineralisation and hydrocarbon formation in peralkaline igneous systems: Insights from the Strange Lake Pluton, Canada. Lithos 91:19–34

Sandford SA (2008) Terrestrial analysis of the organic component of comet dust. Ann Rev Anal Chem 1:549–578

Sandford SA, Aléon J, Alexander CMO et al (2006) Organics captured from Comet 81P/Wild 2 by the Stardust spacecraft. Science 314:1720–1724

Schidlowski M (1988) A 3,800-million-year isotopic record of life from carbon in sedimentary rocks. Nature 333:313–318

Schidlowski M (1993) The beginnings of life on Earth: evidence from the geological record. In: Greenburg JM, Mendoza-Gómez CX, Pirronello V (eds) The chemistry of life's origins. Kluwer, Dordrecht, The Netherlands

Schidlowski M (2001) Carbon isotopes as biogeochemical recorders of life over 3.8 Ga of Earth history: evolution of a concept. Precambr Res 106:117–134

Schidlowski M, Appel PWU, Eichmann R, Junge CE (1979) Carbon isotope geochemistry of the 3.7×10^9-year-old Isua sediments, West Greenland: implications for the Archaean carbon and oxygen cycles. Geochim Cosmochim Acta 43:189–199

Schopf JW (1993) Microfossils of the early Archean apex chert: new evidence of the antiquity of life. Science 260:640–646

Schopf JW, Packer BM (1987) Early Archean (3.3-billion to 3.5-billion-year-old) microfossils from Warrawoona Group, Australia. Science 237:70–73

Sephton MA, Verchovsky AB, Bland PA et al (2003) Investigating the variations in carbon and nitrogen isotopes in carbonaceous chondrites. Geochim Cosmochim Acta 67:2093–2108

Sherwood Lollar B, Frape SK, Weise SM et al (1993) Abiogenic methanogenesis in crystalline rocks. Geochim Cosmochim Acta 57:5087–5097

Sherwood Lollar B, Westgate TD, Ward JA et al (2002) Abiogenic formation of alkanes in the Earth's crust as a minor source for global hydrocarbon reservoirs. Nature 416:522–524

Sherwood Lollar B, Lacrampe-Couloume G, Slater GF et al (2006) Unravelling abiogenic and biogenic sources of methane in the Earth's deep subsurface. Chem Geol 226:328–339

Sherwood Lollar B, Lacrampe-Couloume G, Voglesonger K et al (2008) Isotopic signatures of CH_4 and higher hydrocarbon gases from Precambrian Shield sites: a model for abiogenic polymerization of hydrocarbons. Geochim Cosmochim Acta 72:4778–4795

Strauss H, Moore TB (1992) Abundances and isotopic compositions of carbon and sulfur species in whole rock and kerogen samples. In: Schopf JW (ed) The Proterozoic biosphere: a multidisciplinary study. Cambridge University Press, Cambridge

Summons RE, Albrecht P, McDonald G, Moldowan JM (2008) Molecular biosignatures. Space Sci Rev 135:133–159

Taran YA, Kliger GA, Sevastianov VS (2007) Carbon isotope effects in the open-system Fischer-Tropsch synthesis. Geochim Cosmochim Acta 71:4474–4487

Thomas KL, Blanford GE, Keller LP et al (1993) Carbon abundance and silicate mineralogy of anhydrous interplanetary dust particles. Geochim Cosmochim Acta 57:1551–1566

Ueno Y, Yurimoto H, Yoshioka H et al (2002) Ion microprobe analysis of graphite from ca. 3.8 Ga metasediments, Isua supracrustal belt, West Greenland: relationship between metamorphism and carbon isotope composition. Geochim Cosmochim Acta 66:1257–1268

Ueno Y, Yoshioka H, Maruyama S, Isozaki Y (2004) Carbon isotopes and petrography of kerogens in 3.5-Ga hydrothermal dikes in the North Pole area, Western Australia. Geochim Cosmochim Acta 68:573–589

van Zuilen MA, Lepland A, Arrhenius G (2002) Reassessing the evidence for the earliest traces of life. Nature 418:627–630

van Zuilen MA, Lepland A, Teranes J (2003) Graphite and carbonates in the 3.8 Ga old Isua Supracrustal Belt, southern West Greenland. Precambr Res 126:331–348

Wickman FE (1941) On a new possibility of calculating the total amount of coal and bitumen. Geol Foren I Stolkholm Förb 63:419

Wickman FE, von Ubisch H (1951) Two notes on the isotopic constitution of carbon in minerals. Geochim Cosmochim Acta 1:119–122

Index

S.D. Golding and M. Glikson (eds.), *Earliest Life on Earth: Habitats, Environments and Methods of Detection*, DOI 10.1007/978-90-481-8794-2,